T0136470

Charles Valentine Riley

Charles Valentine Riley

Founder of Modern Entomology

W. CONNER SORENSEN, EDWARD H. SMITH,
AND JANET R. SMITH, WITH DONALD C. WEBER

THE UNIVERSITY OF ALABAMA PRESS TUSCALOOSA

The University of Alabama Press
Tuscaloosa, Alabama 35487-0380
uapress.ua.edu

Typeface: Adobe Caslon Pro

Manufactured in Korea

Frontispiece: Charles V. Riley examining insect with hand lens (Courtesy of the National Agricultural Library Special Collections, Beltsville, Maryland).

Cover image: Charles V. Riley examining insect with hand lens; courtesy of the National Agricultural Library Special Collections, Beltsville, Maryland

Cover design: Michele Myatt Quinn

Publication made possible in part by a generous contribution from The Charles Valentine Riley Memorial Foundation, established in 1985 to build on the legacy of C. V. Riley by promoting a broader and more complete understanding of agriculture and to enhance agriculture through increased scientific knowledge. http://rileymemorial.org/

Publication is also made possible in part by a generous contribution from the Entomological Society of America (www.entsoc.org), founded by C. V. Riley with a mission to promote and advance scientific inquiry, discovery, and communication of entomology. A key component of ESA's mission is to enable our members to share their science globally, such as with this book, as well as with all of our activities.

Library of Congress Cataloging-in-Publication Data
Names: Sorensen, Willis Conner, author.
Title: Charles Valentine Riley : founder of modern entomology / W. Conner Sorensen, Edward H. Smith, and Janet R. Smith, with Donald C. Weber.
Description: Tuscaloosa : The University of Alabama Press, [2019] | Includes bibliographical references and index.
Identifiers: LCCN 2018032195| ISBN 9780817320096 (cloth) | ISBN 9780817392222 (ebook)
Subjects: LCSH: Riley, Charles V. (Charles Valentine), 1843 -1895. | Entomologists —Biography.
Classification: LCC QL31.R5 S67 2019 | DDC 595.7092 —dc23
LC record available at https://lccn.loc.gov/2018032195

To Ed Smith (1916–2012) and Janet Smith (1922–2018),
we dedicate this biography
which they initiated and pursued through many years.

Contents

Illustrations

Preface

It was 1875, and Charles Valentine Riley was describing the difference between the Snout Beetle, the Plum Curculio, and the Apple Curculio to a gathering of horticulturists. A reporter in the audience took note of Riley's spirited presentation. "[His manner was enthusiastic], his face beaming with animation, his eyes sparkling, his manner eager . . . intense, and all about the crooked snout of a beetle."[1] It is Riley's enduring passion for insects, juxtaposed with his pioneering mission to aid farmers in their efforts to control insect pests in the expanding American agricultural environment, that make him such an interesting subject. This duality of fascination with insects and the need to control pest species has typified modern entomology from Riley's time to the present. Just as Riley was drawn to the fascinating beauty, staggering diversity, complex adaptability, and amazing persistence of insects, so too are today's entomologists. Regarding pests, however, the stakes today are much higher: since Riley's time, the world's population has quintupled to nearly eight billion people. The need for wise and environmentally friendly insect management practices is more urgent than ever. We believe that Riley's story uniquely embodies the evolution of entomological science, broadening from its natural history origins to its application to practical insect pest management. If he were with us today, Riley would, no doubt, have many animated comments to share with us on both the wonder and the challenges of the insect world.

Conner Sorensen, Eschbach, Germany
Don Weber, Arlington, Virginia
March 2018

Acknowledgments

WE WOULD LIKE to thank Emilie Wenban-Smith Brash, Riley's granddaughter, for making the Riley diaries, personal correspondence, and family pictures available. Emilie died in 2015. Recognition is also made to the Cathryn Vedalia Riley Trust for early substantial support of the book project.

We thank Lester Stephens, Paul Farber, Janet Moore, and Michael Sparks, for reading the manuscript and helping us to improve it. Carol Anelli researched the National Archives holdings of the Bureau of Entomology and, with Ed Smith, researched local records in Kankakee, Illinois. She also commented on Yucca Moth pollination and Benjamin D. Walsh. Carol recruited Conner, and later Don, to join the project to its completion, and is a true collaborator on this biography.

We thank Jeffrey Lockwood for comments on the Rocky Mountain Locust, Jeffrey Granett for comments on phylloxera, John Klasey for comments on Kankakee history, Pamela Henson for comments on John H. Comstock, Gene Kritsky for comments on Darwinism, Vern LaGesse for supplying historical maps of Kankakee, and Yves Carton for his collaboration on phylloxera.

We thank Elizabeth Harvey for secretarial and organizational assistance over many years and Donna Kolwalski, Ed Smith's secretary at Cornell University, for assistance during the early stage of the project. Helen L. Grove kindly translated Riley's Civil War diary from French to English.

Our special thanks to Marty Schlabach at the Cornell University Library for providing interlibrary and internet support over many years. We thank Sara Lee, Susan Fugate, and Diane Wunsch at the National Agricultural Library Special Collections, Tad Bennicoff at the Smithsonian Institution Archives, Nan Meyers at Wichita State University, and other unnamed librarians and archivists for their support.

We also thank Robert and Mary Margaret Wallace for providing Conner Sorensen with hospitality during several research visits to Washington, DC, and Conner's parents, Howard and Claire Sorensen (posthumously) for their hospitality during research trips to Wichita, Kansas.

Portions of the text based on two articles—W. Conner Sorensen, Edward H. Smith, Janet Smith, and Yves Carton, "Charles V. Riley, France, and *Phylloxera*," *American Entomologist* 54 (Fall 2008): 134–49; and Conner Sorensen and Edward H. Smith, "Charles Valentine Riley: Art Training in Bonn, 1858–1860," *American Entomologist* 43 (Summer 1997): 92–104—are used here with permission.

We are grateful to the Charles Valentine Riley Memorial Foundation, and to the Entomological Society of America, for their generous financial support of this work, resulting in the high quality book in your hands.

The entire team of the University of Alabama Press has been essential as well as gracious in their smoothing over of the bumps in the road on the way to publication. Claire Lewis Evans, Jon Berry, our indexer Bonny McLaughlin, and especially recently retired acquisitions editor Beth Motherwell, have all had a big hand in the gratifying completion of this long project.

Lastly, Ed and Jan Smith's children, especially Janet Moore, have contributed substantially to the completion of the book, including the proper disposition and curation of all of the primary sources, archives, and notes from their parents' research.

Abbreviations

AAAS	American Association for the Advancement of Science
AE	*American Entomologist*
Am. Ent. and Bot.	*American Entomologist and Botanist*
AgH	*Agricultural History*
AN	*American Naturalist*
Ann. Rept. Mo. Bot. Garden	*Annual Report of the Missouri Botanical Garden*
Ann. Rev. Ent.	*Annual Review of Entomology*
ANSP	Academy of Natural Sciences of Philadelphia
APS	American Philosophical Society
ARCA	*Annual Report of the Commissioner of Agriculture*
ARSA	*Annual Report of the Secretary of Agriculture*
ASUSNM	Assistant Secretary, United States National Museum
CAS	California Academy of Sciences
CE	*Canadian Entomologist*
Corr.	Correspondence
CRW	*Colman's Rural World*
CVR	Charles Valentine Riley Papers
DSB	*Dictionary of Scientific Biography*
E	Entry (for National Archives citations)
Ent. News	*Entomological News*
GPO	Government Printing Office
IL	*Insect Life*
JHB	*Journal of the History of Biology*
JHI	*Journal of the History of Ideas*
MC	Manuscript Collection

MCZ	Museum of Comparative Zoology
MSB	Museum of Science, Boston
MSBA	Missouri State Board of Agriculture
NA	National Archives
NAL	National Agricultural Library
NYRB	*New York Review of Books*
NYSM	New York State Museum
OSIC	Office of the Secretary, Incoming Correspondence
OSOC	Office of the Secretary, Outgoing Correspondence
PF	*Prairie Farmer*
Proc. AAAS	*Proceedings of the American Association for the Advancement of Science*
Proc. BSNH	*Proceedings of the Boston Society of Natural History*
Proc. BSW	*Proceedings of the Biological Society of Washington, DC*
Proc. ESP	*Proceedings of the Entomological Society of Pennsylvania*
Proc. ESW	*Proceedings of the Entomological Society of Washington*
Proc. MSHS	*Proceedings of the Missouri State Horticultural Society*
RFR	Riley Family Records
RG	Record Group (for National Archives citations)
RU	Record Unit (for Smithsonian Institution Archives)
SciAm	*Scientific American*
SIA	Smithsonian Institution Archives
Trans. ISAS	*Transactions of the Illinois State Agricultural Society*
Trans. ISHS	*Transactions of the Illinois State Horticultural Society*
Trans. SLAS	*Transactions of the Academy of Science of St. Louis*
UNA	University of Nebraska Archives
USEC	United States Entomological Commission
USNM	United States National Museum

I

The Thames, the Channel, the Rhine, 1843–1859

I was born at Caroline Cottage, Queen Street, Chelsea, September 18, 1843. I was taken down in the country with my brother George, when I was 3 years old, stayed with Mrs. Miller, Walton-on-Thames, Surrey about 2 years, after which we went to Mrs. Bissett's who lived North of Walton near the river in a large red-brick house in which two other families lived, Mrs. Greathead and Mr. Groom.

—Charles V. Riley, first journal entry, undated, ca. 1850, RFR

On Monday, September 18, 1843, in Caroline Cottage in genteel Chelsea, West London, Mary Louisa Cannon gave birth to a boy whom she named Charles Valentine Riley. A year and a half later, on January 20, 1845, again in Caroline Cottage, she gave birth to a second boy whom she named George Riley. The father of the two boys, Charles Edmund Wylde, a middle-aged married Anglican clergyman who sired the two boys out of wedlock, was likely not present at their births. His sexual affair with an unmarried woman in her twenties might raise some eyebrows in middle-class London circles; however, still more attention might be focused on the surprising fact that Wylde's mother, Emma Prichard Wylde, assisted with the births of her son's two illegitimate offspring. It was well known that Emma's father, Robert Prichard, and his wife owned Caroline Cottage. It seems altogether probable that Wylde's parents knew about, perhaps even supported their son's extramarital affair.[1]

If so, Charles Wylde's behavior in his mid-thirties may reflect a nonconformist streak in his upbringing. Born and raised in a clergy house, Wylde studied theology at Cambridge University with the intention of following in his father's footsteps. In 1835, while a student, he married Jane Derby Knox, a young widow. In 1837, he graduated with a degree of Bachelor of Arts. That year Jane gave birth to a baby girl. The baby was given no name and apparently died soon thereafter, possibly causing stress in the Wyldes' marriage. At the time Wylde became involved with Mary Cannon, he was a "Clerk in Holy Orders and Incumbent" [Rector] at Trinity Church in Lambeth, on the south bank of the Thames, across from Chelsea.[2]

Mary seems to have had a defiant streak that likely came by way of her father, George Cannon, an associate of William Cobbett, who led the radical dissenters, a heterogeneous group of reformers who opposed social inequities they blamed on the Industrial Revolution.[3] It is not clear to what extent Cannon supported Cobbett's program of returning England to an idealized preindustrial commonwealth when English gentry and commoners supposedly enjoyed equality and self-sufficiency; however, it is clear that he agreed with Cobbett's denunciation of the established church. Cannon and Cobbett objected to the church's dogmatism, intolerance, and its alliance with the monarchy, made explicit in the Anti-Sedition Act (1817). Under the sedition laws, a person who questioned the literal truth of the Bible could be jailed for libelous blasphemy. Many of Cannon's fellow radicals, like the poet Leigh Hunt, William Benbow (a non-conformist preacher and publisher) and Daniel Isaac Eaton (a political activist), were imprisoned under the anti-sedition laws. In 1815–16, Cannon published the *Theological Inquirer; or Polemical Magazine*, a periodical in which he wrote under the pseudonyms Reverend Erasmus Perkins or Philosemus, in order to avoid charges of blasphemy. There, he published portions of Percy Bysshe Shelley's poem *Queen Mab*, a utopian fantasy advocating peaceful revolution of a commonwealth through moral transformation of individuals. *Queen Mab*'s thinly disguised Jacobin leanings infuriated the Society for the Prevention of Vice and led to the prosecution of several publishers of the poem, probably including Cannon.[4] In 1815, perhaps in response to Cannon's publication of the *Queen Mab* excerpts, the mayor of London demanded that Cannon sign an oath of allegiance to the crown, an oath to the supremacy of the Anglican Church, a declaration against popery, and a declaration of his Christian faith.[5] Although Cannon avoided imprisonment on this occasion, he soon experienced prison conditions firsthand. In 1821–22, when married but without children, he served a one-year sentence in debtor's prison in place of his father who was also named George Cannon.[6]

About 1820, while participating in a Unitarian-style meeting, George

Cannon met Mary Miller, a woman considerably younger than himself. Reportedly "electrified" by Cannon's argument that the truth of religion should stand separate from the personality of its teachers, Mary married him soon thereafter, although for several years they kept their marriage secret. Following several miscarriages, Mary gave birth to three daughters: Mary (Charles and George's mother), Laura, and Mira. Mary also gave birth to a son who died when a child. Shortly after the son's death, a woman appeared at the Cannon house with a two- or three-year-old girl she said was George Cannon's illegitimate daughter. Overcoming her initial shock at this revelation, Mary Cannon accepted the girl, Amelia, thus making a family of four daughters.[7]

Cannon's restless, innovative, and crusading nature led him to law, medicine, and finally to theology and polemics. His many interests, however, distracted him from the practicalities of family finances and his daughters' futures. Eventually, his brother-in-law, William Miller, seeing his sister's plight, urged Cannon to capitalize on his knowledge of old books. With start-up capital furnished by William, Cannon achieved moderate success as a dealer in antique books and manuscripts. When his eldest daughter, Mary, reached age eighteen, Cannon sent her to a Catholic finishing school in France.[8]

Mary returned in 1842, an attractive, educated, impetuous, and strong-willed woman in her early twenties, and promptly began her affair with thirty-five-year-old Wylde. When Wylde's wife died in 1844, between the births of Charles and George, Mary probably expected Wylde to marry her, but he declined. What exactly occurred between the two is not clear, but Wylde may have envisioned an "open marriage," not uncommon in London's upper-class circles, with Mary as his extra-marital partner. Mary, for her part, may have expressed her sexual freedom and independence in the style of Flaubert's *Madam Bovary*. Whatever their motivations, by 1846 the union between Charles Wylde and Mary Cannon that had produced Charles and George Riley was unraveling.[9] That year, Mary married Antonio Hipolito Lafargue, a ship's agent.[10]

In 1851 Wylde left London to join his uncle, John Fewtrell Wylde, a widower and retired military surgeon, at Uplands House, Bridgenorth, Shropshire, his substantial estate in the west of England. According to family tradition, John Wylde, in his early seventies, in failing health and having no heirs, proposed that his nephew add the name of Fewtrell to his own, in return for being named John Wylde's heir. Charles Edmund Wylde thus became Charles Edmund Fewtrell Wylde and at his uncle's death in the mid-1850s, he inherited the estate.[11] In 1852, Wylde married Cecilia Elizabeth Bell, daughter of

Captain Charles William Bell, a ship captain in the Merchant Navy of the East India Company.[12] At that time Wylde was enjoying his inheritance, but he soon drifted into financial duress. Tradesmen and retailers customarily extended credit to clients of apparent social standing like Wylde, but his debt soon rose to unacceptable levels. On May 21, 1855, he was committed to King's Bench (later Queen's Bench) Prison, in East London, where he remained (with several short intervals on parole) until he died, December 15, 1859.[13]

Meanwhile, Mary settled with her husband in Rotherhithe, East London. She placed her two small sons (now ages three and one) under the care of Mrs. Miller, a family friend they referred to as "Auntie Miller," but who was not related to Mary, at Walton-on-Thames, about eighteen miles west of London on the south side of the Thames (see figure 1.1). Lafargue adored Mary and provided her with a good home, but Mary nevertheless maintained her correspondence with Wylde for three years, though never mentioning her marriage to him. She also maintained a relationship with a man whom the family referred to as "the Captain."[14]

Walton's rivers and woodlands gave Charles his first experience of nature. Nestled in a semi-rural agricultural district eighteen miles from Hyde Park Corner, Walton supplied fresh garden produce to the estates of Oatlands, Ashley Park, Sunbury, and Weybridge. Mary visited her sons regularly, investing their outings with a sense of adventure.

In 1848, Mary gave birth to a daughter, Louisa Josephine (nicknamed Nina).[15] That year, she moved Charles and George, now five and three, from Mrs. Miller's care to the home of Andrew Bissett, a Scottish gardener, and Annie, his English wife. The Bissetts lived north of Walton and had two grown children, George and Annie. The boys were fortunate, once again, to find a haven with a woman who treated them as her own.[16]

Mary waited at least five years before introducing her two boys to her husband. Until then she had not told Charles and George that she was married.[17] Mary apparently never told her boys who their real father was, and it is not clear when they might have found out about their biological father.

Despite uncertain relations between Mary's two families, she demonstrated affection for her sons. She took Charles and George to the theater in London, to the bazaar in the Thames Tunnel, and to Greenwich Fair.[18] In June 1851, Mary, Antonio, Charles, and George visited the Crystal Palace Exhibition where the boys had their portraits taken.[19]

Having been reared in a literary family, Mary enrolled her sons at ages seven and five in the Children's School in Walton. Later they went "down the alley" to Mr. Edward's school. Both schools stressed spelling, reading, and arithmetic as well as expert penmanship; the latter skill Charles maintained throughout his life.[20] In the afternoons and on weekends, the boys went swimming, fishing, and boating, often with the Wheatley boys. Nearby parks and estates offered playgrounds and collecting grounds where Riley dug snowdrops, violets, and blue bottles to transplant to his little garden at home.[21]

With her literary background, Mary likely subscribed to the blend of natural theology and natural history characteristic of the pre-Darwinian era where the moral and the useful were intertwined, and leisure activities were infused with industriousness. Naturalists in that romantic era often wrote about birds whose fluttering wings symbolized dynamism and insects whose metamorphosis dramatized a vital living force. In the British Isles, insect collecting and study became so popular that it threatened to swamp natural history's other branches.[22] Two Cambridge classmates who later had a profound influence on Charles, Benjamin Dann Walsh and Charles Darwin, became enthusiastic insect collectors.

While Mary encouraged Charles' early fascination with nature, William Chapman Hewitson, a wealthy retired neighbor living in Oatlands Park near the Bissett residence, directed his attention toward insects. Curiously, Charles does not mention Hewitson in his journal, but it is clear that during the years 1845 to 1853, when he lived at Walton or at boarding schools in Kensington and Chelsea, Hewitson played a key role in Charles' choice of entomology as his life's vocation.[23]

Hewitson belonged to a closely knit circle of entomological enthusiasts in and around London that included Henry Tibbats Stainton, editor of the *Entomological Intelligencer* and *Entomologist's Annual*; John Curtis, author of *British Entomology*; William Kirby, lead author of the *Introduction to Entomology*; Henry Doubleday, lepidopterist; and Oxford professor John Obadiah Westwood, England's most prominent entomologist. Members of the entomological fraternity referred to themselves as "brethren of the net," and they maintained a set of rituals specifying equipment, dress, and demeanor. Meeting regularly in "At Home" meetings where neophytes were welcomed, they communicated through notices in the *Entomological Intelligencer* where they announced discoveries, offered specimens for exchange, and posted notices of specimens they desired. Whether Charles met these men as a child is not documented, but it seems likely because he collaborated with many of them once he became established in America.

Through Hewitson's travels, and through purchase, he amassed one of the finest collections of British and exotic Lepidoptera, the favorite insect group of Victorian collectors. During Charles's childhood, Hewitson was issuing serial installments of *The Genera of Diurnal Lepidoptera* (1846–52) and *Illustrations of . . . Exotic Butterflies* (1851–78) that became classics of Victorian entomology.[24]

The surge of interest in entomology in Hewitson's generation was largely inspired by the serial publication of the *Introduction to Entomology* (4 vols., 1815–26) by William Kirby and William Spence.[25] Its engrossing literary style inspired readers to take to the field with insect net and collecting bottle, where alone, or in the company of others, they sought spiritual uplift, practical knowledge, and prizes for the cabinet.[26] The popularity of this publication sprang not only from its engaging literary style but also from its mix of intellectual, philosophical, and political viewpoints that appealed to various groups in pre-Victorian England. Its natural theology, primarily from Kirby, a clergyman in Burnham who echoed the views of William Paley's natural theology, appealed to defenders of a hierarchical society, royal supremacy, and Anglican orthodoxy. Its treatment of insect pest control and agricultural improvement, primarily from Spence, appealed to Whig promoters of economic growth but also to radicals like Cobbett, since it was grounded in Spence's call for an agrarian English economy, independent of capitalism and industrialism. It is likely that Charles read portions of the *Introduction* as a child because its concept of insects as key participants in the balance of nature was central to his thinking. Riley also shared Kirby and Spence's view that insects were sentient creatures, with thinking, reasoning, and emotional capacities.[27]

In some respects Hewitson was ideally suited as a father figure and mentor for Charles. He shared with Charles his entomological cabinet and library, among the finest in England, and he demonstrated his skill in illustrating the form and color of pinned butterflies. Hewitson's expertise in lithography made a lasting impression on Charles. This technique involved drawing figures with wax or other water-repellant substance on limestone after which the surrounding stone was etched away with acid, leaving the raised drawing on a plate ready for printing. From its invention in the 1790s until photography replaced it in the late nineteenth century, lithography constituted a central medium of natural history illustration.

Hewitson was essentially an armchair naturalist and an iconographer. He directed Charles's attention to the beauty of insect forms and introduced him to the fundamentals of natural history in the British tradition, with its fraternal spirit, disciplined procedures, and attractive publications. Young Charles, by contrast, was fascinated by biological phenomena. In a probable reference

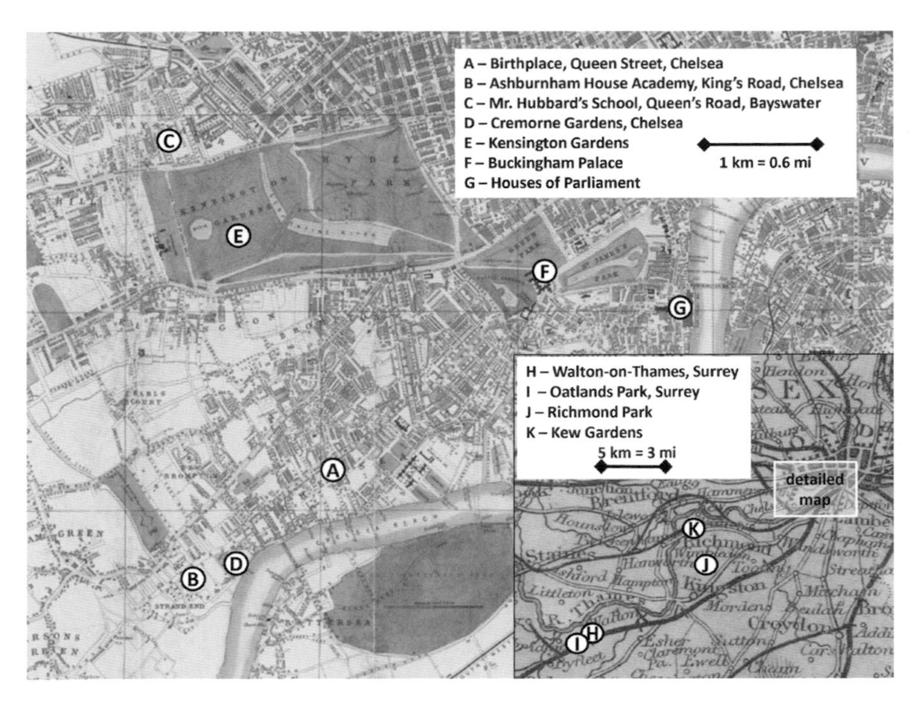

FIGURE I.1 Composite map of Riley's childhood and youth in London and surroundings (prepared by Don Weber from historical sources, 2016, and used by permission).

to his explorations on Hewitson's estate, Riley later recalled that he observed a spanworm (geometrid caterpillar), an allied species of the American Gooseberry Span-Worm, that was "very common in a dearly loved garden at Walton, England, where in watching its metamorphosis I first, as a child, became interested in insect life."[28]

Charles's Walton idyll came to an end at age ten. In September 1853, Mary enrolled Charles and George in Mr. Hubbard's boarding school, Queen's Road in Bayswater, Kensington, on the north bank of the Thames. In July the next year, they transferred a short distance to Ashburnham House Academy, Kings Road, Chelsea, adjacent to Cremorne, a popular resort park near their birthplace.[29]

The stern discipline of the boarding school now replaced the comfort of the Bissett home. In the rough, competitive atmosphere of their new school-home, Charles, like other newcomers, had to fight upper classmen to establish his place in the pecking order, and he learned early that discipline was severely enforced.[30] Despite these constraints, the boys enjoyed considerable freedom to explore beyond the school grounds. In Kensington Gardens they ran, hid, and played games of "Hunt the Stag" and "Prisoner's Base." On visits to Cremorne

Park, they enjoyed treats of curds and whey with brandy snaps. At the Royal Polytechnic they viewed the science demonstrations and the panorama of "Symbad [*sic*] the Sailor."[31] One summer evening, Charles and George viewed an open-air production of the *Siege of Sebastopol* from their quarters above.[32]

In late November 1855, Mary took the boys to Moses and Sons in London to be fitted with school uniforms in preparation for their enrollment in the College Sainte Paul in Dieppe, France. At Dieppe, Rouen, and Caen, sixty miles across the channel, English families had established enclaves complete with finishing schools for young men and women.[33] Charles and George, like their mother before them, left their childhood surroundings to study in a Catholic school in France. Following a rough passage aboard the steamship *Paris* accompanied by August Ryder, the founder and principal of the school, the boys joined the Ryder family for dinner. Charles was so miserable with seasickness that he could not eat. As he listened to the lively repartee of the eleven Ryder children, he despaired of ever learning French.[34]

With a population of sixteen thousand, Dieppe featured a busy harbor and superb bathing beaches, rimmed with two-hundred-foot white chalk cliffs. Charles and George took to the beaches at every opportunity. High tides delivered flotsam and jetsam and low tides revealed a marine world populated with crabs, periwinkles, and other tidewater organisms new to lads who had grown up along the tranquil Thames.[35]

The town's docks, shipyards, townhouse, theater, hospital, and tourist shops possibly struck Charles and George as a bit humdrum compared to the hustle and bustle of London. The plain, gray stone school building directly abutting the street contrasted to their boarding school on the Thames, surrounded by parks and gardens; however, out in the countryside the boys often encountered kindly French farmers who offered them fresh milk with biscuits or fruit.[36]

St. Paul offered a classical education with instruction in Latin, Greek, French, classical literature and history, arithmetic, composition, penmanship, and art. Charles's teachers applauded his rapid mastery of French, Latin, and Greek, and they encouraged his collecting of insects and his experiments with silkworms. In a small book he entitled *Natural History of Insects* (1857), Charles sketched and painted silkworms, butterflies, and moths in various stages of development.[37]

On July 4, 1857, having completed two years at Dieppe, Charles and George celebrated their graduation. Their mother, who traveled from London, shone like a queen bee amidst the conservative, predominantly male assembly of priests and parents gathered for the ceremony. Charles's half-sister, Nina, wrote years later that when "Momma" assisted in conferring on Charles the laurel wreath for the highest scholastic achievement, "she could not have felt more proud of her son than he did of his mother; she was so young and so pretty and such a center of admiration."[38]

The following week, Mary and the boys shopped, swam, strolled on the promenade, and took long walks in the country. After a week of excitement and laughter, Mary departed on the steamer *Paris*. Watching the ship disappear into the gray mist of the Channel, the boys must have sensed the impenetrable curtain that separated them from their mother's life in London.[39] Were they aware, for example, that the year before they entered St. Paul, Mary (to the consternation of her husband, her mother-in-law, and her own family) set up her own room on Milton Street, where, their Aunt Mira reported later, she entertained gentlemen visitors?[40] Mira described this as the beginning of Mary's descent into "moral depravity," but how much Charles knew, or wanted to know, and how such knowledge may have affected him is unclear.

During the fall or winter of 1857, Charles's educational path took a decisive turn.[41] By 1858 he was enrolled in a private art school in Bonn, Prussia, now without George.[42] Nestled on the left bank of the Rhine, where the river emerges from confining hills onto the broad plain surrounding Cologne, Bonn featured its own enclave of British residents, attracted by its picturesque setting, intellectual atmosphere, and inexpensive goods and services.[43] Bonn's Frederick Wilhelm University, founded in 1818, attracted an outstanding faculty that included the zoologist Georg August Goldfuss and the botanist Christian Gottfried Daniel Nees von Esenbeck. In 1853, Bonn professor Hermann Schaaffhausen proposed a radical theory of the transformation of species along genealogical lines, a theory Darwin later cited in the *Origin of Species*. In 1857, two years prior to publication of the *Origin*, Schaaffhausen was called to examine fossils from a quarry in the Neanderthal valley near Düsseldorf. His judgment that these fossils represented a prehistoric "Neanderthal Man" supported the (soon to be published) Darwinian thesis that present-day human species had developed gradually over long periods of time.[44]

Charles may have known of Bonn's professors; however, the primary reason for his move was to study art under Nicolaus Christian Hohe.[45] Riley's talent in drawing and painting apparently prompted his teachers at Dieppe to suggest that he continue his art training under Hohe. His new teacher descended from a family of artists in Bayreuth, Bavaria, that included Christian's father, Johann Hohe, a well-known master painter, and his brother, Friedrich Hohe, a skilled painter and lithographer. When called to Bonn in

FIGURE 1.2 “Silkworm,” cover page of Charles V. Riley, *Natural History of Insects* (1857) (courtesy of the National Agricultural Library Special Collections, Beltsville, Maryland).

1824, Christian Hohe’s primary emphasis was drawing lithographic illustrations for naturalists like Goldfuss, Esenbeck, botanist Carl Friedrich Philipp von Martius, zoologist Johann Baptist Ritter von Spix, naturalist Johann Lukas Schoenlein, and Louis Agassiz, in Neuchâtel, Switzerland. Hohe’s talent in natural history illustration prompted Bonn’s scientists to propose his appointment to a university chair; however, the dominating liberal arts professors insisted on more “philosophical” academic credentials and rejected the proposal. By the time Riley arrived, Hohe was engaged in tutoring private students, portrait painting, renovating medieval wall paintings in cathedrals, and painting scenes of Bonn and its surroundings. Hohe mass-produced his

Rhine paintings by lithography for sale to local patrons and tourists. He also produced a popular series of art instruction books.[46]

Although Hohe's eclectic background hampered his academic aspirations, he had much to offer the young Riley. Rejecting the prevailing custom of using plaster models in his instruction, Hohe led his students on field trips along the Rhine where they sketched and painted natural objects, historic structures, and landscapes. Hohe's depiction of objects and scenery from nature no doubt reinforced Riley's conviction that the "Book of Nature" was the best teacher. In sketching and painting, Hohe preferred simple lines rather than fanciful romantic impressions favored by many contemporary artists. His clear, straightforward style is reflected in Riley's Bonn art book and later works.[47] Hohe's promotion of Bonn and the Rhine through mass-produced lithographic scenes reinforced Charles's predilection for addressing public audiences. Under Hohe's direction, Charles filled his sketchbook with landscapes, natural history subjects, portraits, and architectural and historical landmarks, including many scenes found in illustrated guidebooks and paintings in England. With schoolboy ardor, Charles inscribed the flyleaf of his sketchbook:

> Please spare the book,
> Cut not a single leaf,
> You dream not of the pains I took
> Or you'd regard my grief.
>
> For many a thoughtful hour
> I've [culled] my fruitful brain
> To draw both scene and flower
> And thus have earnt my fame.
>
> Friend spare the book
> It is my fancy's pet,
> Turn gently o'er its leaves, and look
> Now everything is "nette". Dec. 1, 1858.[48]

As in Dieppe, Charles found camaraderie and friendship in his teacher's large family. In his art book is a sketch by Hohe's son, Rudolf, six years Riley's senior, depicting two young men in school uniform sketching the Drachenfels (Dragon's Peak).

The scene very likely depicts Charles and a fellow student absorbed in their art studies. It captures an important stage in Riley's development, as he perfected his artistic talents under his teacher surrounded by his surrogate

FIGURE 1.3 Sketch entitled, "Drachenfels, 1859," by R. Hohe. From C. V. Riley, Bonn art book, 1858–59. One of the two students figured is likely Riley (courtesy of the National Agricultural Library Special Collections, Beltsville, Maryland).

family. Although Hohe urged Charles to continue his art training in Paris, an unknown family crisis in 1859 prompted him to return to England.[49]

At age sixteen, Riley faced an uncertain future. From his mother's family he acquired a penchant for literary expression in the tradition of radical dissenters who strove to improve society through exposing exploitation, blind adherence to tradition, and superstition. In his home and at school he acquired a gentleman's classical education, with fluency in French, German, Latin, and Greek, and familiarity with British, French, and German literature and cultures. Coached by Hewitson and his teachers in Dieppe and Bonn, he developed a distinctive artistic style and became familiar with the techniques of lithography. While the stages of his education as an artist and illustrator are well documented, the record of his development as a naturalist and scientist is not so clear. He expressed his growing interest in living insects in his natural history book written in Dieppe and to some extent in his Bonn art book. The cover page of the Bonn book illustrates the four stages of a butterfly common around Bonn. On subsequent pages are butterflies in association with flowers (as well as some flowers alone), in most cases with scientific names.[50]

FIGURE 1.4 "Butterfly," Charles V. Riley, *Natural History of Insects* (1857) (Courtesy of the National Agricultural Library Special Collections, Beltsville, Maryland).

Riley later referred to his observations of insects during his youth. In 1873, for example, he wrote, "In searching for insects in the winter time in England and other parts of Europe, I recollect very well, when a lad, how common the naked and suspended chrysalides [of the Cabbage Butterfly] were along the ledges of palings, and in other sheltered situations; and how a large percentage of them were always parasitized, and generally distinguishable . . . by their discolored look."[51]

Fortune suddenly forced Riley to make his own way in the world. Like many of his countrymen, he responded to the promise of America.[52] As he recalled years later, he was ripe for adventure and infatuated with America's free, democratic society. It is also possible that he wished to escape the stigma of his illegitimate birth. Although he never referred to this and, within his family, the subject remained taboo, he may have welcomed the anonymity of a new social setting where no one questioned or made insinuations about his birth. Whatever his thoughts and feelings, Riley determined that his future lay in America.

2

Along the Kankakee River, 1860–1863

The people about here are much more inventive and do much more for themselves which is natural as they have not the convenience they have in England, we make our own candles, soap, butter and lots of other things which I never knew how they were made, nor what with before. In fact since I have been here I have learnt a great many things much more useful than anything I learnt before.

—Riley to Grandmamma [Mary Miller Cannon] and Aunt [possibly Mary Miller Cannon's sister, Riley's great aunt], December 12, 1860, Folder 7C, RFR

On August 27, 1860, Charles boarded the *American Eagle* in London. He listed his occupation as "farmer." With normal weather, he and his 107 fellow passengers would reach New York in three to four weeks; however, a gale off Portsmouth and rough seas on the open Atlantic extended their voyage to five weeks and two days. Midway on the Atlantic, on September 18, Charles celebrated his seventeenth birthday.[1]

Upon his arrival in New York on October 3, the city brimmed with "Lincoln for President" placards in anticipation of the election on November 4. He quickly boarded a train bound for Chicago, the first leg of his journey to Kankakee, Illinois, fifty-six miles south of Chicago, where he would join Charles and Mary Edwards, friends of his family who had taken up livestock farming. Boarding the southbound Illinois Central train in Chicago, he soon found himself gazing at a panorama of grass and farmland extending to the horizon.[2] At Kankakee, Charles and Mary Edwards and George Edwards, a younger relative who operated an adjoining farm, greeted him. Charles

FIGURE 2.1 "Court Street, Kankakee, 1860." Left to right, Boot and Shoe Dealer, Ehrich Grocery Store, Sibley General Merchandise Store, and the three-story Hobbie Block, built in 1859 as one of the first brick buildings in Kankakee (courtesy of the Kankakee County Museum Photo Archive).

Edwards pointed out that, although Kankakee County had only been established in 1853, the county seat now had four hotels, two weekly newspapers, and a new courthouse building.[3]

Heading south and east past the French Canadian settlement of Aroma, they crossed farmland, prairie, and woodlands.[4] An early frost had produced a riot of color with yellow, orange, and crimson maples bordering the pale prairie grass. To the south, Mount Langham rose a hundred feet above the horizon.[5] To the north was the ridge bordering the Kankakee Basin that stretched some one hundred and thirty miles across Indiana and Illinois. Since the eighteenth century, French, English, and American settlers had broken sections of tall-grass prairie and cultivated crops on some of the most fertile soil on earth. Settlement accelerated in the 1850s with the construction of the Illinois Central line. The Edwards family joined immigrants from eastern and southern states and Europe to farm and ranch near the Kankakee railroad station. Charles wrote to his family that when he arrived about half the land was cultivated and half was still open range.[6] Charles soon learned that the two Edwards operations, emphasizing cattle and corn, were among the more prosperous in the area. Their combined holdings amounted to more than two hundred acres of land, more than two hundred head of cattle, plus swine, chickens, draft animals, houses, barns, and farm implements.[7]

Charles Edwards, age fifty, and Mary, age thirty-five, may have felt like substitute parents toward Charles, as they apparently had no children. George, age thirty-five, was called "uncle," though the exact kinship of the two Edwards men is unclear.

In the evenings, Riley shared the London *Illustrated Times* his mother and aunt sent him, and he and Mary and Charles Edwards took turns reading aloud the latest installments of Charles Dickens's *A Tale of Two Cities*.[8] Although churches played an important role in the Kankakee community, the Edwards clan and Charles seldom attended services.[9]

Despite the familial atmosphere, Charles was also a hired hand. He was paid a daily wage (the exact amount is unclear) and furnished with room and board. On the few occasions when Charles was sick, he forfeited his wages. Hired hands like Charles provided critical labor in the farm economy, and they could look forward to becoming independent farmers, a process described as climbing the agricultural ladder.[10]

A few months after his arrival, Charles recounted his daily round of activity: "I get up at about day-break, light the fire, milk the cows, have breakfast, mind the cattle 'till noon, have dinner, then feed and water the cattle, then have supper and spend the evening in reading or writing. . . . [I]t is just what I like, in fact, I could not have come to a place more suited for me for although the food is so different, it agrees with me and I have enjoyed first rate health since I have been here."[11] Charles reported that they had meat at every meal (beef, pork, buffalo jerky, prairie chicken, rabbit) plus buckwheat cakes, vegetables in season, and the ever-present cornbread dipped in molasses.[12]

During Charles's first month, prairie fires lit the skies nightly. Many fires were set intentionally to encourage new grass the following spring. Charles reported that the saying "he runs like wild fire" came from the fact that a fire out of control spread "faster than a horse can gallop." He and the Edwards men spent one night fighting a fire that threatened their haystack.[13]

Charles marveled at the riding skills of local boys who bounded onto their mounts and dashed off at breakneck speed, so different from the staid riding routine at the Bissett stable in Walton. With practice and a number of falls, Charles honed his own riding skills so that by November he was ready to join the fall roundup.[14] For eight months, the cattle had been roaming free. It was now time to pen them up for the winter. Neighbors who knew the terrain, the owners' brands, and the foraging patterns of the cattle joined forces for the roundup. Once the cattle were corralled, the men began harvesting corn, spurred on by signs of an early winter. Charles noted that winter on the prairie, where snow and ice accumulated to several feet and wind chilled one to the bone, was much more severe than winter in England. As though preparing for siege, the men secured roofs and siding, insulated foundations with straw, and drained water away from the cattle enclosures.[15]

In January and February, the Edwards men traded cattle at Saint Anne. The Edwards men sold their stock to local buyers rather than driving or shipping the animals to Chicago.[16] Charles fetched animals to be sold, learning how to cope with stubborn cows—some responded to a lead rope, others to a whip to the flank. When trading involved French- and English-speaking farmers, Charles helped with translation. In February, Charles helped butcher two hogs and twelve chickens. One hog they kept for home consumption; the other—weighing in at 323 pounds—they sold to a shop in Kankakee.[17]

During that first winter, George and Charles Edwards' talk of rising land prices fueled Charles's desire to own land. In his native England, where tenants and day laborers tended crops and animals on large estates, there was practically no opportunity to become a landowner.[18] In Illinois, they assured him, things were different. On February 2, 1861, they took Charles with them to Thomas P. Bonfield, the Kankakee County Superintendent of Schools, who was authorized to sell school lands. Charles Edwards bought forty acres that lay between the two Edwards farms.[19] Charles Riley purchased forty acres located a quarter of a mile east of the Edwards holdings at six dollars per acre. Charles paid for the land by signing a five-year note for $241.90 to a neighbor, Frank Coyer.[20] His status as a landowner was short lived. Eight months later he sold his parcel to Coyer for eight dollars per acre, two dollars per acre more than he paid. His reason for selling is unclear.[21] Charles's dreams of landowning may have suffered a setback but not so his entrepreneurial spirit. In May and June 1861, he bought a cow from Coyer, built a beehive, and bought another hive of bees.[22]

On April Fools' Day 1861, Charles noted, "We dug a ditch," signaling that the land had thawed sufficiently to drain marshes and bogs.[23] Mud, mud, and more mud, black and sticky, represented a hazard for livestock that sometimes foundered in the mire. Charles recorded the casualties: "Uncle's mare got down . . . Mr. [Edwards] killed her . . . A red heifer got down . . . and died."[24]

In mid-April, the men debated when to turn the cattle out. Hay was running short and spring rains transformed the barnyard into a mire of mud and manure. Charles's observation of a bull snake on April 22 convinced them that spring had arrived. Two days later they turned the cattle out to pasture.[25]

The community's preoccupation with weather and cattle almost obscured news that war had begun. On April 12, 1861, as Charles began his early morning chores, Confederate batteries fired on Fort Sumter. At day's end, he noted, "The war between N and S commenced." Earlier entries in Charles's diary indicate his awareness of the looming conflict. On March 4, 1861, he noted "President took the chair," referring to the inauguration of Abraham Lincoln, the Illinois native son who was elected on a platform opposing the extension of slavery into the territories.[26]

Having been given responsibility for breeding and birthing livestock, Charles gathered information on the subject from books and local authorities. In March and April, barnyard births came thick and fast: "Mooley had a bull calf . . . Rockcreak had a calf . . . Fanny had a mule colt . . . the old sow had twelve pigs."[27] The Edwards family owned neither jack nor stallion so Charles rode or led brood mares to sires at the Van De Karr farm nearby.[28] He kept meticulous notes of each breeding, often discreetly couched in French. On April 18, 1861, he noted that Fanny "prit de Saureau [took the bull]."[29]

With the cattle now ranging free, the men commenced planting corn, the main staple for livestock and humans. Folk wisdom prescribed that planting should begin when oak leaves reached the size of squirrels' ears.[30] On April 25, the day after they released the cattle, Charles and Charles Edwards began planting corn.[31] Unfortunately, that year folk wisdom led them astray. A late frost in early May killed the corn seedlings. On May 11, they replanted, this time with success.[32]

In early June, Charles helped Mary Edwards plant the family garden. With cash in short supply, the family depended on homegrown beans, carrots, melons, onions, potatoes, pumpkins, and squash, supplemented by meat, milk, cream, and cheese.[33] Charles's energies seemed limitless. Once the garden was planted, he turned to cultivating corn. As the summer progressed, he learned that two to three rounds of cultivation were necessary to give the corn seedlings a head start over the weeds.[34]

In June, Charles assisted in castrating the colt, Nick. Not having the Morgan bloodline, Nick was destined to remain a workhorse under bridle and harness. A year later, Nick exacted his revenge by throwing Charles, leaving him stunned.[35]

Cattle breaking through fences caused more quarrels in frontier settlements than any other, whiskey not excepted.[36] By the time Charles arrived, local law decreed that fenced-in crops were off limits to cattle. The worst happened on July 9 when the Edwards' cattle broke through neighbor Cole's fence and trampled his corn.[37]

On July 24, 1861, Charles's brother, George, arrived from England, no doubt enticed by Charles's glowing accounts of farm life.[38] Lacking his elder brother's self-confidence and adaptability, George soon drifted from one job to another. The day George arrived, the Edwards team was making prairie hay. Charles carefully studied the art of building compact, wind resistant, and water repellent stacks before building his first one.[39]

In September, between summer cultivation and fall roundup, Kankakee folks gathered at the county fair to socialize and exhibit their jams, jellies, preserves, needlework, quilts, vegetables, and livestock.[40] Charles entered their classiest horse, Fanny, a sow, a pen of chickens, and some artwork. The livestock failed to place, but he won prizes for his painting, "The Turk," an Ottoman warrior astride his galloping steed that he had drawn in Bonn, and for his pencil drawings.[41]

On September 18, 1861, Charles came of age. Apparently there was no special celebration, only a note in his diary, "My birthday, 18 years old."[42] On October 9, noting, "I have been here one year," he took an inventory of his economic status.[43] Tallying up purchases, trades, and loans, he figured his net worth was $77.69.[44]

While his earnings were modest, so were his expenses. The Edwards family provided room and board so he needed cash primarily for clothing, entertainment, and special items like beehives and a microscope. Shortly after his arrival, he paid George Edwards's housekeeper $1.25 to sew him some rugged overalls and shirts, and later he bought a coat for five dollars and a hat for one dollar. Community dancing, skating, and candy pulling were free with the exception that benefit dances cost fifty cents.[45] Many of his possessions were obtained through barter. As he informed his family, "the people here do not use money very much, they exchange things and call it trading."[46] In October 1861, he traded one of his heifers for a microscope.[47] In December 1862, he traded a colt to Joseph Legg for a better colt named Fly, plus ten dollars and a hive of bees.[48] At the end of his second year, he was confident that his speculative ventures (amounting to the equivalent of twenty-five dollars) would pay off.[49]

Charles was maturing intellectually and physically. He noted in February 1862, "I shaved for the first time." Later that year, Charles Edwards gave him his first bottle of after-shave cologne.[50] He was appointed secretary of the debate society that met at the school. Friends and neighbors bought his paintings, and French-speaking neighbors asked him to help write and translate letters.[51]

Following the April rains in his second spring (1862), Charles helped Mary Edwards plant flowers, vegetables, and fruit trees.[52] Recalling groomed gardens in England, they hoped to enliven the monotonous flat terrain with flowers and trees reminiscent of their homeland. Charles noted that "the same fruits grow here as in England, some things are more plentiful than they are there."[53] In June, they harvested an abundance of strawberries.[54]

In summer 1862, the Seventy-Sixth Illinois Volunteer Infantry Regiment was recruited in Kankakee and Iroquois counties. A number of Charles's friends volunteered, reporting for duty at the oak-shaded fairgrounds a mile north of Kankakee, where a fine water supply created an ideal campsite. Companies D, F, G, H, and I were drawn from Kankakee; Company D was composed largely of French-speaking recruits.[55] The regiment was ordered to proceed by rail to Columbus, Kentucky. The night before their departure, Charles noted in his diary, "The railway track was pulled up between Kankakee and Shebance [Chebanse] by Rebels."[56] The sabotage failed, and the troops proceeded to Columbus. During three years of duty, the regiment suffered heavy casualties, in particular at the Union siege of Vicksburg.[57]

With the departure of young recruits, Illinois farmers relied more heavily on farm machinery like reapers, mowers, and threshers. Mobilization of the Union Army called for increased production of corn, wheat, beef, pork, and draft animals. Illinois, already the leading producer of farm commodities and machinery, increased its lead. Whether the Edwards men acquired new machinery is not recorded, but they increased their production of cattle and other livestock, particularly draft animals.[58]

When time allowed, Charles painted landscapes, portraits, and livestock.[59] Winter was also the season for séances, popular gatherings where participants claimed to communicate with deceased relatives and friends. Charles considered this nonsense. To Uncle Tim, who believed in spiritualism, he wrote, "[W]hen winter comes and people sit by the fireside . . . the Spirits return. . . . [W]hen there is plenty of work [during the spring and summer] . . . one seldom sees or hears any of the [spirits]."[60]

Charles's second full year on the farm, 1862, ended on a note of uncertainty and growing restlessness. Torrential rains in the summer and fall transformed the prairie and fields into swampland, delaying planting and harvest. Charles spent day after day herding cattle through slush and mud into their winter quarters. In mid-December, he lay sick in bed for three days and by year's end he had not fully recovered. In addition, the year had brought several disquieting events. One of his beehives was stolen. Equally unsettling was the discovery of a foundling baby early one morning on a neighbor's doorstep. Just before Christmas, a little girl from a neighboring farm died of scarlet fever.[61]

The monotony of farm chores and the concerns of the local community suddenly seemed too confining to a young bachelor with ideas, talent, and ambition. Chicago, not far away, beckoned. Despite newspaper editors' rosy picture of farm life, young Americans knew, from novels, word of mouth, and common sense that the future lay in the city. Romantically inclined newspaper editors might portray farm boys as potential independent yeomen, but many young men suspected that those who stayed on the farm might be regarded as hicks, hayseeds, or country bumpkins.

As 1863 began, Charles lay in bed, probably entertaining such thoughts. Still weak from his illness, by the fifth of January, after husking corn and performing chores in rain and mud for four days, he decided to move to Chicago.[62] Charles had only limited experience of Chicago, based on his transfer in the train station on his way to Kankakee, and a three-day visit in September 1861.[63] Years later, Charles's colleague, George Brown Goode, explained his decision in terms of frustrated ambition and overwork: "Young Charles was simply too enthusiastic and too bent on excelling in everything. He took no rest. Often he would be up . . . getting breakfast . . . and milk half a dozen cows before the others were about. When others were resting at noon . . . he would be working at his flowers under a July sun . . . He kept a lot of bees, got hold of the best bred colts and some of the best heifers in the county, secured a good quarter section [*sic*] and spent his Sundays reading, sketching, and studying insects. Three years [*sic*] of this increasing effort . . . broke [his] health."[64]

Having made his decision, Charles moved expeditiously. He collected $20.00 of the $81.50 Charles Edwards owed him and got a note for the remainder payable in six months. He settled his debt to Mr. Legg by transferring his cow Effie, a hive of bees, three sacks of corn, and eight dollars. He left his colt, Fly, in Legg's care. Another calf, Jenny, he left with George Edwards, along with his boots and his kerosene can.[65] Two items among his possessions indicated his future as an entomologist: his microscope, worth eighteen dollars (close to a month's wage), and his insect collection that he left with Charles Edwards who later sent it to him in Chicago.[66] With mixed feelings he succumbed to whatever restlessness called him to Chicago.

Throughout his life, Charles referred with pride to his years on the farm. Like his references to insects he observed as a boy in England, his references to insects he observed during his Kankakee years indicate a sophistication not evident in his diary and letters. In 1873, as Missouri state entomologist, he wrote: "[The Buck Moth] was one of the first acquisitions to my cabinet. . . . During a farmer's life . . . in Kankakee County Ils., it was my fortune to spend many a day in the so-called 'oak-ridges' lying along the Indiana line. Here, late in the months of October and November—when the still and hazy atmosphere, and the somber brown of thc craggy oaks, boded so eloquently the coming of cold . . . when the rustling leaf under the horse's tread, or the modulated echoes of the woodman's ax were the only sounds of life, and animated nature seemed to have been wooed to Lethean slumber—this crape-winged moth would often flutter by as though loth [*sic*] to follow in the general sleep."[67]

Charles's years on the farm represented a rite of passage. Coming as a boy with a classical education and talents in writing, sketching, painting, and collecting and rearing insects, in Kankakee he came of age, mastered the skills of farming, won a place in the community, and acquired a deep sense of loyalty to rural people struggling to cope with the vicissitudes of farm life. Now he turned to Chicago and the future.

3

Chicago and the *Prairie Farmer*, 1863–1868

Wanted: A situation as Clerk or Salesman in some store or office, either in this city or in some country town by a young man between nineteen and twenty years of age. Understands thoroughly both the French and German languages and can give good references. Address: C. V. Riley, Post Office Box 819, Chicago, Ill.

—*Chicago Tribune*, March 23, 1863

At four o'clock Tuesday morning, January 12, 1863, Riley stepped off the train at Chicago's gas-illuminated Central Station. His departure from Kankakee having been delayed five hours by a train accident, he slept in his seat until arriving in predawn darkness. After arranging for room and board with Mr. Buck at the Union Hotel, he made the rounds of advertised openings on South Water Street, only to be told time and again to come back the next week or that a young man had just been hired.[1]

After two weeks of pounding the pavement in Chicago's bitter cold, he found a job collecting subscription bills for the *Evening Journal* at $3.50 per week, exactly what he paid for room and board.[2] Several weeks later, still waiting for money owed him by Charles Edwards and others in Kankakee, he quit his job at the *Journal*, sold his hat and two razors to pay back a loan from a friend, and resolved to live on "hope."[3] Placing an advertisement to cure corns in the *Chicago Tribune* under the pseudonym, L. Lafargue, he received one response at 50 cents, not enough to cover the $1.50 he spent for the advertisement.[4] He placed another advertisement in the *Tribune*: "Wanted: A situation as Clerk or Salesman . . . by a young man . . . Understands . . . French and

German."[5] Eight days later, on March 31, 1863, Henry D. Emery, editor of the *Prairie Farmer*, offered him a job beginning April 1. At the time, Emery was planning French and German editions of the *Prairie Farmer* and was probably intrigued by Riley's fluency in languages.[6] With a weekly salary of five dollars and having received forty dollars from Charles Edwards, Riley spent thirty dollars—the equivalent of ten weeks of room and board—for a suit of clothes, a linen overcoat, a hat, and refurbished boots.[7] The farm lad turned dashing man-about-town was now equipped for adventure in the city.

Though dwarfed in comparison to London's three million inhabitants, Chicago's mushrooming population of one hundred twelve thousand offered a fully developed, albeit frontier-style, urban ambience. Families of diverse ethnic backgrounds worked and lived in wooden houses and factories rather than in buildings of brick and stone. Uneven boardwalks paved with pine blocks rather than cobblestones did little to guard against ankle-deep slush in winter or pulverized dust in summer. Wooden ships jammed the docks and wooden bridges spanned the Chicago River that served as sewer for Chicago's slaughterhouses. Chicago, now the foremost rail junction in the United States, boasted a total of fourteen theaters, opera houses, and concert halls; forty-nine fraternal lodges; dozens of churches; eighty-four newspapers; three colleges; and an academy of science.[8]

With his roommate, Alan Bowersack, Riley attended shows, often with free admittance, at Kinsbury Hall and Canterbury Hall. After attending the minstrel show by Arlington, Kelly, Leon & Donniker in Kinsbury Hall, Riley reported being "very much amused."[9] One day, he saw elephants from Mabie's Menagerie, an animal show originating in London, meandering down the street. Another day he attended the German theater. He bought two Dickens novels, *The Old Curiosity Shop* and *Nicholas Nickleby*, for ten cents each.[10] He attended church for the first time in two years. He also took time to paint portraits, landscapes, and buildings.[11]

Emery quickly recognized that his new employee was a precocious linguist, a talented artist, an experienced all-around hand, and a budding entomological genius. Emery had bought the *Union Agriculturist and Western Prairie Farmer* in 1858 and merged it with *Emery's Journal of Agriculture*, shortening the title to the *Prairie Farmer*. Through skillful combination of reader input and professional discussion of farm issues, he set the standard for farm journalism in the Midwest for decades to come. Emery assigned his new employee to rearranging the office, helping Mrs. Emery with their vegetable garden, drawing house and barn plans for publication, and designing a circular for distribution to subscribers.[12] Emery encouraged Riley's entomological ambitions by assigning him to write articles on the housefly, the Pine-leaf Scale, and the May Beetle, complete with illustrations. He purchased a microscope for forty dollars, the equivalent of two month's pay, for Riley's use.[13] After two months he raised his weekly salary to six dollars.[14]

A week after joining the *Prairie Farmer*, Riley met Isaac A. Pool, curator of insects at the Chicago Academy of Sciences. Through Pool he met other academy entomologists including George Hathaway, Charles Sonne, Andrew Bolter, and Robert A. Kennicott.[15] Riley sold the insect collection he had begun in Kankakee to Pool and commenced making another. The new collection grew rapidly, enriched by weekly field trips with fellow collectors.[16] From dealers in Chicago, Philadelphia, and London, Riley acquired specimen boxes, pins, a collecting net, and other accessories. He constructed breeding cases to rear insects in captivity, a practice he had begun at Dieppe, and recorded the results in an entomological journal.[17] In February 1864, academy president Eliphalet W. Blatchford invited him to a dinner in honor of Louis Agassiz, who was in the city to deliver a series of lectures.[18] After meeting Agassiz that evening, Riley attended two lectures, noting in his journal "if everyone went home as satisfied as I did, they had their money's worth."[19]

Emery's new entomologist-artist soon came to the attention of Benjamin D. Walsh, the former Cambridge classmate of Darwin who immigrated to Illinois in 1838 and settled in Rock Island near the Iowa border in 1851. In 1857, Walsh retired from farming to devote himself to the study of insect pests of crops and orchards. In June 1863, he asked Riley to draw some insects for a publication, initiating a partnership that lasted six years, until Walsh's untimely death.[20] Curiously, during his Chicago years, Riley seems to have had no contact with William LeBaron and Cyrus Thomas, two other Illinois agricultural entomologists. LeBaron had written on injurious insects since the 1850s. In 1860, Walsh and Thomas presented reports on injurious insects to the Illinois State Agricultural Society. In 1861, Walsh, Thomas, and John Hancock Klippart, another Illinois naturalist, debated the causes of the Armyworm infestation.[21]

Riley also contacted John L. LeConte, America's leading coleopterist, and other members of the Entomological Society of Philadelphia, as well as Asa Fitch, New York state entomologist.[22] LeConte responded immediately, but Fitch, a notoriously poor correspondent, failed to answer his letters.[23]

Encouraged in his entomological endeavors by his employer and his scientific associates, Riley wrote to his uncle Tim: "Had I a fortune, I know I should become a great naturalist and that most of my time would be spent in painting nature[,] for I have an inherent love for it." He concluded, "If I live long enough, I shall someday have a very interesting and useful book."[24]

Three months after starting at the *Prairie Farmer*, in July 1863, Riley joined the Independent Order of Good Templars. This fraternal order, based on the principle of temperance, originated in New York State in 1851 and spread rapidly to Chicago and other midwestern cities. During his five years in Chicago, the Templars was the center of Riley's social life.[25] Riley attended various lodge meetings before joining Houston Lodge, where he passed through the first, second, and third degrees to become a full member. At Houston he served as secretary for two years, then as lodge chaplain. Templar lodge meetings combined business with socializing. Initiation involved a pledge to abstain from the consumption, production, and distribution of alcoholic beverages. At the meetings, singing, prayer, and planning future events promoted fraternal and social cohesion. Badges and secret handshakes facilitated recognition among members.[26] The Templar's central tenet of abstinence may have struck Riley, of British upbringing, as a bit strange, but he honored his vow.[27] At Templar meetings, picnics, dances, fundraising, and essay nights, Riley met many eligible young women and was especially attracted to Emily St. John, whom he often accompanied home after Templar events.

During his first year and a half in Chicago, Riley was so caught up in his work, his social activities, and his budding romance with Emily that he rarely noted the war in his diary. On April 6, 1863, he reported the news that Charleston had fallen.[28] In July 1863, when news reached Chicago that Union forces had taken Vicksburg, he joined the celebration that was highlighted by bonfires and fireworks.[29]

With the Mississippi River now under Union control, General Ulysses S. Grant transferred the bulk of Union forces to eastern Tennessee and Georgia to challenge the Confederate plan for an offensive in 1864.[30] Meanwhile, the governors of Illinois, Ohio, Indiana, and Iowa called for volunteer infantry troops for one hundred-day service to control areas that were still subject to raids by Confederate guerrillas.[31] When in early April 1864 Illinois governor Richard Yates called for twenty thousand volunteers, Riley vouched for his Kankakee friend Charlie Boswell so that Charlie could collect his enlistment bounty.[32] Ten days later, on May 8, 1864, Riley left his job at the *Prairie Farmer* and, together with several Templar friends, enlisted for one hundred days in Company C, 134th Regiment of the Hancock Guards, Illinois Volunteers.[33] Posing in uniform, he sent one picture to his mother and gave another

FIGURE 3.1 Charles V. Riley portrait, taken upon his enlistment in Company C, 134th Regiment of the Hancock Guards, Illinois Volunteers, May 1864 (Riley Family Records, National Agricultural Library Special Collections, Beltsville, Maryland).

to Emily. When Emily in turn gave him her picture, he confided in his diary, "she promised to be faithful and true to me."[34]

On May 30, with thirty dollars enlistment pay in hand, Riley joined a raucous crowd for a night in the city.[35] Stationed for two months near Chicago, he thrived on army rations of beef, salt pork, bean soup, bread, potatoes, and strong black coffee. With overnight passes, he and his friends attended Templar meetings. Visits from Emery, Corbett, Emily, and brother George kept his spirits high. On one visit from George, Riley forthrightly advised him not to join the army.[36]

On June 3, 1864, Riley's regiment departed for garrison duty in Columbus, Kentucky, four hundred miles south on the Mississippi River. During the two-day train trip to Cairo, Illinois, at the confluence of the Mississippi and Ohio Rivers, Riley saw the first signs of war. At a deserted farm between Rantoul and Urbana, four teams of steers pulled a plow with no driver in sight and at Cairo stores were closed and the streets largely deserted.[37] As the steamship *Nevada* rounded a big bend in the Mississippi, they arrived two hours later at Columbus. One evening from their camp on a bluff overlooking the river, Riley described in his journal the panorama of seven steamboats docked in the moonlight like brightly lit floating palaces with smoke stacks belching sparks in the afterglow of the sunset. Music from a calliope melded with voices singing a familiar refrain.[38]

During Riley's approximately three-and-a-half-month enlistment, camp life assumed a familiar routine. Although signs of hostilities were visible, the volunteers encountered only sporadic contact with guerrillas. One day in Columbus, Riley noted bullet-ridden trees that he learned resulted from a battle in 1862 when Union forces captured Columbus and the fortified stronghold of Island 10 that blocked Union navigation downriver.[39] On patrol one day, Riley and his companions found only blacks working the fields, the masters having fled.[40] One evening the troops responded to cries of "Rebels, Rebels!" Within five minutes the regiment formed into battle lines, only to learn that the colonel had sounded an alarm to see how quickly they could assemble. In the excitement some soldiers appeared without trousers, while others discharged their weapons by mistake.[41]

For recreation, Riley and his companions took early morning dips in the Mississippi, on occasion swimming from the Kentucky to the Missouri shore and back.[42] At the outset, when the regiment was near Chicago, Riley and fellow Templars organized a regimental lodge. Although Riley was pleased with the new lodge, it lacked one vital ingredient: young ladies. When the worthy chaplain read from the order of service "here one sees males and females," Riley and his companions hooted with laughter.[43]

During his enlistment, Riley corresponded with Pool, Hathaway, and Walsh and also with officials at the Smithsonian Institution.[44] While stationed in Columbus he met an entomologist named Botter, who, Riley noted, knew all the entomologists in Chicago. He also met an avid insect collector named Sherman. Later in Chicago, he met General Hathaway, who invited Riley to his quarters and helped him identify insects Riley had captured.[45] Shortly after enlisting, he bought a bottle for his captures, and later, in Columbus, he purchased brass wire to make a net. Riley collected insects near Columbus and he later recalled observing the beetle and larva of *Doryphora juncta*, a relative of the Colorado Potato Beetle, feeding on a species of wild potato near town.[46]

In the heat of July and August, Riley took ill, contracting diarrhea, then stomach and intestinal flu, then dysentery. Unable to walk, he was carried to the post hospital in Columbus where he contracted typhoid. An open abscess on his left ear became so swollen with pus that he had to sit in one position and couldn't open his mouth to eat or talk.[47] Riley recorded, "I don't know how I lived[,] but I never lost courage."[48] After a week in the hospital, and fortified by dosages of "Drakes Plantation Bitters" and "Ayers Pills," he began to recover.[49]

While Riley was confined to the hospital, his regiment departed for Mayfield, Kentucky, on the south bank of the Ohio River. On August 12, he and other discharged patients followed them on the steamer *Columbus*, arriving in Cairo that afternoon. Fearful of missing the steamer's departure at daybreak, they spent the night on ship, plagued with clouds of mosquitoes. At Paducah, the army doctor was astonished to see him alive, but by the time he rejoined his regiment at Mayfield in mid-August, he was fit for service.[50]

At the beginning of their duty in Mayfield (August 21 to September 19), guerrillas frequently fired on pickets; however, after the cavalry captured and executed a number of guerrillas, sentry duty was quiet.[51] One afternoon, when Riley had watch from three to six p.m., he spent the hours reading in the shade before returning to his boarding house for supper. Resuming guard duty from midnight to three a.m., he slept soundly at his post, apparently with no fear of being surprised by enemy combatants or his superiors.[52] Evidently he viewed military regulations as guidelines to be interpreted in light of current circumstances, rather than as God-given commandments. At Mayfield, Riley's regiment worked with pickaxes and spades, side by side with conscripted secessionists, to erect a fort with emplacement of a cannon. Riley enclosed a detailed sketch of the fortification in the pocket of his war diary.[53]

As the end of their hundred-day enlistment approached, morale sagged and discipline weakened. Riley went foraging for mushrooms.[54] When a rumor circulated that the tour of duty would be extended for thirty days, unrest bordering on mutiny broke out among the troops. The commanding officers urged them to join a fifteen-day campaign to rout twenty-five hundred Confederate troops near Dresden, Tennessee. Most of the officers were willing, but the privates voted overwhelmingly to go home. General Meredith attempted to intimidate Riley's regiment by ordering it to march without firearms between the armed New Jersey 34th and a regiment of African Americans.[55] With discontent among the troops mounting, regimental officers gained permission from Governor Yates to return to Chicago. On September 19, they departed by train, and after travelling two days and nights in open cars with no protection from hot cinders, rain, and hail, they arrived in

Chicago.[56] Having been granted a five-day pass, Riley made a short trip to Kankakee and Aroma.[57] Returning to Chicago, he found that his company was joining a brigade of volunteers from four other Illinois regiments for duty in St. Louis. Upon his request, the company doctor extended his sick leave until October 15, and with two weeks of free time he resumed working at the *Prairie Farmer*.[58] The week he returned to work, Emery and Corbett departed for a fair in Indianapolis, Indiana, leaving Riley to manage the office and issue the payroll, and they raised his pay to twelve dollars per week.[59]

Riley spent evenings at the theater or at lodge meetings, often in Emily's company. Riley soon learned that the course of love sometimes encounters rough waters. Emily was indignant that Riley had not visited her immediately upon his return. As the month of October wore on, he concluded that brother Cook, "smitten by Emily's charms," was taking his place.[60]

Riley's regiment was formally discharged on November 3 but recalled for emergency duty on November 7.[61] Responding to reports that Confederate sympathizers planned to free five thousand Confederate prisoners at Camp Douglas outside Chicago and another seven thousand prisoners at Rock Island, Illinois, Colonel Bigelow ordered the regiment to assemble for duty. Rumors circulated that Confederates planned to burn Chicago after releasing the prisoners. Although Confederate sympathizers claimed eighty-five thousand supporters in southern Illinois, a much smaller number joined the raid. In the end, Union forces captured the leader, General Marmaduke, and fifty guerrillas, along with twenty-nine hundred pistols, many carbines, and several hundred pounds of powder.[62]

Riley strongly supported the Union.[63] Like many immigrants, he felt called to defend the American Union and its Constitution. In July 1864, in a report sent to the *Prairie Farmer* from Columbus, Kentucky, he deplored the fact that southern Union supporters had to flee their homes, and he criticized war profiteers.[64] In a letter to his aunt and uncle, he praised the sacrifice of friends who had died. To those who claimed the war effort had failed, he replied, "let the fathers, mothers and sisters in the country who have seen those dearest to them pour out their life['s] blood . . . decide whether it is a failure. Let the freed Negro tell!"[65]

Riley returned to the *Prairie Farmer* as a full-fledged member of the editorial staff with responsibility to develop the entomology section. This involved advising farmers and fruit growers across the Mississippi Valley who were plagued with insect pests. Following Emery's tested formula of combining reader and staff input, he organized the entomological section as a series of questions and answers. Riley addressed all members of farm families in his column: housewives like Mary Edwards who were trying to overcome the bleak farm surroundings by cultivating flowers, young people whom he encouraged to study natural history, and farmers who combated insect pests that threatened the family's livelihood. Variously entitled "Entomological Notes" or "Queries Answered," his column centered on practical advice based on his own investigations. For example, his "Queries Answered" for September 7, 1867, contained clear, original, and entertaining commentary (slightly edited here):

> Q: What can I do to save my crop of plums from worms? [specimens enclosed]
>
> A: The "worms" are larvae of the curculio. Check back issues of the *Prairie Farmer* for its description and life cycle. It is probably too late to save this year's crop. Try jarring trees and destroying the adult curculios next spring.
>
> Q: In reply to your request for samples of hop vine caterpillars, I am sending eggs and larvae of several insects found on my hop vines.
>
> A: The eggs are *[M]antis carolina*. They prove that the mantis breeds as far north as Ogle County. The largest worm is the common Hop Vine Hypena, of the second brood. Have placed them in breeding cages where they have spun cocoons, and will emerge as moths in about two weeks. As remedy, I recommend the application of soapsuds and hand picking. The other worm is unknown to me, but I have placed it in a breeding cage where it has spun a cocoon. I will be able to inform you of its identity in a few weeks.
>
> Q: Am sending two boxes of various worms found in my granary. What are they and what can I do to rid my granary of them?
>
> A: The two from the small box are very different, one producing a beetle, the other a butterfly. The worms from the large box are larvae of the "meal moth" which I have learned this year will feed on clover as well as grain. As remedy, I recommend heating the wheat to 200 degrees Fahrenheit.
>
> Q: Am sending small grubs that are destructive to young plants. They eat the roots of trees, corn, etc.
>
> A: The insect is unknown to me, but it is closely allied to the apple tree borer. I will breed the adult insect from the grubs and give you an answer later.[66]

Like Fitch and Walsh, Riley grounded his philosophy of insect control on the balance of nature in the tradition of Kirby and Spence. According to

Fitch, Divine Providence would not have placed an insect species in the world without endowing man with sufficient intelligence to discover its vulnerabilities, exploit them, and thereby maintain man's ordained dominion.[67] Walsh described how the balance operated: Spiders that preyed on insects were stung by mud wasps that placed their paralyzed victims in their nests as food for their larvae. Mud wasp larvae in turn became food for ichneumon flies that deposited their eggs in their larvae. Walsh, the apostate churchman, agreed with natural theologians that: "the whole system [is so skillfully] balanced . . . that in a state of nature . . . it is only in . . . special seasons, and in certain localities, that a particular insect becomes unduly numerous."[68]

Riley urged farmers and horticulturists to help restore nature's balance through four methods: biological (primarily through encouragement of predator insects), cultural (crop rotation, planting resistant varieties, and planting early or late to avoid insect pests), mechanical (plowing, rolling logs over fields to crush hoppers, and erecting barriers to prevent the migration of Chinch Bugs), and chemical (hot lime, limewater, brine, tobacco water, wine, turpentine, benzene and sulfur, salt, kerosene, and other caustic substances). The advent of Paris Green and other arsenic insecticides came later, about the time Riley moved from Chicago to St. Louis in 1868. Riley recognized that the recommendations of Harris and others that were developed for agriculture in the northeastern United States were poorly suited to large-scale market farming in the new West. Hand picking of caterpillars, while appropriate to home gardens, was impractical on large acreages. Furthermore, farmers in the west encountered insect pests unknown to eastern agriculture. The Chinch Bug, for example, although originating in the South-Atlantic states, became a major pest in the extensive grain fields of the Mississippi Valley.

Riley also assisted farmers by exposing entomological nonsense in the agricultural press. In response to a proposed advertisement for "Smith's Patent Curculio Trap" that claimed that the Curculio had no wings and could not fly, a fallacy anyone could refute by observing the winged insects, Riley denounced the inventor as a "quack" and his machine as "humbug," and refused to print it. Riley concluded, "We have on hand half a dozen circulars of the same stamp, believing them not worth the trouble of . . . denouncing."[69]

Following his return to the *Prairie Farmer*, Riley turned out articles and illustrations in ever-increasing tempo. Prior to his enlistment in the army, he published six articles. In 1865, the first full year after his volunteer service, he published nineteen articles. The following year, 1866, he published forty-three articles, seven of which were in other journals. In 1867, his last full year in

Chicago, he published fifty-eight articles, two of which were in other publications. In the first three months of 1868, before his move to St. Louis, he published seventeen articles and his total output for that year reached sixty-four. On average, during his last two and a half years in Chicago, Riley published more than one article per week, an astounding output for a neophyte author. Some of these publications were quite extensive. In January 1867, for example, in the *Prairie Farmer*'s first annual *Agricultural and Horticultural Advertiser*, Riley treated the Sweet Potato Beetle, the Grape Leaf Folder, the Tomato Fruit Worm, the Stalk Borer, the Phlox Worm, and the Clover Worm.[70]

Riley's insect illustrations became his hallmark. He believed that "Written descriptions . . . are of little use to the Horticultural and Agricultural community; whereas, a good illustration is at once a substitute for the text, impressing those unversed in technical terms, far more correctly."[71] During Riley's Chicago years, he published approximately 125 drawings, as accurate and attractive today as they were over a century and a half ago.[72] Riley's articles also included systematic reviews of entomological literature relating to a species, based on books at the *Prairie Farmer*, the library of the Chicago Academy of Sciences, Smithsonian publications, and literature furnished by John L. LeConte and others at the Entomological Society of Philadelphia.

During the war, demand for food, fiber, and draft animals for the Union Army led to an enormous expansion of acreage and extensive cultivation of crops that in turn produced what one historian has called an "insect emergency."[73] In 1862, the president of the Horticultural Society announced that "an . . . innumerable multitude of . . . insect . . . enemies must be met and vanquished."[74] Five years later an Illinois horticulturist announced, "[T]he great questions . . . are not whether to plant, or what to plant, but how to protect what has been . . . planted from . . . insect enemies . . . and diseases."[75] Following the end of the Civil War, Illinois horticulturists spearheaded a drive to establish an office of state entomologist and to name Walsh as state entomologist.

Horticulturists were familiar with Walsh through his appearances at society meetings, his reports on insects to the society, his articles in the *Prairie Farmer* and elsewhere, and through his entomological exhibits at fairs. In 1865–66, Walsh served first as western editor then as editor-in-chief of the *Practical Entomologist*, published by the Entomological Society of Philadelphia. In 1866, the Illinois State Horticultural Society and the Illinois State Agricultural Society called for Walsh's appointment as state entomologist.[76] A bill authorizing Walsh's appointment was passed in 1867 but had to await senate confirmation that came a year later. The horticultural society, unwilling to wait, authorized payment of $500 in May 1867, so that Walsh could begin work. He began preparing his first annual report, devoted to insects of fruits, which was issued in 1868. The following year, the Illinois legislature confirmed Walsh's appointment and paid his salary retroactively.[77]

Whereas Walsh was the chief spokesman for the Illinois State Horticultural Society, his protégé Riley became the primary advisor for fruit growers in southern Illinois and neighboring Missouri. In 1866 when Walsh and Riley exhibited collections at the Illinois state fair, Walsh displayed one case of insects while Riley, with neophyte enthusiasm, displayed five cases, complete with oil paintings depicting the metamorphosis of harmful Lepidoptera on their host plants.[78] Whereas Walsh and the Illinois State Horticultural Society addressed the concerns of horticulturists and agriculturists statewide, Riley spoke to the special concerns of fruit growers in "Egypt," the moist, fertile, and mild southern tip of Illinois.

Geographically and horticulturally, "Egypt" comprised the Illinois sector of a large fruit-producing region that extended across the Mississippi River to St. Louis and its hinterland. The horticulturists east and west of the river knew each other and participated regularly in meetings of their sister societies.[79] During the 1860s, Illinois and Missouri growers produced about the same quantity of fruit. Although Missouri's climate and geography offered a better environment for fruit growing, Illinois soon took the lead in the transportation and marketing of fruit, and Chicago surpassed St. Louis as a center of fruit distribution.[80]

In the early 1860s horticulturists in "Egypt" organized as the Southern Illinois Fruit Growers Association. At the same time, fruit growers at Alton, Illinois (across the river from St. Louis), organized as the Alton Horticultural Society. In the years 1866–68, Riley met frequently with growers of both organizations, as well as with the fruit growers in St. Louis.[81] During a memorable three-day meeting of the Southern Illinois Fruit Growers Association in Cobden, Illinois, in December 1866, Riley discussed insects attacking fruit trees in that region, including the Apple Tree Borer, Codling Moth, Curculio, and aphids. At the close of the meeting, the fruit growers adopted a resolution requesting Riley to continue his investigation of insect pests in their region and to report his findings the next year. They passed a second resolution thanking Riley for his participation in the meeting.[82]

In 1867, the Missouri State Horticultural Society petitioned the Missouri General Assembly to create an office of state entomologist.[83] Although unsuccessful that year, with additional support from the Missouri State Board of Agriculture and the St. Louis Academy of Science, and with the urging of Walsh and Norman J. Colman, editor of *Colman's Rural World* (St. Louis), the bill establishing the office of state entomologist was passed in early 1868.[84] Riley was the candidate of choice. A Riley supporter wrote in *Colman's Rural World* in December 1867: "Chas. V. Riley, a young and enthusiastic entomologist is a candidate . . . for the office of state entomologist. . . . If he weren't so occupied with his studies as a naturalist, his linguistic talents would soon secure him a professorship in almost any institution of learning. . . . He is an honorary member of all the leading societies of naturalists, and has contributed considerable [*sic*] to science by his researches, experiments, and studies."[85] On April 1, 1868, exactly five years from the beginning of his employment at the *Prairie Farmer*, Riley took up his duties as the first Missouri state entomologist.

4

Missouri State Entomologist, 1868–1877

The prosperity of a State does not depend solely on its material wealth, but to a greater extent on its mental wealth. Knowledge—that great interpreter of oracles—moves the world!

—C. V. Riley, *First Annual Report* (1869), 172

Having worked with fruit growers in and around St. Louis, Riley was familiar with his new hometown. With its large foreign-born contingent and industrial-urban character, St. Louis contrasted sharply with the Missouri hinterland. In the 1850s, railroads and immigrants had transformed the former French fur-trading center into the leading metropolis in the Mississippi Valley. At that time, 60 percent of St. Louis citizens were born outside the United States, the highest proportion in any American city, and 25 percent were German immigrants. Steamboats along the mile-long waterfront discharged bulk cotton to be pressed into bales for shipment to New York textile mills. Texas cattle and Arkansas hogs crowded stockyards for shipment to Chicago. In south St. Louis, blast furnaces with flames suggestive of Dante's inferno reduced lead, iron, and copper ore to molten metal. St. Louis had paved streets, a progressive school system, colleges and trade schools, and Shaw's Garden (later the Missouri Botanical Garden). Prior to the Civil War, few St. Louis citizens owned slaves, whereas the majority of farmers in rural Missouri had owned one or more slaves. St. Louis overwhelmingly supported the Union, whereas rural Missourians favored the Confederacy. In matters of religion, St. Louis, with its prominent German-Irish Catholic and large Unitarian congregations, contrasted with Baptist-Methodist rural Missouri.[1] St. Louis even

Featured Insect

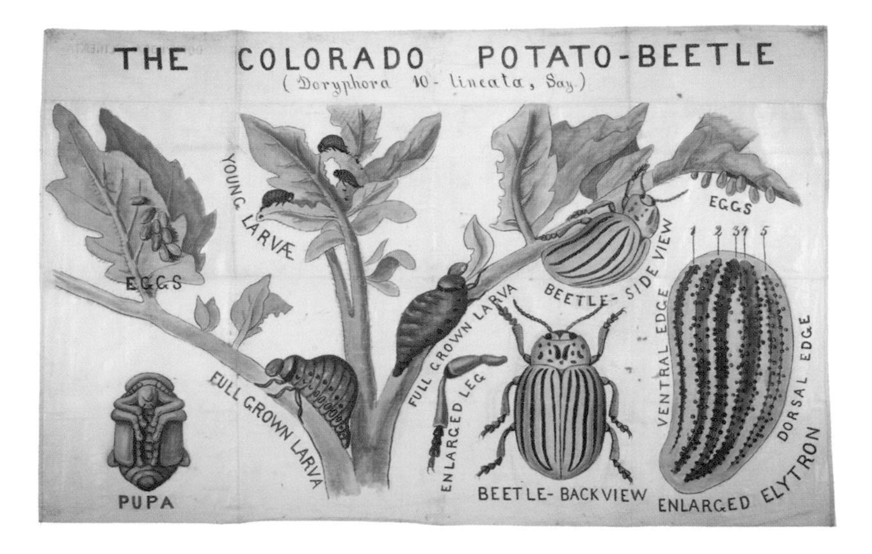

Colorado Potato Beetle [current scientific name *Leptinotarsa decemlineata* (Say) (Coleoptera: Chrysomelidae)]. Riley Insect Wall Chart (courtesy of Morse Department of Special Collections, Kansas State University Libraries). In 1859, with its acquisition of potato as a new plant host in Nebraska Territory, Colorado Potato Beetle spread east in a spectacular invasion, reaching the Atlantic coast before 1880 and resulting in catastrophic damage to potatoes, a staple crop in North America. It remains a major pest of potatoes (and also of eggplant and, in some locations, tomatoes) in North America, and in Europe and Asia because of its introduction to France in 1922 and its subsequent spread eastward to China through the end of the twentieth century. Its evolution of resistance to each insecticide used against it is perhaps the best example of pesticide resistance known to entomology.*

*Donald C. Weber, "Colorado beetle: pest on the move," *Pesticide Outlook* 14 (2003): 256–59.

had a prosperous African American elite, dating back to pre–Civil War days.[2]

Riley may have wondered how farmers in backcountry Missouri, where Jesse James's Confederate band still roamed, might greet an English immigrant and Union Army volunteer. He need not have worried. Western Missouri grain and cattle farmers welcomed him with the same enthusiasm as fruit growers around St. Louis in their battle with their "tiny but mighty insect foes."[3] Their cooperation boded well for future campaigns against the Colorado Potato Beetle, the Rocky Mountain Locust, and other insect pests.

With his appointment in 1868, Riley became the nation's third state entomologist. Fitch, who had served in New York since 1854, provided some precedent, but the responsibilities and activities associated with the office of state entomologist were largely up to the person who assumed the office (Walsh in Illinois had been in office only one year). Fitch was thirty-four years Riley's senior, declining in health, and a semi-recluse at "Fitch's Point," surrounded by his books, microscope, and breeding cages. He rarely answered letters and only reluctantly journeyed to agricultural meetings. Riley, on the other hand, steadily expanded his circle of correspondents and participated in so many horticultural, agricultural, and scientific meetings that some must have wondered if there were multiple Rileys. Fitch never traveled west of New York state and he confined his investigations to insect species in the East; Riley traveled extensively and investigated species across the continent and beyond.

The enabling legislation for Missouri state entomologist, written with the help of Walsh, Colman, and possibly Riley, directed the state entomologist to "study . . . insects . . . injurious or beneficial to . . . farmers and fruit growers."[4] Five years after assuming office, Riley explained how he had implemented this mandate. He had produced annual reports, answered questions by letter or through his publications, exposed false claims regarding insect control, lectured on injurious insects in Missouri and outside the state, investigated and experimented to develop controls for injurious species, discovered the habits and transformations of many species, and assembled a collection of injurious insects, including many new species.[5]

Riley considered his nine Missouri Reports (1869–78) his most enduring legacy.[6] He submitted his first report on December 2, 1868, eight months after he assumed office, commenting that it was "neither so full nor so valuable as I had hoped." The nine reports totaled approximately fifteen hundred pages, with information on the habits of hundreds of insect species, many of them new to science, in a unique blend of prose, poetry, and illustrations. Like Fitch and Walsh, he arranged his reports in categories of "noxious," [harmful] "beneficial," and "innoxious" [neutral] species and cast descriptions of species in small print that the non-specialist could skip. Although primarily addressing growers with practical needs, he insisted that descriptions of new species in agricultural publications (e.g., his reports) were as scientifically valid as those published in strictly scientific journals. He thus took issue with academic purists like Agassiz who held that "the man of science who follows his studies into their practical application is false to [his] calling."[7] Riley noted with pride that scientists like Darwin, Fritz Müller, and Asa Gray read and cited his reports.

Riley typically interwove scientific writing with personal commentary, often embellished with poetry. When discussing a new parasite species, he urged that it be named in honor of his Chicago collecting partner, Andrew Bolter, "an entomologist as enthusiastic as he is modest." Recalling the "many happy hours we have spent together," Riley quoted a poem in memory of their outings:

> I long to walk by the meadow's brook,
> To visit the fields and the woods once more . . .
> Or, under the spreading branches to lie
> And watch the clouds in the azure sky.[8]

Riley often ascribed human characteristics to insects. While gardeners normally considered Striped Blister-Beetles and Ash Gray Blister-Beetles as enemies because they consumed potato foliage, Riley discovered that they also fed on the larvae of the Colorado Potato Beetle, making them, on balance, friends of the gardener: "'When dog eats dog, then comes the tug of war,' and now that certain potato-beetles have taken to feeding on other potato beetles, the American farmer may . . . shout for joy."[9] Noting that the Convergent Ladybird, a predator of the Colorado Potato Beetle, sometimes devoured its own siblings, Riley concluded, "It is . . . cruel . . . to . . . take advantage of a helpless brother; but in consideration of its good services, we must overlook these unpleasant traits in our little hero's character!"[10]

Riley frequently expressed his fascination with insects as objects of beauty and wonder. With regard to the caterpillar's transformation to butterfly or moth, he wrote: "I am moved to admiration and wonder as thoroughly today as in early boyhood, every time I contemplate that within each of these varied and fantastic caterpillars . . . is locked up the future butterfly . . . destined, fairy like, to ride the air on its gauzy wings, so totally unlike its former self."[11] In his account of the seventeen-year cicada, one of the most distinctive American insects, he interwove natural history with American history: "We are moved to admiration in contemplating the fact that [the Cicada] has appeared in some part . . . of the United States at regular intervals of 17 years for . . . ages in the past. Long ere Columbus trod on American soil this lowly insect . . . filled the woods with its rattling song, when none but wild beasts and savages were present to hear it. . . . [I]n the month of June . . . 1738, for instance, . . . they appeared in the southern part of Missouri, and . . . six years previously they . . . appeared in the northeastern corner of the same state!"[12] Riley's essay on Katydids opened with the poetry of Oliver Wendell Holmes:

> I love to hear thine earnest voice
> Wherever thou art hid,
> Thou testy little dogmatist,
> Thou pretty Katydid.[13]

He then launched into a long diversion in which he compared spring (and summer and autumn) in England with spring in the Mississippi Valley: "The worshiper at Nature's shrine in this country must miss . . . the transport which the European may experience as he goes forth at morn . . . to view our Mother Earth decked in her brightest and most pleasing garments . . . Spring, with us, is apt to be but a narrow leap from winter to summer . . . Nor will the bright colors of our birds and flowers fully compensate for the enchanting song and sweet fragrance of those which add to the sylvan attractions of the more southern portions of England and the Continent of Europe . . . But in autumn, when the leaves are turned by the . . . touch of the Great Artist . . . the American may read and enjoy the Book of Nature to most advantage, and need envy no one on any other part of our terraqueous globe."[14] Following his comparison of seasons, he explained the evolutionary significance of the male Katydid's song: "The male Katydid doubtless feels something of the same satisfaction in playing to his . . . Katy, as a prima donna does in singing to an audience. . . . [T]he fact that the males are principally the players . . . shows that the . . . rivalry among the males . . . is . . . [important] to the species. The best player wins his coveted love, while the feeble and cripple stand no chance to impair the vigor of the race."[15]

Riley then combined natural history and human history in a gem of nature writing: "These insects [grasshoppers, crickets, and Katydids] are true fiddlers. . . . [L]ong even before birds had been fashioned to pour forth their vocal melody, there is good paleontological evidence that grasshoppers, not greatly different from present forms, fiddled away among the carboniferous ferns, and enlivened . . . those preadamic times."[16]

The Missouri Reports were distinguished above all by their illustrations. Fitch, Walsh, and Townend Glover (at the Agricultural Department) were all competent illustrators, but none had Riley's natural ability or artistic training and none drew with his rapidity, exactness, and flair. Riley declared that "illustration . . . conveys to the mind, in an instant, what the ear would fail to

Featured Insect

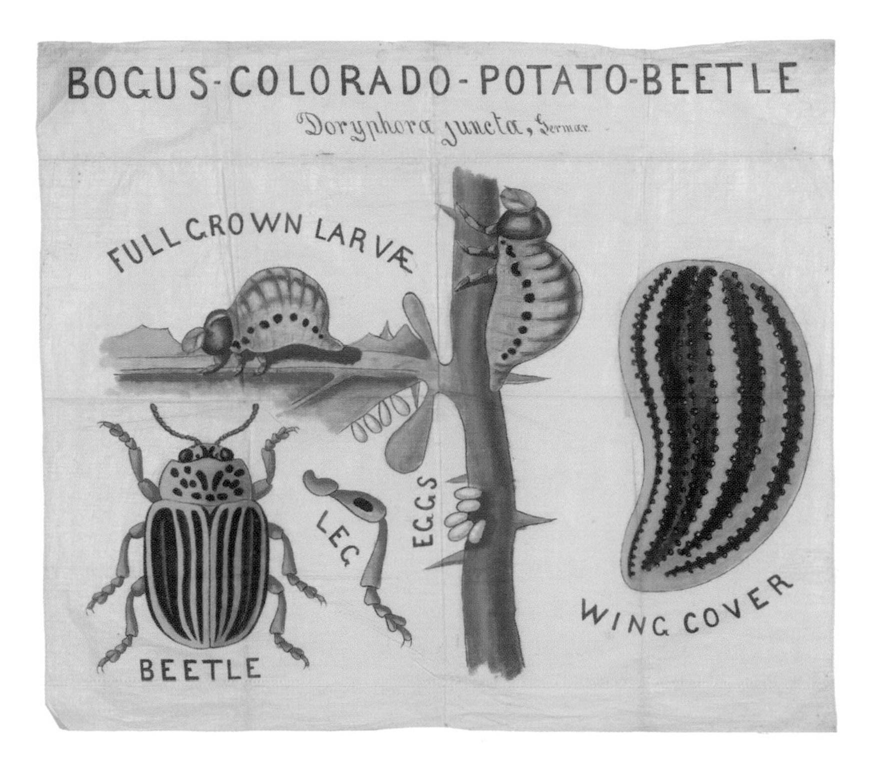

False Potato Beetle [current scientific name *Leptinotarsa juncta* (Germar) (Coleoptera: Chrysomelidae)] (Riley Insect Wall Chart; courtesy of Morse Department of Special Collections, Kansas State University Libraries) was well-known in Riley's time as a close relative of Colorado Potato Beetle and was sometimes mistaken for it. Feeding on the native perennial weed horsenettle, the False Potato Beetle never adopted potato as a food plant, occurring only as an occasional pest of eggplant, first documented by Riley in 1882.*

*C. V. Riley, "Change of habit; two new enemies of the eggplant," *American Naturalist* 16 (1882): 678–79.

do in an hour."[17] He therefore insisted that the Missouri Reports be copiously illustrated, and he drew each illustration from specimens, never from other illustrations, as did Glover on occasion.[18] The index to the Missouri Reports lists a total of 524 illustrations, the vast majority being his own.

Riley was able to produce high-quality illustrations at low cost because engraving and printing techniques were rapidly evolving and improving in the mid-nineteenth century. For the *Prairie Farmer* illustrations, he utilized the technique of wood engraving that Thomas Bewick, ornithologist and wood engraver, had developed around the turn of the century, whereby engravers carved the image into the end grain of a wood block with an engraver's burin (a finer instrument than knives). The completed wood block image was then inserted alongside movable type in the press. By the 1860s, engravers often coated the wood block surface with linoleum, which was more durable than wood. A further improvement was the electrolyte, a copper plate produced from a wax mold of the wood block and then coated with copper by a process of electrolysis, whereby an image could be printed multiple times with no loss of quality. Using these techniques, Riley produced illustrations for the *Prairie Farmer* and the Missouri Reports that rivaled the best stone and copper plate engravings he knew from his youth.[19] While Riley invested his life's blood in the scientific, literary, and visual material that graced the reports, he was dissatisfied with their poor printing and distribution. The entomologist's reports were bound with the Board of Agriculture Reports that appeared late in the summer, too late to be of use to growers and too bulky for wide distribution. In 1874, in the process of gathering information about the Chinch Bug, Riley asked farmers whether they knew of his Chinch Bug report published in 1870. To his chagrin, respondents estimated that only one in ten, perhaps one in one hundred, knew of his earlier report.[20] Some farmers complained that county officials "injudiciously distributed" reports to "landless lawyers and impecunious county editors."[21] Thereafter, Riley regularly paid to have about three hundred copies of his reports printed on good quality paper. It was principally through Riley's extras that the reports became known worldwide.[22]

In order to reach farmers and orchardists with timely information, Riley proposed to Walsh that they launch a separate publication that they named the *American Entomologist*. The first issue appeared in September 1868, five months after Riley assumed office. Published in St. Louis and distributed to subscribers at one dollar per year, the *American Entomologist* was in every respect a remarkable publication. Printed in large newspaper format, lavishly illustrated by Riley, with informative and entertaining articles by two gifted entomological journalists in their prime, the *American Entomologist* holds the reader's attention today, even as it did a century and half ago. It was the kind of publication Riley and Walsh wanted to issue as state entomologists. Riley spearheaded the project.[23] The *American Entomologist* replaced the *Practical*

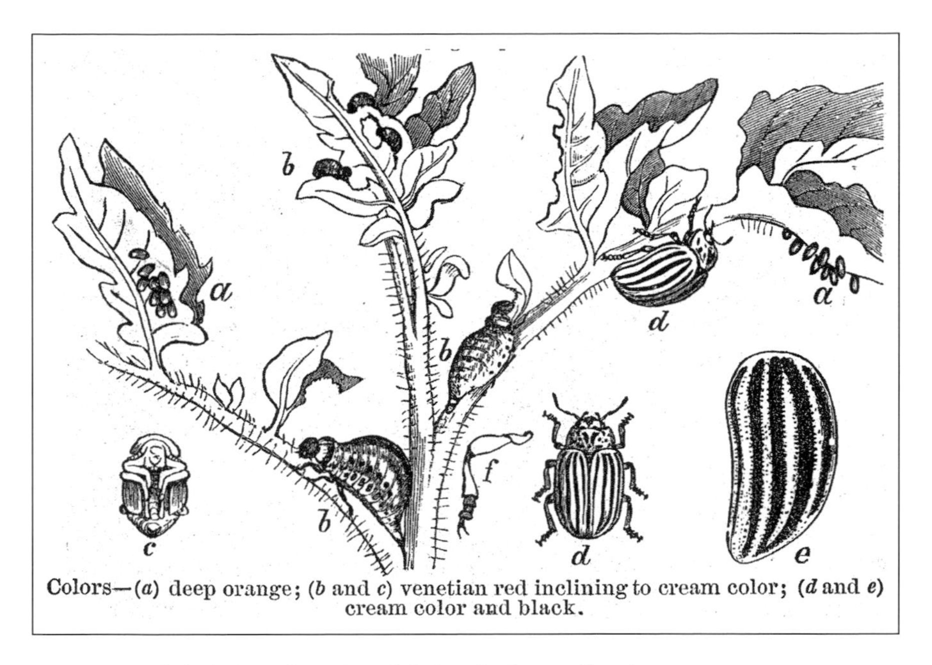

FIGURE 4.1 Riley's line drawing of Colorado Potato Beetle, *American Entomologist* 1 (1868): 41. Image scanned by Donald C. Weber from original, October 2016.

Entomologist that ceased publication in 1867. Following Walsh's death in November 1869, Riley continued the *American Entomologist*, but within a year he had to suspend publication.

Beginning in 1870, Riley began lecturing on economic entomology before agricultural and horticultural society gatherings and at Washington University in St. Louis. For his talks before lay audiences, he prepared large colored "Wall Charts" of insects, their life cycles, and their food plants that simplified the insects' anatomic features but highlighted their destructive habits.[24] Riley's most frequent out-of-state appearances were at Manhattan, Kansas, where in 1871 he was the featured speaker at the Fifth Annual Farmers Institute. In 1871–72 and subsequent years, he taught a course on "Insects Injurious to Vegetation" at Kansas State College at Manhattan.[25] In addition to lecturing and teaching at Manhattan, Riley wrote reports on injurious insects for the Kansas Academy of Science, and he advised Kansas farmers, functioning, in effect, as Kansas state entomologist.[26]

In October 1872, at the invitation of Burt G. Wilder, professor of zoology and physiology at Cornell University, Riley delivered twelve lectures at Cornell. The Natural History Society and the Farmer's Institute co-sponsored his visit. In anticipation of Riley's appearance, John Henry Comstock, Wilder's assistant, offered a preparatory course in entomology. In his university lectures, Riley spoke on the classification of insects with special reference to species of economic importance. At the Natural History Society, Riley discussed plant and insect co-evolution (pollination of the yucca) and mimicry in insects. At the Farmer's Institute, Riley talked about injurious insects. Leland O. Howard, a student of Comstock's, recalled that Riley, "tall, slender, romantic . . . with long wavy hair and a luxuriant moustache," attracted large audiences.[27] Riley's teaching and lecturing influenced the inclusion of agricultural entomology in the college curriculum.[28]

Unfortunately, Riley's attempt to promote economic entomology at his own state university got off to a rough start. In 1868, the catalog of the University of Missouri at Columbia listed him as lecturer, even though the university at the time had no agricultural program.[29] In 1872 and 1873, civic and educational leaders founded an agricultural college as a program of the university to be financed through the sale of Morrill Act lands.[30] Riley apparently delivered some lectures in 1873, but in 1874, when the board of the agricultural college proposed to pay Riley $300 to deliver ten lectures, Dean George C. Swallow cancelled Riley's courses, citing lack of funds and students.[31] Swallow's action may have been in retaliation for Riley's critical remarks about Swallow delivered at the state horticultural society. Riley charged that the agricultural college was inferior to those in other states and characterized Swallow, who taught most of the courses, as "[A] professor of agriculture, who rattles . . . dry bones of natural history . . . out of poor textbooks [who] never reads . . . the book of nature . . . or . . . lays before [his students] . . . new . . . agricultural truths"[32] Riley's remarks may have reflected St. Louis snobbishness with regard to the new university in Columbia. In 1876, twenty-one St. Louis boys attended Washington University in St. Louis and nineteen attended the St. Louis Polytechnic Trade School, but only five were enrolled in the university.[33]

The projected insect collection for the university further strained relations between Riley and the university. The legislative mandate for state entomologist required that the state entomologist furnish the agricultural college "when established" with an insect collection. In 1874, the University Board of Curators summoned Riley to produce a collection.

In December, with the help of his assistant, Otto Lugger, Riley delivered a cabinet with sixty drawers containing an assortment of important insects, including many immature stages.[34] Because the collection was housed in unsuitable facilities with no qualified curator, it soon disintegrated. Despite prickly relations between Riley and university officials, the university awarded him an honorary degree of Doctor of Philosophy in 1873. Based on this title and

FIGURE 4.2 "Bug Hunting," illustration from *Rural New Yorker* (March 28, 1874), Charles Valentine Riley Papers, 1866–1895, Record Unit 7076, Box 1, Scrapbook 8, p. 91 (courtesy of Smithsonian Institution Archives, Washington, DC).

an honorary degree of Master of Arts from Kansas State Agricultural College (1872), Riley insisted on being called "Professor Riley" from that time on.

Like Walsh, Riley raged against unscrupulous vendors of bogus insect cures and the naïveté of farm folk who believed them. In 1872, when discussing an insect trap developed by Thomas Wier of Lacon, Illinois, Riley objected to the claim in the advertisement that it destroyed "every species of insect infesting fruit."[35] To make his point, Riley constructed a fictional dialog between "Agent Gaingreedy" and "Farmer Glauball," a dialog based on an actual encounter. Agent Gaingreedy first demonstrated his patented trap, consisting of two shingles to be attached to the tree. He then showed onlookers how the worms gnawed their way in between the shingles and how easily one could press the shingles together to destroy them. "'Ach,'" cries the credulous German farmer, "und is it true das die wurm rader eat de shindel als de apfel?" "Oh yes," says Gaingreedy, "Screw one of the traps to this tree, and in a week I will come back, and you will see." A week later Gaingreedy returns, and finds Farmer Glauball has killed dozens of worms with the trap, not noticing that they had already eaten his apples.[36]

Riley's position as state entomologist underwent a severe test in 1873, when Missouri and the entire nation experienced a drastic economic contraction known to history as the "Panic of 1873." Governor Silas Woodson, elected in 1872 as the first Democratic governor after the Civil War, appointed a Committee on Retrenchment and Reform. The state entomologist office, established by Radical Republicans four years previously, presented an obvious target. Riley, Colman, and Charles W. Murtfeldt, secretary of the State Board of Agriculture, defended his position before the General Assembly.[37] When Senator Pope of Wright County introduced a bill to abolish the state entomologist, Senator Palmer, also of Wright County, came to Riley's defense. Palmer testified that Riley had identified the pest infesting his apple orchard as the Apple Tree Bark Louse and had prescribed effective controls. Abolishing the office, he said, would be "penny wise and pound-foolish."[38]

Nevertheless, Woodson, in his message to the Missouri General Assembly in 1874, proposed that the state entomologist office be abolished.[39] Alert to a crowd-pleasing debate, the *St. Louis Dispatch* reported the afternoon proceedings under the headline, "Torturing the State Entomologist, Riley as a Sacrifice."[40] When coached witnesses ridiculed the state "Bug Master," the debate degenerated into afternoon public entertainment.[41] When Riley's turn before the assembly came, he cited letters of support from Missouri horticulturists and farmers. George Hussman, a leading Missouri vintner, attested to Riley's service to the state's wine growers, and Thomas Meehan, editor of the *Gardeners' Monthly*, wrote that Riley's phylloxera discoveries alone more than paid for his position. Other witnesses expressed pride in having an internationally esteemed scientist as Missouri state entomologist.[42]

A reporter from the *St. Louis Dispatch* who interviewed Riley concluded that the state entomologist was of vital importance to Missouri farmers and fruit growers, who lost millions of dollars to insects each year.[43] A German correspondent for the St. Louis *Fortschritt* referred to the "Dummheit" and "Arroganz" of those who failed to recognize the value of the entomologist.[44] Despite this show of support, the House voted to abolish the office of state entomologist. Following more political maneuvering, the bill was "laid over informally" and the session ended without a vote on the bill.[45]

In 1875, Governor Woodson and budget-cutting legislators again sought to abolish the office. Some legislators opposed anything that smacked of science, learning, or education, such as the State Board of Agriculture, the "mineralogist, geologist, conchologist, entomologist, and other scientists." Some of these offices, like the conchologist, didn't even exist. A few extreme legislators proposed the abolishment of the state's normal school system.[46] Again, the horticultural and agricultural societies, the Grange, and other supporters rallied and saved the day. In 1876, newly elected governor Charles H. Hardin set about reducing debt by carefully examining expenditures rather than sacrificing educational, agricultural, and other essential programs.[47] The value of the state entomologist office was confirmed in the mid-1870s when locusts swarmed over the prairie states and territories and invaded western Missouri. Riley, who emerged as the central figure in the locust emergency, received solid support in Missouri and beyond, insuring the continuation of his office, at least until the end of the locust threat.[48]

Shortly after his arrival in St. Louis, Riley fell in love with Emilie Conzelman. Her father, Gottlieb Conzelman, was a master upholsterer who had emigrated from Germany in the late 1840s. He tested business opportunities in Cincinnati and Louisville before settling in St. Louis. There he prospered as an upholsterer and paperhanger, investing profitably in the brisk trade between St. Louis and New Orleans and augmenting his income with real estate transactions and rental property. In 1852, he built a handsome brick house on Clark Avenue in Mill Creek Valley. In 1861 his wife, of Danish descent, died of tuberculosis, leaving him with the care of his three children, Emilie (age eleven), William (age five), and Theophilus (age three). His wife's sister, Mary, came from Denmark to assist with the family, but six years later she married and moved to California. Emilie, now a young woman of seventeen, substituted as a mother for her two younger brothers. These circumstances help explain why she and Charles waited until June 1878 to marry. From all accounts, Emilie was demure and retiring, in striking contrast to her flamboyant husband.

Riley's mother died on April 18, 1877, in London. Riley's contact with her had decreased during his Chicago and St. Louis years. On his first return to England in 1871, Riley observed the obvious estrangement between his mother and his stepfather. When Antonio Lafarge died three years later, Riley's Aunt Mira wrote, "I shall always rejoice . . . that you came while Antonio was living, that you show[ed] him kind attention and that he liked you."[49] Following his mother's death, Riley asked his Aunt Mira to share what she knew about her. Mira revealed how Mary had betrayed her husband for some fifteen years (from about the time Riley left for America) and how she drifted ever further into what she called the "vile license" of adultery and debauchery before dying at age fifty-two.[50] What effect these revelations had on Riley is not clear. Following Mary's death, Riley's half-sister Nina moved to St. Louis, where she found refuge and security with her prominent brother.[51]

Riley was barely established in St. Louis when he was called to northern Michigan, where brother George was laid up with a broken leg.[52] Four years later, Riley confided to his colleague Joseph A. Lintner that he had spent several weeks in Kansas nursing George through a serious illness, lamenting that George was "not well; nor have I hope that he ever will recover."[53] In January 1875, Riley learned to his astonishment that the preceding October George had been committed to the Insane Asylum in Dickinson County, Kansas. An irate Riley wrote to the doctor who belatedly informed him of this action, "I am perfectly astounded . . . that this action should have been taken without my being notified, or given an opportunity of taking better care of my brother."[54] Riley arranged to have George transferred to the Insane Department of the Poor House in St. Louis. There he vegetated for over a decade, during which time the Rileys moved from St. Louis to Washington, DC. George died at age thirty-eight on May 29, 1888. Listed as "farmer" who died of "entero colitis," George was buried in the Gottlieb Conzelman plot in Bellefontaine Cemetery, St. Louis. On his gravestone was engraved: "The spirit bound in life, in death released from strife."

In St. Louis, Riley found companionship among a group of artist-bohemians known as the "sketch club," who met, and drank, in waterfront inns on Olive Street. Prominent among them was John Martin Tracy, widely known for his paintings of hunting dogs. Other members of the group sketched and painted "pot boiler" panoramas of St. Louis and the Mississippi that they sold at local auctions.[55] Through contact with members of the sketch club, with German immigrants who frequented beer gardens, and with clients who grew hops and grapes for beer and wine production, Riley modified his strict Templar abstinence, now approving of moderate social drinking.[56] He also met frequently with the St. Louis Philosophical Society, a "little coterie of metaphysicians" that William T. Harris, superintendent of the St. Louis public schools, and later US commissioner of education, initiated in 1866. The metaphysicians met regularly at a restaurant in St. Louis to drink beer, smoke their pipes, and debate philosophy. In 1867, Harris, upon the recommendation of Ralph Waldo Emerson, hired Thomas Davidson, a Scottish-American philosopher, as a teacher of classics. Davidson, who had participated in Emerson's circle, was a valuable addition to the St. Louis gathering. The St. Louis "metaphysicians" represented a leading outpost of American transcendentalism in the Midwest. They supported educational opportunity for all members of society, regardless of race, sex, and religion, as the practical basis of a working democracy. Under Harris's leadership, the St. Louis school system was transformed into the most advanced and egalitarian system in the nation. Joseph Pulitzer, at that time a struggling journalist in St. Louis, listened with rapt attention to the conversations of the metaphysicians. Later, as his newspaper empire flourished, he featured articles by Davidson. Among the topics debated by the metaphysicians was the relevance of ancient and modern philosophers to contemporary issues.[57] When the group discussed the question, "Was modern art superior or inferior to that of ancient Greece?" Riley defended the moderns, protesting against the "unqualified praise of the antique."[58]

Riley worked with Harris to introduce science into the school curriculum in St. Louis.[59] In January 1872, in a speech before the St. Louis district school convention, Riley urged a "revolution" in teaching methods that would place more emphasis on science. Advocating what Anna B. Comstock later championed as nature study, Riley asserted, "School children are now blind to the beauties of nature . . . for want of [someone] to tell them what those beauties are."[60] Under Harris, the St. Louis school system revised the elementary school curriculum to devote one hour per week to science.[61] Riley contributed an article on entomology with instructions on capturing insects and organizing an insect collection as one component of the expanded science curriculum. The article appeared in his fifth annual report and in *Campbell's New Atlas of Missouri*.[62] Riley also supported trade schools in St. Louis, a special concern of his future father-in-law, who was the primary financier of the St. Louis Polytechnic School.[63]

Riley's prodigious output during his Missouri years was made possible with the help of assistants he recruited. In his report for 1873 he noted, "[T]here is a limit to one man's capabilities" and in 1874, he threatened to resign if no assistants were funded.[64] Despite the legislature's refusal, he assembled a corps of loyal helpers, including Otto Lugger, Mary E. Murtfeldt, Theodore Pergande, Moritz Schuster, and Joseph T. Monell.

Riley met Lugger in Chicago about 1864, shortly after Lugger emigrated from Germany. When Riley was appointed state entomologist in 1868, he called Lugger to serve as his part-time assistant. Lugger commuted from Chicago until 1871, when he moved to St. Louis to work on a regular basis.[65] Quiet and methodical, Lugger was the ideal assistant to the visionary Riley. In 1875, Lugger married and moved to Baltimore, Maryland, where he served as curator for the Maryland Academy of Science, and naturalist for the city parks.[66]

Theodore Pergande emigrated from Germany in 1860 and immediately enlisted in the Union army. Discharged four years later in St. Louis, he worked in a gun factory and married a young German Fräulein. One Sunday, the two Germans met while collecting insects. At that time, Lugger was preparing to move to Maryland and he suggested that Riley hire Pergande as his replacement. Pergande served as Riley's assistant in St. Louis until Riley and his family moved to Washington, DC. Some years after Riley's move, Pergande joined Riley's team in Washington, DC, where he proved indispensable in rearing insects.[67]

Murtfeldt, born in New York, moved with her family to Rockford, Illinois (Walsh's hometown), where she studied botany and other natural history subjects at Rockford College. In 1868 or 1869, her father, Charles W. Murtfeldt, an agricultural writer and publisher, became editor of *Colman's Rural World*. The family moved to St. Louis, and in 1871, to Kirkwood, Missouri, just west of St. Louis. Riley encouraged her entomological studies, and she soon became an expert in the micro-Lepidoptera. She helped Riley unravel the role of the Yucca Moth in Yucca pollination. Riley listed her as his assistant in three reports (1876, 1877, and 1878). In 1880, after his move to Washington, DC, Riley appointed her as a field agent for the Division of Entomology, a function she continued intermittently until 1893. From 1888 to 1896, she also served as second Missouri state entomologist.[68]

Schuster, a native of Germany, was in his forties when he became associated with Riley in St. Louis. Listed in the 1877 *Record of American Entomology*, he was an enthusiastic student of entomology in St. Louis until his death in 1894.[69]

Monell was the teenager among Riley's coterie. Born in St. Louis and orphaned at age four, he grew up under the care of Henry Shaw, the founder of Shaw's Garden. Beginning in 1881, he worked as a "volunteer student" under Riley for two years and with the physician-botanist George Engelmann for one year, while at the same time studying at Smith's Academy and at Washington University for a degree in mining engineering. Specializing in aphids, he co-authored a report on this group with Riley, had one genus and several species named after him, and continued his entomological studies (though without publishing) throughout his career as a mining engineer.[70]

Because the position of state entomologist was funded by lump sum appropriation with no provision for pay for assistants, Riley paid Lugger, Pergande, Murtfeldt, and perhaps others from his appropriation.[71] Riley estimated that his expenses, including assistants, materials, and services (books, cabinets, chemicals, railroad travel, engraving, etc.), averaged $2,350 per year. Of his $3,000 annual appropriation there remained approximately $650 per year, or around $54 per month for room and board, clothing, recreation, and miscellaneous expenses. Considering that the average Missouri farm family in the 1870s realized a net profit of $700 per year, Riley's personal income of $650 seems reasonable. Whatever the exact figures were, Riley paid out approximately three quarters of his annual appropriation for assistants and other expenses.[72]

Viewed from a modern perspective, such financial arrangements appear highly irregular, but they were standard practice in the nineteenth century. The same difference in perspective applies to Riley's supplemental sources of income. In 1874, a St. Louis newspaper announced, "[O]ur popular and scholarly State Entomologist has been named to the board of the Missouri Gas-Light Company, thus guaranteeing its success."[73] The company evidently had contracts with municipalities and private consumers across the state to supply gas lighting, a major utility before electrical lighting. The company also marketed a device called a gas "utilizer" that was advertised to save up to 50 percent on gas consumption.[74] It is not known how much Riley was paid as a board member, nor whether he was involved in other entrepreneurial ventures. Riley accepted the Gas-Light position because it did not directly involve his position as state entomologist. He resolutely declined a number of other potentially lucrative offers, such as marketing grapevine rootstock to French growers or endorsing patented insecticides, because he would profit directly or indirectly from his position as state entomologist.[75] In the nineteenth century, and well into the twentieth century, such apparent conflicts of interest were accepted. Public officials were expected to supplement their incomes from outside sources.

During his Missouri years, Riley became internationally prominent through his phylloxera and Rocky Mountain Locust investigations, which are discussed in separate chapters.[76] His investigations of other species that are treated here, such as the Colorado Potato Beetle, Periodical Cicada, Oyster Shell Bark Louse, Plum Curculio, and Army Worm, further enhanced his reputation.

Walsh, who first reported on the "new" potato beetle, explained that Thomas Say had discovered the small, bright yellow beetle with black wing stripes in 1823 and named it *Doryphora 10-lineata*. It was first reported as a pest of potato plants in eastern Nebraska in 1859. Walsh speculated that around that time a population had switched from feeding on buffalobur, a solanaceous weed related to the potato, to cultivated potatoes grown by miners in Colorado. He speculated further that the beetle spread from one potato patch to the next, reaching settled regions of Nebraska, Iowa, and Wisconsin by the early 1860s.[77] Riley, whom Walsh credited with tracing its life cycle by breeding several generations, coined the name "Colorado Potato Beetle" in 1867, based on Walsh's speculations regarding its origin.[78] In 1864, the Colorado Potato Beetle reached central Illinois, by which time it had become a serious pest in the entire region west of Chicago and St. Louis.[79] By 1870, it had become so destructive that potato prices in the Midwest rose to two dollars per bushel, a prohibitive price for thousands of Americans, who then had to do without this staple. Spreading eastward, it became the most notorious insect pest of the time nationwide. Gardeners and growers appealed to agricultural editors, state entomologists, and the Agricultural Department for help.[80] Walsh, Riley, Fitch, LeBaron, and Glover investigated its life cycle but found no weak point in the insect's life history nor any habits that might lead to its control. Hand picking of beetles, larvae, and eggs was the only known method of control. In 1867, farmers in Illinois discovered that dust containing Paris Green, or copper(II) acetoarsenite, a compound used to color paint green, when sprinkled on potato plants, killed beetles that fed on the dust-covered leaves. Walsh, Riley, Glover, and LeBaron conducted experiments to determine the effectiveness of Paris Green as an insecticide. Following Walsh's death in 1869, Riley was the leading authority on the Colorado Potato Beetle. After extensive testing, Riley published his formula for mixing Paris Green with other ingredients. He revised Walsh's prediction of the beetle's advance eastward at a rate of fifty miles per year upward to seventy miles per year. Even Riley's revised estimate fell short. By hopping trains and ships, the beetle reached the Atlantic in 1874, four years earlier than Riley predicted.[81]

When Europeans became alarmed that the beetle might cross the Atlantic, lawmakers in Belgium, France, Switzerland, and Germany responded by prohibiting the importation of American potatoes, and Great Britain, Italy, and the Netherlands debated doing the same. Riley assured Europeans that because adult beetles consumed only the leaves of the plant and not the tubers, restricting the importation of potatoes was pointless. He recommended that posters with illustrations of the adult beetle be placed on ships destined for Europe with instructions for passengers and crew to report any sightings of the beetle so that ship's officers could destroy the unwanted passengers. The German government implemented Riley's recommendation; others followed suit more informally, and soon the alarm subsided in Europe.[82] Riley's writings on the Colorado Potato Beetle, which he collected and published in book form, spread his growing reputation.[83]

Riley's first summer in St. Louis (1868) was a "locust year," when seventeen-year Periodical Cicada emergence was widespread throughout the East and Midwest. That summer Riley discovered a thirteen-year brood whose range extended to the south of the seventeen-year broods, the two broods overlapping along the 37th and 38th parallels, the latter bisecting the state south of St. Louis. In 1868, both seventeen- and thirteen-year broods appeared simultaneously in Missouri, an event that Riley calculated had last occurred in 1647 and would next occur in 2089.[84] Prominent scientists including Walsh, Darwin, Gray, and Joseph D. Hooker (director of the Royal Botanical Gardens) praised Riley's mapping of twenty-three broods across the eastern and central states. Although the Cicada attracted attention primarily because of its unusually long larval stage underground, its sudden appearance in overwhelming numbers, and its raucous chorus, Riley pointed out that it also had some economic importance. When Cicadas appeared in outbreak numbers, the females damaged twigs of fruit trees when laying eggs. Riley reported that he spent $200 on experiments searching for a remedy, without success.[85]

During his first year in St. Louis, Riley conducted experiments on the Oyster Shell Bark Louse (now known as Oystershell Scale), a European species that, since its introduction seventy years earlier, had become a serious pest in apple orchards throughout the upper Midwest. To test its adaptability to local conditions, Riley placed eggs of this scale insect on apple trees in St. Louis. When only a few eggs hatched and the larvae soon died, he concluded that the Oyster Shell Bark Louse presented no threat to apple growers in Missouri or further south.[86]

In 1870, Riley challenged Walsh and others regarding the number of broods of the Plum Curculio, known as the "Little Turk," the most notorious destroyer of plums and other stone fruits. Walsh asserted that the Curculio was double brooded, producing two generations in the same season. Riley contended that the Curculio was single brooded. To settle the question, Riley constructed a frame over a large peach tree, and, after the first peaches had been infested by Curculios, enclosed it with stout gauze. At the end of the season, the adults of the first generation were still alive, but no second generation had appeared. Riley cited this as proof that, with minor exceptions, the Curculio was single brooded. In the latitude of St. Louis, he concluded, only one out of one hundred Curculio females that hatched out in summer deposited eggs the same season. He also concluded that the Curculio hibernated as an adult beetle.[87] While the overwintering stage is always the adult, later research has determined that Plum Curculio populations fall into two distinct types that cannot interbreed: a northern race which has one generation per year, and a southern race which is multivoltine (with two or more races per year); both of these are now found in Missouri and Illinois.[88]

In 1861, a massive outbreak of the Army Worm occurred in the upper Midwest. Walsh, Cyrus Thomas, and others responded by investigating the life history and habits of the Army Worm.[89] When Riley began his own investigations, central points in the life cycle of the Army worm remained unresolved: When and where did the moth deposit the eggs? In what state did the insect hibernate? Were there one or two broods each year? In 1876, in a lengthy article on the Army Worm, Riley offered answers to these questions. Like other investigators, Riley had failed to find a female in the process of laying eggs, but, based on the structure of the female's ovipositor and farmers' reports regarding the times and places the worms appeared, he concluded that the female laid her eggs on the stalks of mature grass and grain, between the stalk and the sheath, or she attached the eggs in rows along the stalk. Some females deposited their eggs in the spring and some in the fall. Riley concluded that Army Worms generally hibernated as adult moths, but that some also hibernated as pupae underground, the proportion of hibernating pupae increasing from south to north. Concerning the most intensely debated question, the number of broods, Riley believed that Walsh and Thomas had erred when they concluded that the Army Worm had two or more broods per season. He concluded that the Army Worm was single brooded. After more than a decade of debate on this issue, Riley considered the question settled. The Army Worm had only one generation annually.[90]

By the mid-1870s, the list of Riley's correspondents read like a who's who of entomologists worldwide: LeBaron and Thomas in Illinois; John L. LeConte and Ezra T. Cresson in Philadelphia; Lintner in New York; Townend

FIGURE 4.3 Cartoon figure of C. V. Riley, 1870s, source unknown (Riley Family Records, National Agricultural Library Special Collections, Beltsville, Maryland).

Glover in Washington, DC; John G. Morris, lepidopterist, in Baltimore; Simon Snyder Rathvon in Lancaster, Pennsylvania; Alpheus Spring Packard Jr. in Massachusetts; Charles Bethune and William Saunders in Canada; John O. Westwood, Henry Tibbats Stainton, Robert McLachlan, John Jenner Weir, and Charles Darwin in England; Fritz Müller and Philip C. Zeller in Germany; and Etienne Mulsant, Victor Antoine Signoret, Jules Émile Planchon, and William Auguste Jules Lichtenstein in France.

In 1872, the American Association for the Advancement of Science (AAAS) met in Dubuque, Iowa. Attendance was low because many scientists in Boston, Cambridge, Philadelphia, and other eastern cities regarded this aspiring outpost of science as too distant; however, the meeting served as Riley's induction into the nation's scientific elite. In Dubuque, Riley, Morris,

Bethune, Saunders, and others organized the Entomological Club of the AAAS.[91] The Entomological Club was distinguished from the American Entomological Society (originally the Entomological Society of Philadelphia) and earlier organizations by its national representation and its emphasis on agricultural applications. It served as the central forum for American entomologists until superseded by the American Association of Economic Entomologists in 1889.[92] At the Dubuque meeting, Riley was elected to the rules committee and to the committee on nomenclature. His paper on fertilization of yucca elicited an animated discussion at the general assembly of the AAAS.[93] In 1872, Riley also helped organize the National Agricultural Congress. At its organizational meeting in St. Louis that year, he spoke on Paris Green as a control of the Colorado Potato Beetle and the Cotton Worm.[94]

Riley's professional relationships suffered at times from his frank, even blunt, statements in public or private, an example being his uncomplimentary remarks about Swallow quoted earlier. With reference to a disagreement with Walsh, aired in the *Prairie Farmer*, Riley remarked that he (Riley) was willing to acknowledge his error but questioned whether Walsh would admit his.[95] With reference to Packard's description of a new species, Riley asserted that it was not new, then added, "If a new species is to be made out of such trifling characteristics . . . what is the science of entomology to come to?"[96] Walsh and Packard overlooked Riley's frankness and remained his friends; others were less forgiving. Regarding reports that the lepidopterist Augustus R. Grote felt offended by something he had written, Riley wrote to Lintner, "I have no sympathy with the surly spirit of self-importance which takes offence at honest criticism or the expression of differences of opinion," adding that he [Riley] was "always open to correction, and willing to stand corrected for any fault committed."[97] In his *Seventh Annual Report*, Riley reported being "amused" by remarks of the mayor of St. Louis who published his reply to a Belgian official regarding the Colorado Potato Beetle in the local newspaper: "[I]nstead of ascertaining the facts [by asking me] he chose to display his [ignorance] by publishing a reply [in which] there is scarcely a statement . . . not opposed to the facts."[98] Years later he confessed to Lintner that his "blunt expressions . . . made . . . with the kindliest of feelings may at times seem harsh to my friends. It is my unfortunate way to write just as I feel without mincing words . . . I am glad to have you point out that my language is liable to be misconstrued. But I never say anything I do not mean though the manner of saying it might be more studied."[99]

In 1871, Riley returned to England to visit family and to investigate the phylloxera outbreak and silk production in France. Eleven years earlier he had sailed to America as an immigrant farm laborer. He returned now as a renowned scientist. The *Rural New Yorker* noted that Riley "does not go abroad for recreation . . . He has in view his entomological work [and he] will visit prominent entomologists in Europe."[100] The highlight of his trip was his visit with Charles Darwin at his residence in Downe.[101]

In 1875, Riley returned to Europe, again citing phylloxera and silk culture investigations as primary goals. Again, he met with Darwin at his home. On this trip Riley also made appearances before the London Entomological Society and the Entomological Society of France.[102]

Through his writings and illustrations in the Missouri Reports, the *American Entomologist*, and other publications; through participation in the AAAS, the National Agricultural Congress, and horticultural and agricultural meetings; through his association with Darwin and other famous scientists; through his role in rescuing French winegrowers from phylloxera; and through his response to the locust invasion, Riley became, by the 1870s, the leading spokesman for the emerging field of economic entomology. In 1871, the *Missouri Western Ruralist* declared, "[Riley] has . . . cast a scientific luster upon . . . Missouri [and drawn] the eyes of the European scientific world."[103] In 1872, the Manhattan, Kansas, *Nationalist* announced, "[A]lthough still young, Professor Riley has already achieved a national reputation."[104] Policy makers in the United States, Belgium, France and Germany sought Riley's advice regarding the Colorado Potato Beetle, phylloxera, and the Rocky Mountain Locust.[105] London papers regularly reported Riley's activities.[106] Scientists in England, Germany, and France, praised his Missouri Reports and published excerpts from them in German and French translations.[107] In 1877, by which time Riley had served almost a decade as Missouri entomologist, an unnamed writer in the *Centennial Record of Missouri* concluded that Riley was "the foremost economic entomologist of the day."[108]

5

The "Book of Nature" According to Darwin

[I]t seems highly probable that [the formation of a species] may some day be traced in insects.

—C. V. Riley, "The Archippus Butterfly," *Third Annual Report* (1871), 173

In November 1859, the year before Riley set sail for America, Charles Darwin's *On the Origin of Species* broke upon the English-speaking world. Darwin's thesis, that present-day living species had been produced by natural selection operating over a vast span of time, shocked the public and altered every aspect of entomology. What were species? How should one classify them? How did variations arise? How did insect behavior relate to their evolution?

Riley's adjustment to Darwinism may be traced in four stages: (1) initial acceptance of Darwinism during his Chicago years; (2) discovery (with Walsh) of mimicry in American butterflies and their Darwinian interpretation of mimicry; (3) discovery (with George Engelmann) of the mutual interdependence of yucca plants and Yucca Moths; and (4) perception of rapid evolution among insect species in the American environment coupled with Riley's attempt to apply this concept in the control of insect pests.

Little is known about how Riley reacted to Darwin during his Kankakee and Chicago years. His diary does not mention the *Origin* or other works by Darwin, nor is there any reference to Darwin in his correspondence. This is surprising when one considers that Walsh read the *Origin* in 1861, opened a correspondence with Darwin the same year, and wrote to Darwin in 1864, "[T]he first perusal staggered me, the second convinced me, and the oftener

Featured Insect

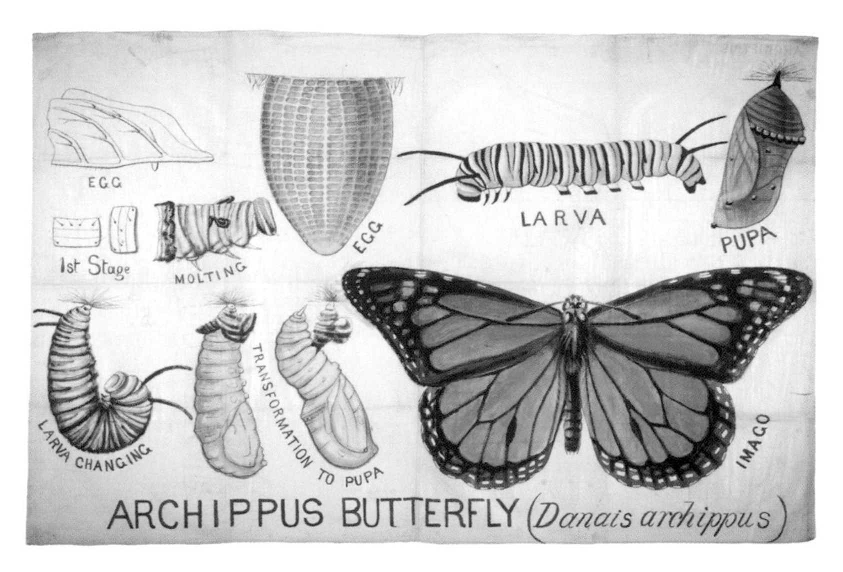

Monarch Butterfly [old scientific name *Danaus archippus*; current scientific name *Danaus plexippus* (L.) (Lepidoptera: Danaidae)]. Riley Insect Wall Chart (Courtesy of Morse Department of Special Collections, Kansas State University Libraries). The Monarch is the best-known butterfly in North America and is renowned both for its amazing migrations and its role as a proposed model in the Monarch-Viceroy mimicry system. As a result of a long period of research, the overwintering grounds for most of the North American Monarch population was discovered in the mountains of southern Mexico in 1975. Although Riley proposed the Monarch as the toxic model in a Batesian mimicry system, more recent work has shown that the Viceroy is also distasteful, indicating instead a Müllerian mimicry duo. In recent decades, Monarch populations have declined drastically, due to decimation of their overwintering habitat in Mexico and to extensive use of herbicide-resistant field crops in the US, resulting in decreased populations of its milkweed food plants.* In 2015, the US Fish and Wildlife Service initiated a review of Monarch status under the Endangered Species Act.**

*Lincoln P. Brower et al., "Decline of monarch butterflies overwintering in Mexico: is the migratory phenomenon at risk?" *Insect Conservation and Diversity* 5 (2012): 95–100; and John M. Pleasants and Karen S. Oberhauser, "Milkweed loss in agricultural fields because of herbicide use: effect on the monarch butterfly population," *Insect Conservation and Diversity* 6 (2013): 135–44.

**Brice X. Semmens et al., "Quasi-extinction risk and population targets for the Eastern, migratory population of monarch butterflies (*Danaus plexippus*)," *Scientific Reports* (2016): 6:23265.

I read it, the more convinced I am of the general soundness of your theory."[1]

That same year (1864), Walsh published a broadside against Agassiz and his students, Alpheus Spring Packard, James Dwight Dana, and Samuel H. Scudder.[2] Drawing on his entomological investigations, Walsh refuted Agassiz's assertion that no insect species were common to both Europe and America, citing 360 identical or closely related species on both continents and arguing that Darwin's theory of common descent explained this phenomenon. He ridiculed Agassiz's theory of special creation because this would require thousands of special insect creations to account for existing species.[3] Walsh charged that Agassiz "approached [the *Origin*] with the same feelings as many men approach a toad or a spider, viz., as something scarcely worthy of his notice."[4] Upon receipt of Walsh's paper, Darwin replied that he was "very much pleased to see how boldly and clearly you speak out on the modification of species."[5]

By breeding insects taken from galls on willows, Walsh demonstrated that fifteen species of cynipid wasps formed galls exclusively on their host species of willow. In many cases the species were so closely allied that they could only be differentiated on the basis of the willow species upon which they formed galls. Walsh postulated that these wasp species had descended from a common ancestor but had formed a succession of new species when the original willow species had split into separate species, thus isolating the larvae that fed on different host plant species. Walsh called these "phytophagic" species because they originated in connection with their food plant.[6]

Walsh's influence on Riley's thinking regarding the evolution of species remains unclear. When Riley met Agassiz in February 1864, he noted approvingly that Agassiz's division of the animal kingdom into four classes beautifully demonstrated "the relative place of all animals."[7] Even more surprising, Riley remained silent on Darwinism for another four years. Then suddenly, in June 1868, shortly after his appointment as Missouri state entomologist, Riley reviewed Darwin's *The Variation of Animals and Plants under Domestication* for the *Prairie Farmer*, where he revealed his familiarity with Darwin's publications and his support of Darwin's theory. "The Origin of Species," Riley wrote, "has been shaking the moral and intellectual world as by an earthquake. . . . [T]he principal cause of the common unpopularity of Darwin's views, is that they did not accord with the 'Law of the Prophets' . . . they are not Mosaic."[8]

When and how did Riley become a convinced Darwinian? Riley provided a hint in 1870 when discussing mimetic butterflies: "[M]y studies of insect life led me several years ago to appreciate the [Darwinian] hypothesis, and the more I became acquainted with these tiny beings in the field, the more I became convinced of its truth."[9] While Riley's reference to "several years ago" provides scant help in pinpointing *when* he became a convinced Darwinian, his reference to field observations of insects definitely points to *how*: Riley became convinced through his observation of insects in their natural habitat.[10] In his review, Riley indicated that he had hesitated to write about Darwinism earlier because "the columns of the P[rairie] F[armer] are not the fit place to discuss these matters."[11] Now, as Missouri state entomologist, Riley felt free to discuss Darwinian theory publicly.

Regardless of when and how Riley became a convinced Darwinian, by 1868 he had arrived. In his first Missouri report (1868), Riley compared the Walnut Tortrix Moth, whose larvae fed on hickory, with an allied species that closely resembled it but whose larvae fed on snowberry, citing these as examples of phytophagic variation.[12] That year he made his first (indirect) contact with Darwin. In August 1868, as Riley and Walsh commenced publishing the *American Entomologist*, Walsh wrote to Darwin: "My partner in this enterprise—C. V. Riley—who is an active and energetic man of 25, without any very deep scientific knowledge, but an excellent hand at breeding insects, and what is most important of all a first-rate entomological draughtsman, is very desirous to get your photograph, and if possible, that of Westwood. . . . He has recently been appointed State Entomologist of Missouri . . . has written a good deal on entomological matters for the Agricultural Press; and not long ago published a tolerably good review of your book."[13] Walsh enclosed copies of two recent articles by Riley. Darwin sent Walsh his photograph and thanked Riley (through Walsh) for the articles but suggested that Riley ask Westwood directly for his photograph.[14] Riley and Darwin remained in contact until Darwin's death in 1882.

Riley's first important contribution to evolutionary theory was his discovery, with Walsh, of the now well-known mimicry of the Monarch Butterfly by the Viceroy. Riley and Walsh were prompted to search for mimetic butterflies by Henry Walter Bates's report of mimicry among butterflies in the Amazon basin, read before the Linnaean Society of London in November 1861 and published in the society's proceedings in November 1862.[15] In the Amazon basin, Bates observed some species of a relatively plainly marked family of butterflies, the Pieridae, that deviated from the normal wing pattern to mimic brightly colored species of a different family, the Danaidae. Bates pointed out that the brightly colored Danaidae "models" had an obnoxious odor and were unpalatable to predators. He reasoned that the mimics, who lacked strong odors and were relished by predators, had arisen through variations that resembled the Danaidae models. Because predators avoided them, they tended to survive and produce more offspring than those that less resembled the model.[16] Bates asserted that the mimetic butterflies offered "better or clearer illustrations of [natural selection] than any other class of animals or plants."[17] Darwin agreed. To Bates he wrote, "the mimetic cases are truly marvelous. . . . I am rejoiced that I passed over the whole subject in the 'Origin' for I should have made a precious mess of it . . . one feels present at the creation of new forms."[18] Bates's report sparked a search for mimetic species worldwide, resulting in similar discoveries by Alfred Russel Wallace in Malaysia, Thomas Belt in Nicaragua, and Roland Trimin in South Africa.[19] In 1865, Darwin arranged to have Bates send a copy of his paper to Walsh, whereupon Walsh and Riley began searching for mimetic butterflies in their neighborhood.[20]

Their search soon bore fruit. In 1868, in the first volume of the *American Entomologist*, Walsh and Riley reported that the milkweed-feeding butterfly *Danaus archippus* was mimicked by the willow-feeding *Nymphalis disippus*, two species now known respectively by the names given them by Scudder in the early 1870s, the Monarch and the Viceroy.[21] The Monarch belongs to the butterfly family Danaidae that Bates found so abundant in the tropics. There are only three species of this family in North America, *Danaus plexippus* L. (the Monarch), which is common in much of the United States and Canada; *D. gilippus* (the Queen), which occurs only in the southern states; and *D. eresimus* Cramer (the Soldier), which is restricted to southern Florida and Texas. Butterflies in the Danaidae family are conspicuous by their bright red, yellow, and orange coloration. They are slow flyers and would normally be easy prey, but they are protected by distasteful, poisonous cardenolide compounds acquired from their milkweed larval food.[22] The Viceroy belongs to the Nymphalidae family whose species are typically colored drab blue-black or black and white. They have short wings facilitating their quick maneuvering to escape predators. In this family, Walsh and Riley found one species, the Viceroy, with brightly colored, longer wings, that was almost an exact replica of the Monarch. The mimicry was so close that some authorities had confused the two.[23]

Walsh and Riley explained this mimicry in Darwinian terms. For one thing, the adult model and the mimic were always found to inhabit the same region during the same season. For another, the mimic Viceroy was abundant in its range, whereas all other species of the family were relatively rare.[24] Since the larval and pupal stages of the Viceroy were almost indistinguishable from other species in the family and would therefore be thinned at about the same rate by predators, Walsh and Riley argued that protection by mimicry in the adult stage explained the abundance of the Viceroy.[25]

Following Walsh's death in November 1869, Riley continued investigating mimicry in butterflies. In his *Third Annual Report* (1871), he cited additional evidence that supported its evolutionary origin. Riley pointed out that the only other species in the same genus as the Viceroy in the Mississippi Valley was *Nymphalis ursula* [the Red-spotted Purple, now known as *Limenitis arthemis* (Drury)], a somber, blue-black butterfly that contrasted in appearance with the bright orange colored Viceroy. Placed side by side, the two species showed little resemblance. The two also differed in their distribution. The Viceroy mimic of the Monarch was plentiful in its range, but the Red-spotted Purple, not a mimic, was rare. How could one account for the relative frequency of the Viceroy and the rarity of the Red-spotted Purple? The larvae and pupae of the two were practically indistinguishable. The only reasonable explanation, Riley insisted, was protection of Viceroy from predators in the adult stage. Even more conclusive was the fact that in the American South, where the Monarch was replaced by a darker relative, the Queen, the Viceroy mimicked this darker species.[26] Ironically, Scudder, at that time probably still an anti-evolutionist, reported the case of the southern relative. Riley stressed the evolutionary implications of this mimicry. The striking resemblance of different species of butterflies, he said, had once been considered curiosities in nature "intended to carry out the general plan of the Creator; but viewed in the light of . . . the Darwinian . . . hypothesis, they have acquired an immense significance."[27]

Riley summarized the debate over *archippus-dissipus* (Monarch-Viceroy) mimicry that ran through several issues of *Nature* (London) in 1870–71. The British-Australian naturalist George Bennett maintained that at least one thousand steps would be necessary to produce a mimic that would warn off predators; therefore, intermediate forms would not have a reproductive advantage. Riley replied that Lepidoptera sometimes exhibited dramatic variation from one generation to the next. When Scudder objected that predation takes place primarily in the larval stage, Riley expressed astonishment that Scudder, who had spent much time in the field, would make such a statement. Riley cited observations of predation of the adult butterfly and pointed out that the larvae and pupae of the Monarch and the Viceroy, though similar in appearance, never occurred in the same habitat, whereas the adults intermingled; therefore mimicry resulting from selective predation could only occur in the adult stage.[28] When Riley visited Darwin in 1871, Darwin congratulated Riley on his discoveries and produced his copy of Riley's third report with the leaves turned down where he had marked Riley's discussion of mimicry.[29]

Riley broadened his investigation of mimicry to include protective coloration in general. In his second report, he remarked that the Goat Weed Butterfly larvae assumed the color of its food plant, enabling it "to escape the sharp eyes of its . . . enemies."[30] In 1871, prompted by experiments of Weir and A. G. Butler demonstrating that brightly colored but distasteful organisms tended to be immune from the attacks of predators, Riley demonstrated that the Monarch possessed a nauseating taste and that "neither turkeys, chickens, toads or snakes would touch it."[31] In his third report, Riley noted that the pupa of the Grape Vine Plume Moth mimicked the colors of the plants to which it clung. Through breeding and experimentation, he demonstrated that pupae attached to the vine stocks assumed a reddish brown coloration, whereas those attached to leaves assumed a green coloration.[32] Riley commented, "The philosophic student of insect life cannot fail to be struck with the wonderful disguises which these little animals . . . assume."[33]

Riley's second important contribution to evolutionary theory was the discovery of the mutual interdependence of the yucca plant and the Yucca Moth.[34] In 1872, George Engelmann, retired medical doctor, founding member of the St. Louis Academy of Science, and leading authority on Western botany, called Riley's attention to the yucca flower's elaborate apparatus for fertilization, which he suggested was accomplished by some nocturnal insect, since the yuccas bloomed only at night.[35] At that time, Engelmann was an anti-Darwinian. Riley interpreted the yucca's elaborate reproductive structure in the context of Darwin's discussion of pollination in the *Origin* and, more particularly, his treatise *On the Various Contrivances by which British and Foreign Orchids are Fertilised by Insects* (1862). Darwin's *Orchid* book was an elaborate debunking of the argument from design. Natural theologians often cited flower's "contrivances" that facilitated their pollination by insects as evidence for the existence of the ultimate Designer. Darwin turned this argument on end by demonstrating that the "contrivances" were in fact adaptations of preexisting organs that had nothing to do with pollination or reproduction. What captured the attention of most entomologists and botanists in the *Orchid* book, however, was its revelation of the complex relationship between flowering plants and the insects that cross-fertilize them. Following Darwin's lead, the German botanist brothers Fritz and Hermann Müller studied such relationships extensively.[36]

Featured Insect

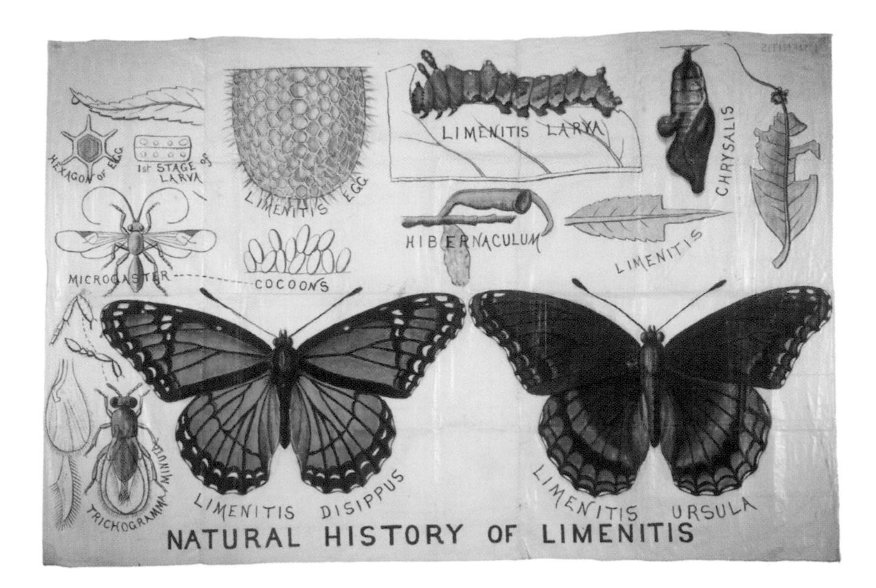

Viceroy, mimic of Monarch [old scientific name *Nymphalis disippus*; current scientific name *Limenitis archippus* (Cramer) (Lepidoptera: Nymphalidae)]. Riley Insect Wall Chart of Viceroy and Red Spotted Purple Butterflies (courtesy of Morse Department of Special Collections, Kansas State University Libraries). The Viceroy feeds on willows instead of milkweeds, and is non-migratory.

Riley soon identified the insects that pollinated the yucca. They belonged to a genus of tineid moth that he named *Pronuba yuccasella*, or the Yucca Moth. (Riley's genus of *Pronuba* has since been revised as *Tegeticula*).[37] The most remarkable aspect of this relationship was that, in contrast to the orchids, which were pollinated by many different insects including butterflies, moths, and bees, yucca was pollinated only by moths of this genus.

The method of pollination was unusual. The stamens of the yucca flower, with their pollen-producing anthers, were located some distance below the top of the pistil and therefore ill suited to promote cross-fertilization. The nectary, located some distance from the pollen-bearing anthers, likewise afforded no particular inducement to insects to visit the flower and cross-fertilize. Given this situation, Riley explained, the female moth acted as a "foster mother" to the yucca. Flying at night from plant to plant, she collected pollen from the anthers, utilizing maxillary tentacles, or mouthparts, wonderfully "contrived" for this purpose, and then rolled the pollen into a ball under her neck. Finally settling on one plant she apparently considered suitable for the rearing of her young, she clung to the top of the pistil and forced the pollen into the stigmatic tube. Only in this way could the yucca be cross-fertilized.

The yucca, in turn, served a vital role in the reproduction of the moth, for the Yucca Moth larvae could only feed on growing fertilized seeds of the plant. During the first season, Riley was unable to determine exactly how the female moth deposited the eggs. He was certain that the eggs developed in the fertilized yucca ovaries, for he had dissected the young fruit and found the larvae there.[38]

Riley reported these findings at the AAAS in Dubuque in August 1872. His paper elicited an animated discussion involving Gray, Darwin's foremost advocate in America; Edward S. Morse, a leading evolutionary theorist and founder of the *American Naturalist*; and others, all of whom marveled at this mutual dependence between an insect and plant species.[39] Riley urged naturalists to report their observations of yucca pollination to help clear up remaining questions. Hermann Müller wrote to Riley that he found the fertilization of the yucca "the most wonderful instance of mutual adaptation yet detected."[40]

With information gathered from naturalists throughout the yucca's range, Riley filled in details of Yucca Moth pollination. After collecting pollen from several yucca flowers, the female sought out flowers that had been open for only one or two nights, the ovules of older flowers not being susceptible to fertilization. Riley initially hypothesized that the female moth first inserted the eggs down the pollen tubes, then applied the pollen to the stigma, because he knew of no Lepidoptera that oviposited by puncturing the ovaries from the outside. After two seasons observing yucca pollination, however, Riley and Engelmann concluded that the female first inserted the eggs by puncturing the ovaries from outside, then proceeded to the top of the pistil where she thrust the pollen down into the stigmatic opening. Riley maintained that the sequence of first ovipositing then pollinating indicated purposeful pollination designed to furnish the larvae with nourishment.

Through the entire sequence, the Yucca Moth did not feed at the nectary, which was located in such a place that it served no apparent "purpose" in pollination. Biologists now speculate that Riley was correct when he suggested that the yucca nectary lures other potential insect pollinators away from the pollen-bearing anthers, thus conserving the pollen for the Yucca Moth.[41] Riley determined that the larva hatched on the fourth or fifth day following ovipositing and began feeding on the ovules. The larva consumed only a portion of the ovules, assuring the survival of the remaining yucca seeds. Upon reaching full size, the larvae dropped to the ground where they burrowed beneath the surface, waiting to emerge as adults prior to next season's blooming. Riley learned later that individuals from a single brood emerged over a period of three years, thus ensuring perpetuation of the species should the yucca fail to bloom in a particular year. About two weeks before the yuccas bloomed, the larvae pupated, emerging as adults just as the yucca came into bloom. Riley attempted to force early maturation of adult moths by submitting larvae to hothouse conditions, but this attempt failed. Riley and Engelmann reported that the yucca stigma closed after the first night of flowering, allowing the female moth a single night to pollinate the yucca and lay her eggs.[42] Riley later found that, on occasion, the yucca flower could be pollinated during two nights.[43] Riley also discovered additional examples of Yucca Moths and the yuccas they pollinated, some pollinating one species and others pollinating several different yuccas.[44]

Having observed yucca pollination night after night without seeing any other species go near the stigma, Riley was convinced that the Yucca Moth was the only insect that fertilized yucca.[45] He cited this as a dramatic demonstration of the co-evolution of plant and insect species. "There is between [the moth] and its food-plant a mutual interdependence which . . . excites our wonder, and is fraught with interesting suggestions . . . Whether we believe, as I certainly do, that this perfect adaptation and adjustment have been brought about by slow degrees through the long course of ages, or whether we believe that they always were so from the beginning, [they are] equally suggestive of that same law and harmony so manifest throughout the realm of Nature."[46] Subsequent research has revealed no non-Yucca Moth pollinators, although two "cheater" Yucca Moths that oviposit into the fruit but do not pollinate have been identified.[47]

Despite praise from naturalists like Gray, Morse, Darwin, and the Müllers, some critics challenged Riley's central finding that Yucca Moths were the exclusive pollinators of yuccas. Riley's primary critics were Zeller, editor of the *Entomologische Zeitung* in Stettin, Prussia; Jacob Boll, a Swiss immigrant in Dallas, Texas, whom Riley engaged as an agent for the US Entomological Commission to report on the Rocky Mountain Locust; Vactor T. Chambers,

Featured Insect

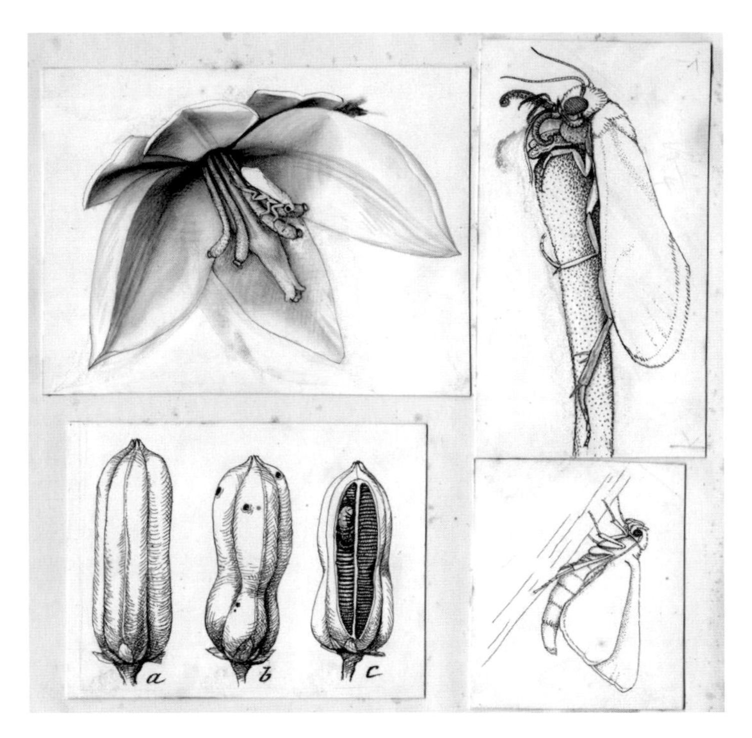

Yucca Moth [old scientific name *Pronuba yuccasella*; current scientific name *Tegeticula yuccasella* (Riley) (Lepidoptera: Prodoxidae)]. Yucca Moth pollination, from C. V. Riley, "The Yucca Moth and Yucca Pollination," *Report of the Missouri Botanical Garden* (1892); *Popular Science Monthly* 41 (June 1892); and elsewhere (courtesy of the National Agricultural Library, Beltsville, Maryland). Yucca moths and their obligate pollination mutualism with the yucca are a continuing source of interest and study for scientists studying evolution and speciation. For instance, in the past twenty-five years, over fifty scientific studies have been published with "yucca moth" or *Tegeticula* in their title.* Based on a "molecular clock" using mitochondrial DNA, Pellmyr and Leebens-Mack in 1999 estimated that this mutualism dates back about forty million years, with a burst of speciation accompanying the development of arid habitats in western North America about three million years ago, and "cheater" moths such as the bogus yucca moth [old and current scientific name *Prodoxus decipiens* (Riley)] evolving about one million years ago.**

*Google Scholar search, February 28, 2018.

**Olle Pellmyr and James Leebens-Mack. "Forty million years of mutualism: Evidence for Eocene origin of the yucca-yucca moth association," *Proceedings of the National Academy of Sciences* 96 (1999): 9178–9183.

attorney in Covington, Kentucky, and specialist on the micro-Lepidoptera, including the Yucca Moth;[48] Thomas Meehan, botanist and editor of the *Gardener's Monthly*; George D. Hulst, editor of *Entomologica Americana*; and (indirectly) Hermann Hagen, head of the department of entomology at the Museum of Comparative Zoology (MCZ).

Three aspects of these objectors stand out. First, the objections of Meehan, Boll, Chambers, and (indirectly) Hagen sprang from a confusion of species. Upon visiting Boll in Dallas in 1879, Riley discovered that Boll's specimens were in fact not Yucca Moths but a closely related species that Riley subsequently named *Prodoxus decipiens*, or the Bogus Yucca Moth.[49] At the AAAS meeting in Saratoga Springs, New York, in August 1879, Riley explained that, despite the close resemblance of the adults, the structure and life history of the two species, particularly in the larval stage, were quite different. The Yucca Moth larvae, which left the yucca plant to burrow into the soil, developed legs, whereas the Bogus Yucca Moth larvae, which remained in the yucca stem, were legless. The female Bogus Yucca Moth lacked the Yucca Moth's characteristic maxillary palps that allowed her to gather and transfer pollen. Yucca Moths emerged as adults just prior to the yucca's blooming, whereas Bogus Yucca Moths appeared before the yucca bloomed, as their larvae fed on the stems as well as the seeds. Above all, the Yucca Moth fertilized yucca, while the Bogus Yucca Moth did not.[50] Riley had, in fact, earlier reported the presence of the then-unnamed Bogus Yucca Moth. At that time, he had pointed out that the legless larvae of the unnamed species were clearly different from Yucca Moth larvae.[51]

Second, Darwin and other leading evolutionists steadfastly supported Riley's evolutionary interpretation of the mutual dependence of Yucca Moths and yucca plants. In 1874, Darwin wrote to Joseph D. Hooker, director of the Royal Botanic Gardens, that Riley's discovery of Yucca Moth pollination was "the most wonderful case of fertilization ever published."[52] Darwin cited Riley's Yucca Moth studies prominently in *The Effects of Cross and Self Fertilization in the Vegetable Kingdom* (1876). Following Riley's rebuttal of Meehan, Boll, Chambers, and Hagen at the Saratoga Springs meeting, Riley sent Darwin a copy of his address. Darwin replied that Riley had effectively answered all this critics, adding that Meehan was "an inaccurate man [whose] Epitaph ought to be 'He retarded natural science in the U. States as much as any one man advanced it.'"[53]

Third, Riley prevailed over his critics because he investigated yucca pollination more thoroughly than any other. In the 1880s and 1890s, he enlisted colleagues in the Division of Entomology, including Howard, Lugger, Pergande, Eugene A. Schwarz, Albert Koebele, and Charles L. Marlatt. He also collaborated with William Trelease at the Missouri Botanical Garden. Throughout Riley's career, the Yucca Moth pollination story was his most popular lecture topic.[54] His explanation of the interdependence of the yucca and the Yucca Moth, coming shortly after the discovery of the Viceroy-Monarch mimicry, reinforced his reputation as an investigator and defender of Darwinian theory. In 1876, Morse cited Riley's discovery of Yucca Moth pollination as a prominent example of American contributions to evolutionary studies.[55]

Whereas Riley's discoveries of butterfly mimicry and yucca pollination sparked widespread interest, his reports of rapid evolution in the habits and structure of various insects received only passing attention. One likely reason for the lack of attention was that Riley published his accounts of new insect species, or new varieties that were harmful, in his annual reports rather than in scientific journals or in presentations at scientific meetings. He typically disregarded the distinction his academically oriented colleagues made between "scientific" and "practical" publication. For example, he often quoted portions of, or entire articles, from the *Proceedings of the St. Louis Academy of Science* in his reports. As noted above, he insisted that discoveries and insect descriptions published in agricultural reports be considered as valid as those published in scientific journals. As state entomologist, he discussed the evolution of pest and beneficial species in the context of control. From a similar, practical perspective, he also suggested that evolution held the key to improving crops and farm animals; however, he felt that experimental breeding of species other than insects lay outside his mandate as state entomologist.

Riley's thinking with regard to the sudden evolution of injurious insects was stimulated by his comparison of insect depredations in America to those of his native England, his observation of rapid ecological change in the American heartland, and his expectation that the Darwinian "struggle for existence" would produce new species in a relatively short time.

As a transplanted Englishman, Riley compared the frequent and severe episodes of insect depredations in America to the relatively mild insect damage in England and on the European continent. Riley and his English mentor Walsh pointed to the fact that most injurious insect species in America came originally from Europe, whereas relatively few American insects had become serious pests there. In 1866, Walsh compiled a list of imported insects that comprised fully half of the most injurious species.[56] In 1870, Riley extended the list of imported injurious species, listing the Hessian Fly, Wheat

Midge, Codling Moth, Cabbage Tinea, Borer of Red Currant, Oyster Shell Bark Louse, Grain Plant Louse, Cabbage Plant Louse, Currant Plant Louse, Apple-Tree Plant-Louse, Pear-Tree Flea-Louse, Cheese Maggot, Common Meal Worm, Grain Weevil, House Fly, Leaf Beetle of Elm, Cockroach, Croton Bug, Carpet-Clothes-and-Fur Moths, Asparagus Beetle, and the Rape Butterfly. Riley noted that Gray's list of imported plants also showed a disproportionate number of European imports flourishing in America.[57] Walsh speculated (and Riley concurred) that the American fauna and flora were much older than those in Europe, and that the newer European species were more vigorous. The European imports thus overwhelmed their weaker American competitors, whereas American species could not prevail against the more vigorous plants, predators, and parasites they encountered in Europe. Walsh (and Riley) postulated further that Australia's still older fauna and flora could not compete with American imports.[58]

A second ground for Riley's belief in the rapid evolution of species was his conviction that the American ecosystem was undergoing a massive transformation. From Illinois settlers in Kankakee and Chicago, he learned that a few decades previously annual prairie fires had thinned out trees and brush to produce a cover of prairie grass. By the time he arrived in the 1860s, Illinois was becoming thickly populated, and Riley witnessed the replacement of former prairie by a dense growth of bitternut hickory and other hardwoods.[59] Riley related these local ecological changes to the massive deforestation of eastern North America. Like many nineteenth century naturalists, Riley assumed that deforestation affected rainfall.[60] At a meeting of the St. Louis Farmer's Club in 1873, Riley spoke on "The Influence of Climate on Trees, Plants, and Animals," and in April 1882, he delivered a paper on injurious insects at the first American Forestry Congress in Cincinnati.[61] In a session devoted to climate at the St. Louis Academy of Science in 1872, Riley announced that forests "equalize the distribution of rain."[62] He believed that profound environmental change like the deforestation of eastern woodlands and the encroachment of trees and brush on mid-western prairies would foster corresponding changes in insect form and behavior.

In 1870, Riley noted that the Rape Butterfly, a harmless insect he remembered from his boyhood, was now inflicting great damage in Canada and in the eastern and central states. "Little did I dream, when, many years ago, I watched this butterfly fluttering slowly along some green lane . . . that I should some day be fearing its presence here."[63] Riley credited the butterfly's new destructiveness to "the immense . . . changes which have come over this broad land during the last two or three centuries."[64] He speculated that insects had become more injurious in settled regions than they had been in pristine, presettlement times. Riley observed that, fifty years previously, "thrifty long-lived [apple] trees were . . . common [because they] had not the vast army of insect enemies to contend with which at the present day make successful fruit-growing a scientific pursuit."[65] The Apple-Tree Borer, for example, known to entomologists since it had been named by Say in 1824, had only recently become a serious threat to apple growers around St. Louis.

In 1877, near the end of his tenure as state entomologist, Riley explained how previously harmless species had recently evolved into serious pests: "There can be no more tangible evidence . . . of . . . evolution, and . . . the constant modification in habit, and . . . in structure and colorational characteristics among animals [of] the lower classes, than the frequent appearance . . . [of pests] that were never reported as injurious before."[66]

Riley's speculations were prompted by the apparent sudden change of habits in *Leucania albilinea* (now *Faronta albilinea* Hübner), a moth species related to the infamous Army Worm moth. The caterpillar of this indigenous, non-pest species, Riley explained, had for centuries fed on native grasses. Beginning in 1874, however, farmers in Maryland and Pennsylvania reported that caterpillars were now feeding on the seed head of wheat and timothy. When depredations in those states increased in 1875–76, Riley asked correspondents to send him eggs, caterpillars, pupae, and adults so that he could breed them in order to study the new race. In 1876, correspondents in eastern Kansas reported destruction by the wheat-head-feeding race. Riley considered it improbable that such ravages could have gone unnoticed by entomologists prior to 1876. Riley speculated that the wheat-head-feeding race had emerged in York County, Pennsylvania, and had subsequently been carried to Kansas, most likely as eggs in grain. Because there were no reports of the injurious race between Pennsylvania and Kansas, Riley concluded that a new race had been transferred by rail from one region to another. He predicted that the Wheat-Head Army Worm would spread and, like the true Army Worm, cause substantial damage in years when conditions favored its increase. As a consequence of the rapid change in the environment, he further expected "an increasing tendency in the [wheat-head-feeding] species to vary, and give rise to still other varieties."[67]

In his Missouri years, Riley struggled to distinguish between factors influencing the abundance or scarcity of a given species from season to season and those factors, including intraspecific competition, that promoted natural selection. In 1870, his explanation of the widespread injury caused to cucumber plants by the previously unknown "Pickle Worm" intermingled balance of nature (interspecific competition) with Darwin's "struggle for existence" (intraspecific competition): "The system of Nature is so complicated, and every animal organism is subject to so many influences . . . that we are not surprised at the fluctuation in the relative numbers of any particular species. The 'Struggle for Life,' as expounded by Darwin, is nowhere more effectual in bringing about changes than in insect life. We are at first a little puzzled to account for the sudden advent, and the equally sudden departure of such insects as the Army Worm, Chinch Bug, Wheat Midge, etc., but when we once acquire a just conception of the tangled web in which every insect is involved, we wonder rather that the balance is so well kept."[68]

In his discussion of the Wheat-Head Army Worm in his final report (1877), Riley ascribed the emergence of new species to evolutionary modification independent of seasonal fluctuations of insect populations. In most cases, he wrote, "new" injurious insects were known species that, because of weather conditions or the introduction of new crops, suddenly multiplied and became injurious. In other cases, however, "in the most [entomologically] well worked-up localities . . . new forms appear, and old forms sometimes disappear, in a manner which can scarcely be explained, except by the extinction of the one and birth of the other through modification. . . . [N]ew forms . . . are thus originating at the present, and we may occasionally get a glance at the process."[69]

In his first four Missouri Reports, Riley reported a number of new insect races that originated when individual insects acquired a new food preference and transmitted this habit (variation) to succeeding generations. In 1869, Riley's colleague William Saunders, in London, Ontario, reported that Codling Moth larvae around London had switched from eating apples (the normal food of this pest species) to eating plums. Riley considered this "new trait in . . . our Codling moth . . . another evidence of the manner in which certain individuals of a species may branch off from the old beaten track of their ancestors."[70] He then speculated that "This change of food sometimes produces a change in the insects themselves, and it would not be at all surprising, if this plum-feeding sect . . . should in time show variations from the normal pip-fruit [apple] feeding type."[71] Three years later, Riley reported that Codling Moth larvae in Missouri had begun feeding on peaches. "Thus, we have a second example in this same species of an insect normally confined to pip-fruit, all at once taking to stone fruit."[72] In his third report (1871), Riley announced that the Apple-Maggot, an indigenous species that had originally fed on wild haws, had in recent years produced a race in the eastern United States with a taste for cultivated apples. Riley considered it more likely that the apple-feeding race of the east would spread west than that western haw-feeding larvae would soon acquire the habit of feeding on cultivated apples.[73]

In that same report (1871), he noted that Glassy-Winged Soldier Bugs, known previously only as plant feeders, were now preying on leaf-hoppers (a potentially beneficial habit);[74] that about 1860 the Bean Weevil, an indigenous species, had began feeding on cultivated beans in eastern states so that it now threatened to become a serious pest in the Mississippi Valley;[75] and that a race of Apple Curculios that normally fed on wild apples had recently begun feeding on domesticated apples.[76]

While the emergence of insect races that fed on cultivated fruits and crops was bad news for cultivators, Riley predicted that Darwinian selection would also produce beneficial races that would act as controls. When the horticulturist W. B. Ransom argued that parasites were ineffective as controls of the Curculio because parasites had always been present but had not prevented Curculio outbreaks, Riley responded that "[T]hose who rightly comprehend the Darwinian hypothesis . . . and who believe that life is slowly undergoing change . . . to day just as it ever has" could take hope because some Curculio parasites "may have acquired the habit of preying upon the Curculio within the past comparatively few years."[77] At that time, Riley was breeding two Curculio parasites for distribution to each county in Missouri.[78] In 1871, he commented that the new race of Glassy-Winged Soldier Bug (noted above) was "an excellent illustration of an insect acquiring a new habit. Some individual or individuals wandering from the oaks (its normal habitat) . . . came upon . . . Spaulding's vineyard and found the leaf hoppers to their taste . . . they soon multiplied . . . and commenced to spread from one vineyard to another."[79] At that time, leaf-hopper–eating bugs were reported within a three-mile radius of the center of their apparent origin in Spaulding's vineyard. He recommended that vintners assist the spread of this beneficial race by transferring them to other vineyards.[80] In 1872–73, Riley colonized a dozen or more larvae of Twice-Stabbed Lady Bird beetles on trees infested by scale insects and was so impressed by their control of scale that he planned to distribute the offspring to fruit growers.[81] About that time, LeBaron introduced a chalcidid parasite around Galena, Illinois, in an effort to control the Oyster Shell Bark Louse. Riley planned to introduce the parasite in two counties in Missouri.[82]

It is difficult to judge the accuracy of Riley's reports of new insect races. In the 1870s there was no consensus regarding the causes and frequency of variation. In fact, Riley, like Darwin, left the cause of variation open. In 1881, Riley cited Darwin in support of his belief that insects sometimes adapted rapidly to changed conditions and that these adaptations could lead to rapid species modification.[83] Riley's emphasis on rapid environmental change in the American interior may have led him to overestimate the speed and frequency with which races, subspecies, and ultimately species emerged. On the other hand, his emphasis on the changing environment informed his valid insight that insect species are indeed plastic and that variations occur more frequently than previously assumed. In the twentieth century, biologists discovered, for example, that local populations of the Codling Moth in the arid American West displayed different behavior patterns than populations in the humid East and advised growers in the West to spray only once, whereas they advised those in the East to spray several times.[84] Similarly, twentieth-century entomologists recommended different control measures for Cotton Boll Weevil varieties in the arid West than they recommended for varieties in the humid Mississippi delta. They even discovered a geographical variety of the Cotton Boll Weevil, the Thurberia Weevil, so thoroughly adapted to arid conditions that it fed on wild cotton rather than on domestic cotton.[85] In the twentieth century, Riley's insight was confirmed through the new concept of "biotypes," that is, subspecies with altered habits. Interestingly, among the first biotypes identified were those in Grape Phylloxera, a species Riley speculated had changed its habits when transferred to Europe. Unlike Riley, twentieth-century biologists were at first reluctant to acknowledge the new forms, although Darwinian theory should have prepared them for this.[86]

Riley's evolutionary views take on sharper focus when compared with his entomological colleague, Alpheus Spring Packard. Born in 1839 and raised in a devout Christian family (his father was professor of theology at Bowdoin College), Packard believed species had been separately created within an orderly balance of nature overseen by a benevolent deity. In 1866, Packard (then Agassiz's assistant), reacted to Walsh's broadside against Agassiz and his followers by insisting that variation among existing species was subject to "a deeper-seated tendency of all . . . organisms towards a perfection of the type."[87] Within the next few years, however, Packard became convinced of the origin of new species through natural agencies. In contrast to Riley, he considered the direct influence of the environment, rather than natural selection, to be the primary mechanism in the evolution of new species. In Packard's view, environmental conditions produced variations. Packard essentially replaced the deity in natural theology with climate and other environmental influences as the originator of new species.[88]

Like Riley, Packard was an entomological generalist. He authored some 579 works, including *A Guide to the Study of Insects* (1869); the *Annual Review of Entomology* (a listing and commentary on entomological publications for each year from 1868 to 1873); three annual reports on the injurious insects of Massachusetts (1871–73); as well as monographs of Lepidoptera and other groups.[89] With Frederick Ward Putnam, he discovered underground crayfish in Mammoth Cave that they believed had descended from aboveground ancestors whose visual organs had atrophied through the absence of light. These crayfish compensated for the loss of sight by an enhanced sense of touch and smell. Packard cited this as evidence for the modification of species through the influence of changed environmental conditions.[90]

Packard elaborated his theory in his monograph on geometrid moths (1876), based on his investigations of moths in the American West under Ferdinand V. Hayden's Geological and Geographical Survey of the Territories, supplemented with material gathered from the eastern United States. Packard argued, for example, that warm, humid conditions on the Pacific slope produced larger moths with fuller development of peripheral body parts and more vivid coloration as compared to moths in dryer, cooler habitats.[91] Packard assigned natural selection a secondary role as the preserver of variations induced by the environment.[92]

Packard's emphasis on environmental causality reflected the dominant trend among American naturalists in the 1870s and 1880s. American ornithologists like Spencer F. Baird, Robert Ridgeway, and Elliott Coues, paleontologists Othniel C. Marsh and Edwin Drinker Cope, and geologists Nathaniel S. Shaler and Joseph LeConte identified climate as the cause of variation and the resulting emergence of new species. American naturalists of this persuasion drew upon the evolutionary theory of Jean-Baptiste Lamarck, who, in his *Philosophie Zoologique* (1809) and *A Natural History of Invertebrate Animals* (1815), proposed that organic change (including the emergence of new species) resulted from the interaction of the "power of life" with the environment. In 1884, Packard proposed the term neo-Lamarckism to describe those, including himself, who held such views.[93] Many American naturalists, including Packard, held that Lamarck's theory was superior to Darwin's theory because it provided an explanation for the origin of variations.

Riley predicted that wholesale changes in the North American environment facilitated the evolution of new insect races that would develop into new species within a few years. Packard agreed that environmental conditions modified existing species, but he believed that new species emerged only during periods of rapid geological and meteorological change, for example during the epoch when geological uplift produced the Appalachian and Rocky Mountains. He viewed present conditions as relatively stable and therefore he did not expect that new species would soon emerge.[94]

Packard briefly considered the theory proposed by Cope and Alpheus Hyatt that species, like individual organisms, had a "lifetime," during which they developed at different rates, sometimes rapidly, other times slowly, a process they called "acceleration and retardation." He soon rejected this thesis because it postulated an internalist mechanism that worked independently of external influences.[95] Riley apparently considered the Cope-Hyatt theory a side issue at best.

Riley and Packard's differences centered on the origin of variations and the transmission of these to subsequent generations through heredity. Riley, following Darwin, held that the origin of variations was not yet known. Packard, following Lamarck, held that variations originate through the direct action of the environment.

Riley and Packard shared the generally accepted view in the 1870s that variation induced change whereas heredity conserved stability. Packard considered heredity to be "soft," allowing variation to make change possible, while Riley thought of heredity as "hard," preserving variations and the integrity of races and species despite environmental influences.[96] In Packard's view, climatic and other influences affected all individuals simultaneously, transforming, for example, entire populations of moths in a given direction. He therefore considered individual variations of the kind Riley cited as insignificant.

Riley believed that spontaneous, random variations that proved to be beneficial were passed on to succeeding generations through natural selection, or perhaps some other agency. In Riley's view, new races (and species) arose when a sufficient number of beneficial variations (e.g., altered food preferences of the larvae) were transmitted to the overall population through natural selection. Riley's conception of heredity was thus more "Darwinian" and Packard's more "Lamarckian."

Riley and Packard also differed about the role of instinct, intelligence, and will in the transformation of animal species. Riley believed that individual organisms played an active role in the modification of habits. His assertion that individual choice is an inheritable variation is clearly different from Packard's emphasis on environmental influences. In discussing the Yucca Moth's apparently purposeful fertilization of the yucca, Riley stated, "I cannot help thinking that instinctive and reasoning faculties are both present, in most animals, in varying proportions, [and that reasoning powers are] called into play."[97] In his account of the Wheat-Head Army Worm, Riley postulated that the sudden preference of some individuals for wheat grain was spread through inheritance, eventually producing a modified race of wheat-head–eating worms. Riley's interpretation of insect transformation was more consistent with Darwinian natural selection than Packard's environmental-induced variations.

In the 1870s, American naturalists (including Riley and Packard) often combined various evolutionary theories depending on their own scientific, religious, and psychological orientation. In the 1880s, however, August Weismann's germ plasm theory challenged Riley, Packard, and naturalists of every persuasion to accept or reject Weismann's strict Darwinism, now defined in terms of natural selection alone.

In a series of essays published from 1883 to 1888 that were combined in book form and published in English translation in 1889, Weismann distinguished between germ (sexual) cells that transmit characteristics unaltered from one generation to the next, and somatic (body) cells that can be altered by use or disuse but cannot transmit these alterations to the next generation. Weismann rejected the inheritance of acquired characteristics. Although he left the question of the origin of variations unanswered, he insisted that they were transmitted, unaltered, through the germ cells to the offspring.[98]

With his insistence that natural selection alone accounted for evolutionary change and his categorical rejection of any Lamarckian agencies like the influence of the environment or the inheritance of acquired characters, Weismann was more purely "Darwinian" than Darwin himself, who, especially in his later books and editions of the *Origin*, left the door open for some such influence. Weismann's ultra-Darwinism found few followers in Germany but attracted a strong following in Great Britain, where, in 1888, George J. Romanes, marine biologist, student, and friend of Darwin, coined the term "neo-Darwinian" to describe Weismann's adherents.[99] American naturalists continued to hold a variety of views. Although not neatly divided into neo-Lamarckians and neo-Darwinians as historians have often held, many felt compelled to respond to Weismann's challenge.[100]

Given Riley's generally "orthodox" Darwinian orientation in the 1870s, it comes as a surprise that, when confronted with Weismann's germ theory, he identified himself as a neo-Lamarckian. When asked by Henry de Varigny, a French physician, journalist, and advocate of Darwinism, for his views on evolution, Riley responded, "In connection with other American naturalists, I am rather neo-Lamarckian."[101] In 1892 he wrote, "My first conviction is that insect life and development give no countenance to the Weismann school, which denies the transmission [inheritance] of functionally acquired characteristics, but that, on the contrary, they furnish the strongest refutation of the views urged by Weismann and his followers."[102]

In the 1880s and 1890s, Riley expressed his revised thinking on evolution in three publications: (1) his tribute to Darwin's work in entomology on the occasion of Darwin's death in 1882; (2) his discussion of the causes of variation in organic forms delivered as his 1887 vice-presidential address before the Zoology Section of the AAAS; and (3) his address on the interrelations of plants and insects before the Biological Society of Washington in 1892.[103] In his 1892 address, Riley explained his Lamarckian position:

> [T]o my view modification has gone on in the past, as it is going on at the present time, primarily through heredity in the insect world. I recognize the physical influence of environment; I recognize the effect of the interrelation of organisms; I recognize, even to a degree that few others do, the psychic influence, especially in the higher organisms—the power of mind, will, effort, or the action of the individual as contradistinguished from the action of the environment; I recognize the influence of natural selection, properly limited; but above all, as making effective and as fixing and accumulating the various modifications due to these or whatever other influences; I recognize the power of heredity, without which only the first of the influences [physical environment] can be permanently operative.[104]

In his address before the AAAS in 1887, Riley paid tribute to both Darwin and Lamarck, but he argued that Lamarck's teaching, that functionally produced modifications (use and disuse) could be inherited, deserved more attention and that Darwin's emphasis on natural selection ignored important Lamarckian factors.[105] In Riley's view, the weaknesses in Darwin's theory were his failure to account for variations, his limiting variation to slight modification, his insistence that new species required long periods of time to develop, and his insistence on the absolute utility of modifications.[106] These deficiencies, Riley maintained, could be rectified through Lamarckian theory. Riley enlisted the support of Lester F. Ward, a paleo-botanist in the US Geological Survey whose primary work was actually studying the dynamics of human society, expounded in his *Dynamic Sociology*. In Ward's presidential address before the Biological Society of Washington in 1891, he defended neo-Lamarckism.[107] The Biological Society, with thirty to fifty members attending bimonthly meetings, seems to have been a stronghold of neo-Lamarckians. Riley, Ward, and others in the society considered inheritance to be the central issue. Ward charged that Weismann and other neo-Darwinians had not disproved the capacity of living organisms to pass on useful adaptations to their offspring. Because scientists lacked the means of observing and measuring variations in the inheritable material at the cellular level, their only recourse was to observe and analyze changes in the whole organism. Since Darwinists and Lamarckians all agreed that organisms had everywhere adapted themselves to their environment, Ward considered the case for inheritance of acquired characters closed.[108] Only radical neo-Darwinists like Weismann and his English followers denied what Ward called "functional" modification as different from "selective" modification.[109] Ward dismissed neo-Darwinian arguments based on artificial, non-environmentally induced mutations like cutting the tails from generations of mice. Such mutilations had not arisen in response to the environment and were of no functional use to the individual organism or its progeny.[110]

Riley supported Packard and other American naturalists who stressed the importance of external factors.[111] Riley asserted that scientists now generally accepted Lamarck's concept of functionally produced modification.[112] Riley was intrigued by Oliver Wendell Holmes's novel *Elsie Venner*, in which the personality of the lead character is formed by experiences in the womb. Riley intended to investigate psychic influences as a biological phenomenon, and he requested that his colleagues send information relating to the subject, but this project was cut short by his death.[113]

Riley was convinced that in higher animals the mind, will, and emotions produced characteristics that were inherited, and he insisted that lower animals had some measure of these characters.[114] He was undecided as to whether plants had similar psychic capabilities.[115] Riley's speculations were not unusual for his time. Darwin, in his final book, concluded that worms had minds and could learn. Such speculations were rejected by later generations, who postulated that the behavior of lower organisms consisted of a collection of automatic (mindless) instincts. In recent years, scientists have reevaluated Darwin's (and Riley's) emphasis on consciousness in lower organisms. Experimentation has demonstrated that certain wasps recognize individual faces of other wasps and that plants register sights, sounds, and have a limited memory.[116]

Riley viewed Lamarckian evolution as the basis of civilization. He agreed with Ward, who asserted that universal improvement on all fronts—in art, literature, and sport—proved Weismann wrong. If every generation had to start from scratch, not having inherited or learned anything from past experience, how could one hope for continued progress?[117] Riley cited Gray, who held that direction in evolution revealed the work of a Creator, even though (for Gray) natural selection governed the process of evolution.[118] Riley speculated that the evolutionary process transformed a primal energy, or vital force, into physical and chemical forces, a process traceable back to an Infinite First Cause.[119] Evolution, Riley insisted, did not detract from the concept of a Creator, lessen the validity of religious beliefs, or diminish the human experience of faith, hope, love, and charity.[120] Riley insisted that "natural law . . . helps to broaden our views and enables us to . . . rise to sublimist contemplation of that unknown Infinity which pervades all."[121] He hoped that theologians would one day revere Darwin as one who enriched the concept of creation. Interestingly, Riley's advocacy of Darwinian evolution in his Missouri Reports apparently elicited no public objections, either from state officials or among his readers.[122] The fundamentalist-creationist reaction to Darwinism came later, surfacing publicly in the 1920s and enduring up to the present.[123]

Riley's belief in the evolutionary progress of society melded with a general intellectual movement later termed Social Darwinism. He believed that a good home environment and right living during childhood would produce improved citizens and an improved society. Linking evolutionary ideas to then-current conceptions of progress, Riley celebrated the rise of humans from a lower to a higher state of development, a process he predicted would lead to unlimited advancement. He even anticipated the emergence of a distinctive American type, not yet fixed, a type that he predicted would influence the future.[124]

Riley was not a major evolutionary theoretician. Furthermore, Riley frequently advocated positions that, from a modern perspective, were wrong, such as the curious Walsh-Riley idea of "old" American vs. "new" European fauna. When he insisted in 1887 that introduced insects have an advantage over indigenous insects, he seems to have momentarily forgotten the lesson from his Grape Phylloxera studies, that plants that co-evolve with insects develop corresponding defense mechanisms, regardless of which continent they inhabit.[125]

On the other hand, Riley's investigation of mimicry in butterflies and his explanation of Yucca Moth pollination strongly influenced the course of evolutionary thinking and continue to be widely cited as compelling demonstrations of evolution.[126] Riley's reports of rapid evolution of insects, though little recognized at the time, foreshadowed biologists' later appreciation of the variability and rapid evolutionary change in insect form , physiology, and behavior.

Above all, Riley was a leading disseminator of evolutionary theory.[127] Richard Billon has argued that the crucial factor in naturalists' acceptance of Darwin's theory was not the general argument in the *Origin* (which seemed at first too speculative and ungrounded in fact) but the demonstration that Darwin's theory produced convincing explanations of phenomena that naturalists investigated day to day. For zoologists, the turning point occurred in 1861–62 with Bates's interpretation of mimicry in butterflies; for botanists the turning point occurred in 1862 with Darwin's *Orchid* book and its evolutionary explanation of various reproductive structures in flowers—structures familiar to all working botanists. Interestingly, Bates presented his paper at the same meeting of the Linnaean Society (November 1, 1861) where Darwin presented his evolutionary botany paper that led to the *Orchid* book.[128] Riley's discoveries involved precisely the phenomena that were central in the triumph of Darwinism among zoologists and botanists, namely mimicry and insect fertilization of plants.

Riley's effectiveness as a disseminator of evolutionary theory sprang both from the subject matter and from his personality. Naturalists and others were impressed not only by Darwin's theoretical argument but also by Darwin's character and by the methods he employed. In the nineteenth century, learned individuals in Great Britain and America presumed an ideal scientist carrying on investigations according to inductive Baconian procedures, patiently examining concrete natural phenomena, "facts," from which he carefully and slowly constructed generalizations or theories. The model scientist thus brought theory into accordance with facts, disciplining his inspiration by patient, systematic work.[129] Riley, as a state and federal entomologist and curator of insects at the National Museum, approached this ideal. His reports of Darwinian evolution grew out of his careful, detailed investigation of insect pest and non-pest species. He combined Baconian attention to facts with the Franklinian search for useful knowledge. His discoveries and views were widely disseminated in government publications, in agricultural publications like the *American Entomologist*, and in newspapers like the *New York Tribune*, the *New York Rural World*, and the *Chicago Tribune*.[130] Furthermore, he frequently discussed evolution before agricultural, scientific, and popular audiences. Riley's evolutionary thinking not only guided his entomological investigations but also informed his optimistic worldview and his steadfast belief in human progress.

6

Subterranean Killers

[H]e who will . . . study [the Grape Phylloxera] and . . . give to the world the results, will . . . render valuable service to the . . . science . . . of economic entomology.

—C. V. Riley, "The Grape-Leaf Gall-Louse—*Phylloxera vitifoliae*, Fitch," *Third Annual Report* (1871), 84–85

In 1868, when Riley assumed office as Missouri state entomologist, French scientists identified a new pest feeding on grapevine roots that they named *Phylloxera vastatrix*. In the next few years, Riley and others demonstrated that the insect was of American origin but known by a different name. Missouri vintners, who represented a substantial segment of the American wine industry, considered the insect of little significance.[1] Nevertheless, Riley sought out Jules Émile Planchon, the leader in French phylloxera research, traveled to France three times, assisted French investigators during their visits to the United States, and featured phylloxera prominently in his reports. Why did Riley spend so much time and energy investigating a tiny insect that was causing problems in faraway France?

One reason was Riley's emotional tie to France, dating back to school days in Dieppe. Another reason was his scientific curiosity, especially the evolutionary implications of its destructive nature in Europe versus its relative harmlessness in America. Above all, Riley saw the opportunity to demonstrate the effectiveness of economic entomology.[2] Although his official position was state entomologist, Riley viewed insects in global context, whether butterfly mimics, migrating locusts, scale insect pests from Australia, or an American plant louse on grape vines in Europe.[3]

In 1868, when vintners in Roquemaure (a village on the right bank of the Rhône River in France) reported vines mysteriously withering and dying, the Hérault Agricultural Society in Montpellier appointed a commission comprised of Gaston Bazille, the society president, Planchon, a physician and professor of pharmacy and botany at the Montpellier Agricultural School, and Felix Sahut, a prominent wine grower in the Midi district around Montpellier, to investigate.[4] They discovered multitudes of aphid-like insects sucking juices from the roots of the vines. Although none of the commission members were entomologists, they described the louse as a new species, *Rhizaphis vastatrix*. Planchon (a botanist) sought entomological advice from his brother-in-law, Jules Lichtenstein, who had studied entomology under his (Lichtenstein's) uncle, Heinrich Lichtenstein, a professor of zoology at the University of Berlin.[5] He also sought advice from Victor Antoine Signoret, in Paris, who specialized in aphids.[6] Signoret suggested that Planchon substitute the genus *Phylloxera* (dry leaf) for *Rhizaphis* (root), on the grounds that the winged form was similar to the winged form of *Phylloxera quercus*, an oak-leaf gall insect that had been described in 1834.[7] Planchon followed Signoret's suggestion, even though his insect had been found only on roots, not leaves. Thus, in 1868, French scientists named the new pest *Phylloxera vastatrix*.[8]

In St. Louis, Riley read French and English accounts of this new vine pest.[9] Of special interest was an article by Riley's countryman, John O. Westwood, in the January 1869 issue of *Gardeners' Chronicle* in which he reported that in 1863 he had received grape vines from Ireland and England containing tiny insects damaging the leaves.[10] In 1867–68, he received additional samples, this time with apparently the same insects on both the leaves and the roots. In a report to the Ashmolean Society of Oxford in 1868, he gave the insect the (unpublished) name *Peritymbia vitisana*. In the 1869 *Gardeners' Chronicle* piece, Westwood referred to Planchon's proposed name and Signoret's opinion that the insect belonged to the genus *Phylloxera*, and he speculated that the insect was identical to the vine louse described by Asa Fitch in 1855 as *Pemphigus vitifoliae*.[11]

Three years earlier, in 1866, Riley, then unaware of an international connection, had published a short note on the "grape louse" in the *Prairie Farmer* in which he informed his readers that the insect that formed galls on Illinois grapevines was the same one described by Fitch.[12] In August 1869, prompted by the crisis in France, Riley published a second article on the grape louse in which he cited Lichtenstein's and Westwood's speculations that the insect was identical to Fitch's grape leaf gall insect. Riley was intrigued by Westwood's suggestion that the root and leaf forms were the same insect, because neither he nor any other American naturalist knew of a root form in America.[13]

Sensing the opportunity to study an insect of potentially immense economic importance, Riley suggested to Lichtenstein in the fall of 1869 that they join forces, and he sent specimens of American leaf forms to Signoret for comparison with European root lice.[14] During the following year and a half, Riley's contact with his French colleagues was interrupted by the outbreak of war between France and Prussia (July 1870), the siege of Paris (September 1870–January 1871), and the bloody suppression of the Paris Commune (March–May 1871).[15] Nevertheless, during the year and a half between Riley's first contact with French investigators in 1869 and his trip to France in 1871, he and his French colleagues made important discoveries about the identity and life history of Grape Phylloxera.

The immediate question was why the European Grape Phylloxera occurred on the roots whereas the American Grape Phylloxera was known only on leaf galls. We now know that Grape Phylloxera, though often fatally infesting the roots of the European grapevine *V. vinifera*, rarely forms galls on leaves of the European vine.[16] Americans had not thought to search for phylloxera on the roots, since the leaf galls on American vines were obvious and the insect caused minimal damage to these vines. Planchon and Lichtenstein's search for gall leaf phylloxera was rewarded when, on July 11, 1869, they discovered galls on the leaves of the European Tinto variety.[17] When they introduced the leaf-galling insects to *V. vinifera* roots, the leaf insects attached themselves to the roots.[18] In 1870, Planchon and Lichtenstein confirmed this by reversing the procedure.[19]

Meanwhile, Riley awaited Signoret's opinion regarding the leaf-galling phylloxera he had sent him in 1869.[20] In September 1870, as Prussian troops laid siege to Paris, Signoret wrote to Riley that he had compared the leaf-galling specimens with the European root phylloxera and he was convinced they were identical.[21] That fall (1870), Riley discovered that American leaf-galling phylloxera wintered on the roots of American grape vines. He confirmed that the root form, like the leaf form, was relatively harmless to the vine.[22] Riley's observation that leaf-galling phylloxera were smooth in appearance, while root-infesting phylloxera were rough, or "tubercled," suggested that they represented separate species, but in 1870 he transferred smooth phylloxera from leaf galls to the roots where they formed colonies of tubercled root phylloxera.[23] He thus

Featured Insect

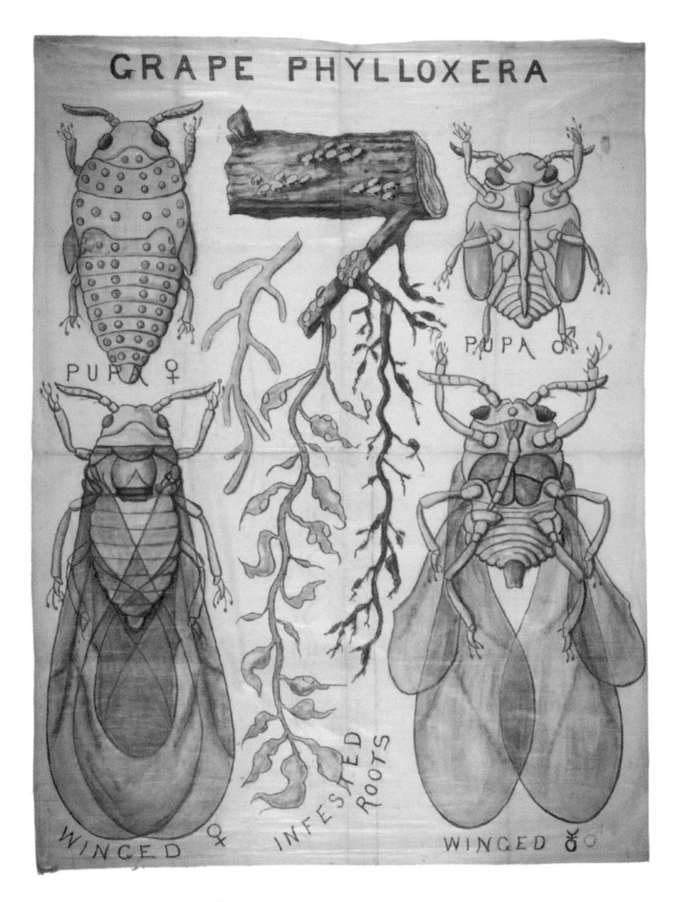

Grape Phylloxera [old scientific name *Phylloxera vastatrix*; current scientific name *Daktulosphaira vitifoliae* (Fitch)(Hemiptera: Phylloxeridae)]. Riley Insect Wall Chart (Courtesy of Morse Department of Special Collections, Kansas State University Libraries). Modern entomologists still puzzle over the

demonstrated the identity of the leaf and root forms in America about a year after Planchon and Lichtenstein had demonstrated their identity in France. The outbreak of the Franco-Prussian War in July 1870 apparently prevented the investigators from learning of these duplicate discoveries.

Convinced that the root and leaf forms represented polymorphic phases of a single insect species, Riley, in 1870, summarized the Grape Phylloxera's life cycle in its American version. The insect wintered in colonies on the roots

Featured Insect continued

complexities of the Grape Phylloxera life cycle and still consider it "the most economically destructive and geographically widespread pest species of commercial grapevines occurring in the most grape-growing countries around the world." Resistance based in American rootstock using conventional breeding is still the most effective management tool against Grape Phylloxera, but many researchers consider reliance on this alone to be risky and favor development of complementary tactics, such as better understanding of soil ecology leading to cultural and conservation biological (including microbial) control, and, since phylloxera is not native over most of its range, the possibility for classical biological control.* Granett et al. have shown the complex interaction among phylloxera galling and fungal pathogens in the soil, which by killing the roots make the phylloxera infestation appear visually less damaging.** This echoes the protracted debates over cause and effect that characterized the early years of the French infestation. Recently, Giblot-Ducray et al. used real-time quantitative PCR, a sensitive molecular method, for detection of small numbers of these tiny pests in soil in Australia, one of only a few wine-growing regions where phylloxera is not yet ubiquitous.*** It is clear that Grape Phylloxera, like several other pests that Riley worked on, is an ongoing challenge to agriculture.

*Kevin S. Powell, "Grape phylloxera: An overview," in *Root Feeders, An Ecosystem Perspective*, ed. Scott N. Johnson and Phillip J. Murray, 96–114 (Wallingford, UK: CAB International, 2008).

**J. Granett, A. D. Omer, P. Pessereau, and M. A. Walker, "Fungal infections of grapevine roots in phylloxera-infested vineyards," *Vitis* 37 (2015): 39.

*** Giblot-Ducray et al., "Detection of grape phylloxera (*Daktulosphaira vitifoliae* Fitch) by real time quantitative PCR: development of a soil sampling protocol," *Australian Journal of Grape and Wine Research* 22 (2016): 469–77.

from which, in the spring, some progeny somehow moved to the leaves, where they formed galls, while others remained on the roots. In both the root and leaf forms, apterous (wingless) females reproduced asexually, giving birth to numerous generations. Many questions remained, especially with respect to the role of winged insects and the manner in which phylloxera spread from one locality to another.[24] Phylloxera's life cycle eventually proved to be one of the most complex in class Insecta.

Featured Insect

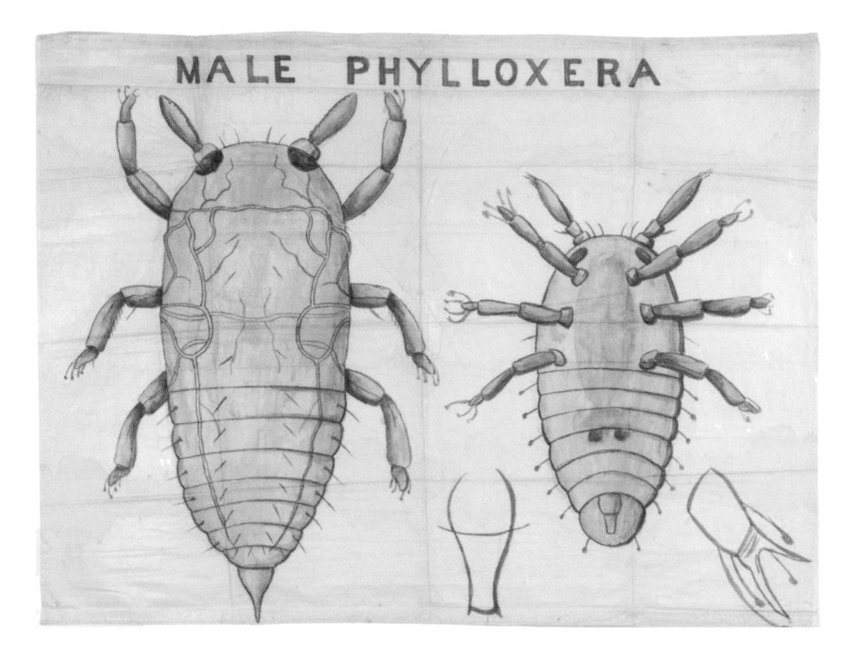

Grape Phylloxera male. Riley Insect Wall Chart (courtesy of Morse Department of Special Collections, Kansas State University Libraries). The wingless, mouthless male's only purpose is copulation with the female, after which she lays a single egg.

Curiously, the initial issue most hotly debated in France was not its life cycle or origin but whether phylloxera was causing the vines to die or whether the insects only attacked vines that were already ailing. Signoret and other leading French scientists subscribed to the dominant physiological theory of disease, according to which some disequilibrium in an organism's internal functions made it susceptible to disease or, in this case, prey to insect pests.[25] Practical and economic concerns played a role as well. Signoret pointed out that if French wine growers admitted that phylloxera was killing French vines, their lucrative export of vines would suffer.[26]

Planchon, Lichtenstein, and Bazille, who were headquartered in Montpellier, a provincial town in the south of France, fourteen hours by train from Paris, maintained that phylloxera was indeed the cause of weak and dying vines. Economic considerations underlay the differences between Planchon in Montpellier and Signoret in Paris. Signoret and the Parisians had close ties to vintners specializing in high quality wines who profited from the sale of vines. Planchon and his associates in Montpellier represented vintners who produced inexpensive table wines, for whom the sale of vines was of minor importance. Parisian savants sneeringly referred to the Montpellier group as "Hérault entomologists," associating them with the cheap table wines of their region.[27] In 1869, Signoret, in a polemical statement of the "effect" theory, proclaimed that phylloxera was the "alleged" cause of vine deterioration.[28] A. H. Trimoulet, secretary of the Linnaean Society of Bordeaux, the wine capital of France, supported Signoret and the "effect" school.[29] Lichtenstein challenged Signoret at meetings of the Entomological Society of France in Paris, while Planchon pressed the "cause" position among vintners in the Midi.[30] Both sides in the "cause-effect" debate sought Riley's support. When, in July 1870, the Entomological Society of France proposed Riley for membership, Signoret supported his election, apparently in hopes of winning him to the "effect" cause. Signoret's overture to Riley failed, because Riley agreed with Lichtenstein and Planchon that phylloxera was the cause of vines dying in French vineyards.[31]

Signoret's adherence to the "effect" position apparently influenced his initial opinion that the American and European species were different. At the Entomological Society of France, June 22, 1870, he quoted a letter from Riley: "from your [Signoret's] descriptions and drawings, I conclude that our insect must be different, mainly because it has never been found . . . on the roots."[32] In fact, Riley's letter argued just the opposite. Directly after the words quoted by Signoret, Riley's summary (as later published in his annual report for 1871) read: "All these apparent differences were rather calculated to give rise to doubts as to the identity of the two insects; but by careful observation and persistency we have been enabled to dispel them all."[33] Apparently Riley remained unaware of Signoret's misuse of his letter. By September 1870, Signoret had finally compared Riley's leaf-galling phylloxera with the European root phylloxera and concluded, however reluctantly, that the two were identical.

In 1871, Riley traveled to England and France, the first of three trips on which he participated in French phylloxera affairs. Following visits with his family and with Darwin in England, Riley crossed the channel to France.[34] Stopping in Paris, Riley met with Signoret, who displayed the living progeny of leaf-galling phylloxera that Riley had sent him and that he had kept alive through the siege of Paris.[35] Riley was pleased to find that Signoret now agreed that the European and American insects were the same species, but he was dismayed that Signoret still held that some unspecified meteorological or other condition weakened the vines, making them susceptible to phylloxera. Riley concluded that Signoret was "too much absorbed in closet studies to make the proper field observations."[36] Riley continued on to Montpellier, where Planchon and Lichtenstein awaited him.[37]

By the time Riley arrived in Montpellier in late July 1871, the phylloxera epidemic was threatening vintners in the Midi region around Montpellier and regions to the north, east, and west.[38] Vintners around Montpellier knew Riley's work from Lichtenstein's translation of his article on the "Grape-Leaf Gall-Louse" that was read at the meeting of the Hérault Agricultural Society in February 1871. They were most interested in his rating of American vines according to their resistance to phylloxera. Fox and Summer grapes, Riley found, proved most resistant, whereas Frost grapes, especially Clintons, were most susceptible.[39] In the spring of 1871, Midi vintners had already begun importing and planting the recommended resistant American vines.[40]

Those who favored the utilization of resistant American vines were at first divided as to whether American vines should be used as direct producers of grapes or as rootstock with European *V. vinifera* topstock grafted to them. Léo Laliman, a vintner and grape vine merchant near Bordeaux, proposed that French vintners substitute American vines for French vines as direct producers of grapes. His suggestion was disregarded in his own district, because "effect" proponents predominated in Bordeaux, and because his fellow vintners blamed him (unfairly) for having introduced phylloxera into France. This charge was ironic, as Laliman was obsessed with the theory that phylloxera was of European, not American, origin.[41] French vintners, especially around Bordeaux and other high quality wine districts, objected to American vines as direct producers on the grounds that American grapes did not produce wine that was up to the standard of French taste.[42] In the Midi district, vintners favored American vines, some preferring direct planting, others grafting.[43] Riley supported both methods, though he hoped that American wine, produced from American vines, would find favor in France.[44]

Advocates of American vines faced stiff opposition from defenders of pure French vineyards, including much of the French scientific elite in Paris, French government officials, and influential growers in Bordeaux and other wine-growing regions. Proponents of "La défense"—the defense of French viticulture against invaders—maintained that French vines should retain their ancient purity and not be "mongrelized" with American stock. As word spread that phylloxera was of American origin, French resistance to American vines stiffened.[45]

Following his return to St. Louis in September 1871, Riley wrote a series of articles for the *Rural New Yorker* in which he explained that phylloxera's subterranean attacks on the roots cleared up the long-standing mystery of why efforts to cultivate European grape vines in the eastern United States had consistently failed. Thirty miles downriver from St. Louis, Isidor Bush had planted European vines in 1867, only to see them wilt and die. When Riley investigated, he had found phylloxera attached to the roots of the ailing vines.[46]

Upon his return, Riley also revised his rating of American vines. In 1870, he had recommended Fox grapes and Summer grapes but warned against Clintons, but now he had second thoughts. In 1871–72, together with Engelmann, Riley conducted a survey of vines around St. Louis from which they identified nine American species of viticultural potential, rated according to their susceptibility or resistance to phylloxera.[47]

It is unclear to what extent American vintners in the eastern United States responded to Riley's message by grafting European vines onto American rootstock; however, in southern France the response was overwhelming. Upon reading translations of his articles, reinforced by illustrations of luxurious clusters of grapes on American vines in the Bush catalogue, growers in the Midi began importing thousands of rootstocks from St. Louis.[48] The demand so outstripped the capacity of the Bush-Meissner nursery in Missouri that the nursery (with the consent of French vintners) sent wild vine cuttings to fulfill orders from the Midi. By the winter of 1872, the Bush-Meissner nursery had sent four hundred thousand dormant cuttings to the Montpellier region. That year the Hérault Agricultural Society appointed a special commission to evaluate the effectiveness of these American imports.[49]

Midi growers, meanwhile, petitioned French authorities to remove restrictions on the importation of American vines. In September 1871, in response to Lichtenstein's pleas, the French minister of agriculture lifted the import ban.[50] Riley approved: "If America has given this plague to Europe, why should she not . . . furnish her with vines which are capable of resisting it?"[51]

Encouraged by these developments, the Hérault Agricultural Society (with support from the Ministry of Agriculture) commissioned Planchon to visit the United States to gather more information about American vines. In August 1873, Riley and Planchon embarked on a survey of vineyards, nurseries, and botanical gardens in New Jersey and Pennsylvania.[52] Riley then returned to St. Louis, while Planchon continued on to Baltimore, Maryland;

Washington, DC; Ridgeway, North Carolina; and Cincinnati, Ohio. Planchon then rejoined Riley in St. Louis, where they exchanged thoughts as to why American vines were resistant to phylloxera.[53]

Riley, following Darwin, maintained that since phylloxera had co-evolved with American grape vines, natural selection had produced vines that were resistant.[54] In the second edition of the *Bush-Meissner Catalogue* (1875), Riley proclaimed that American vines provided "a remarkable verification of the law which Darwin has . . . expressed as 'THE SURVIVAL OF THE FITTEST.'"[55] Planchon, like the majority of French naturalists, was at first skeptical of Darwin's theory, but under Riley's influence he gradually accepted a Darwinian explanation for the resistance of American vines.[56] Following his return to France, he wrote, "[The] hypothesis proposed . . . by Mr. Riley would explain why some American vines survive, while others are only mildly resistant, and others have disappeared. It appears to be a struggle for life where [the] strong . . . survive and the weak die."[57]

In 1875, Riley returned to Europe. Onboard the ship, Riley had the good fortune to visit with Darwin's son, Leonard, and upon arrival in England, he visited Darwin for the second time.[58] When Riley arrived in Montpellier in July 1875, the scene contrasted starkly with 1871: "[W]here four years before the whole country was one vast vineyard . . . the ground was [now] devoted . . . entirely or partly to other crops . . . Yet right in the midst of this [desolation] the American vines were . . . flourishing."[59] The scene confirmed Riley's message that economic entomology could prevent such disasters: "Needs there [be] any more forcible illustration of the importance of economic entomology!"[60]

In the summer of 1875, Riley, Planchon, and Lichtenstein, attired in fine clothes appropriate to scientific dignitaries, made an extended tour of vineyards in the Hérault and Vaucluse and quite likely other regions. Riley sported a white sombrero that contrasted with Planchon and Lichtenstein's black top hats. A sketch of the trio inspecting a vineyard appeared on the cover of *Le Journal illustré*, a widely circulated popular French magazine.[61]

In July 1875, speaking at the Entomological Society in Paris, Riley reported that replanting with American stock since 1871 was rescuing French vineyards. That year, imports of American vines peaked at 14 million cuttings.[62] Unfortunately the celebration was premature. By the late 1870s it became clear that some of the "miracle" American vines had failed to live up to their promise. Some did well, but others didn't take to Midi soils. Also, some American vines were susceptible. Of 7 million Concord cuttings shipped from Missouri before 1875, most succumbed to phylloxera.[63] In addition,

Le Journal illustré

QUINZIEME ANNEE. — N° 39

Gravures :

Visite dans les vignobles suspects de phylloxera, par **Claverie**. — *Les grandes manœuvres*, par **Henri Meyer**. — *Les fêtes de Boulogne*, par **Jules Desprès**. —*Le tonneau hongrois, le trophée canadien*, par **A. Denis**. — *Nos illustrations de* LE MÉDECIN DES FOLLES, par **Henri Meyer**.

DIMANCHE 22 SEPTEMBRE 1878

Le Journal illustré est mis en vente dès le vendredi matin

ABONNEMENT	UN AN	SIX MOIS
Paris.	6 50	3 50
Départements	7 50	4 »

Administration & Rédaction, à Paris, hôtel du **Petit Journal** *rue de Lafayette, 61*

PRIX DU NUMERO : **15** CENTIMES

Texte

Chronique de la semaine, par **Aristide Roger**. — *Beaux-Arts et Théâtres*, par **Charles Darcours**. — *Exposition universelle*, par **Adolphe Kubly**. — *Nos gravures*. — *Le passage du bac*, par **V. Vattier**. — *La Quinzaine judiciaire* par **Chicaneau**. — *Mot en losange inédit*.

La Commission officielle visitant les vignobles suspects de phylloxera!

Dessin de CLAVERIE. — Voir l'article, page 307.

FIGURE 6.1 Front page of the French newspaper, *Le Journal illustré* (September 22, 1878). The two gentlemen in top hats are presumably J. É. Planchon and J. Lichtenstein, and the gentleman in the sombrero is most likely C. V. Riley (courtesy of Yves Carton, Paris, France).

French soils with high lime content—those that produced France's high-quality wines—proved inhospitable to American transplants.[64] The confusion among French advocates of American vines emboldened "La défense" patriots to promote the injection of sulphocarbonates into phylloxera-infested soils as an alternative. In 1878–79, they pushed through new restrictions on the importation of "foreign" (American) vines.[65] Midi growers, seeing no alternative, continued their illegal importation, selection, and experimentation until, by the early 1880s, the principle of inherent resistance of American rootstock was accepted as the best solution.[66]

Following their demonstration in 1870–71 that the leaf and root forms of phylloxera were identical in America and Europe and their conclusion that phylloxera had originated in America, Riley and his French colleagues made only halting progress in unraveling remaining mysteries of phylloxera. In 1873, after three years of effort, Riley succeeded in breeding leaf-galling phylloxera from the root form, thus duplicating his success in the opposite direction (leaf to root) in 1870.[67] The search for the male phylloxera proved fruitless, and Riley concluded none probably existed.[68] Riley also found that many American vines were infested only with root forms, and he concluded that the leaf form was not necessary to the perpetuation of the species. In France, Planchon, Lichtenstein, Signoret, and Maxime Cornu (head of the Phylloxera Commission of the French Academy of Sciences) conducted intense investigations into the life history of phylloxera but reported no major breakthroughs.

The winged insects remained a mystery. Riley's observations of winged insects around Montpellier and the specimens he compared with those in Missouri confirmed the identity of the winged forms in Europe and America but shed no further light on their function.[69] By 1873, Riley and his French colleagues had established that the root nymphs could develop into two different forms. Most often they developed into egg-laying apterous (wingless) females that reproduced asexually for several generations; however, each year in August and September, some nymphs developed into winged individuals. These winged insects in turn developed into two types: (1) females that displayed two to five eggs in the abdomen or (2) winged individuals without eggs that Riley, Planchon, and Lichtenstein initially assumed were males.[70] Yet despite intensive breeding and extensive field study, observers on both sides of the Atlantic had yet to report copulation between the two forms, or any egg laying. The dispersal of phylloxera outward from distinct centers of infestation suggested that they spread by winged individuals, yet years of observation and breeding had failed to confirm this. When Planchon was in St. Louis, Riley showed him two different winged forms hatched in his insectary that he presumed were male and female. When the two forms were brought together, however, they did not copulate. A microscopic examination revealed that the "male" carried a maturing egg. Both winged forms were female![71]

Having been repeatedly frustrated in their attempts to unravel the role of the winged insects, researchers in France and America turned to the study of related species. In 1873, Edouard Gérard Balbiani, a professor at the College de France, began investigating *Phylloxera quercus*, the gall-forming species inhabiting the European oak that Signoret had cited when he recommended that Planchon assign the new grape pest to the genus *Phylloxera*. In contrast to the Grape Phylloxera, the life cycle of *P. quercus* was entirely above ground. *P. quercus* also proved easier to breed than the Grape Phylloxera.[72]

That season (1873), Balbiani found that winged *P. quercus* females dispersed in August and deposited eggs of two different sizes on oak leaves. The smaller eggs produced males and the larger females. Remarkably, neither males nor females had mouthparts or digestive organs. Their only function was to copulate and reproduce. Four to five days following copulation, the female laid a solitary egg, after which both female and male died.[73] Following Balbiani's lead, Cornu soon found sexual Grape Phylloxera without mouthparts. Prompted by these reports, Riley reported that some time earlier he had also observed a female with one large egg, although at that time he had not recognized its significance. In 1873, Riley predicted that, given Balbiani's discoveries regarding *P. quercus*, the life history of the Grape Phylloxera would be completed during the coming year.

By the fall of 1874, Riley's prediction proved true.[74] Balbiani discovered that the winged Grape Phylloxera laid eggs of two sizes that produced wingless and mouthless males and females. Like *P. quercus*, these sexual individuals mated, after which the female laid a single egg that developed into a wingless mother who began the cycle again.[75] Meanwhile, Riley conducted investigations on the American oak phylloxera, *Phylloxera rileyi*, a species Lichtenstein named in his honor from specimens Riley sent him. The life cycle of *P. rileyi* conformed in its essential pattern to that of both *P. quercus* and the Grape Phylloxera.[76]

While clarification of the sexual forms satisfied the investigators' scientific curiosity, it provided little help toward effective control.[77] Following Balbiani's discovery that *P. quercus* overwintered as a solitary egg, based on the theory that the egg stage represented a weak link in the life cycle of the Grape Phylloxera, Balbiani and others initiated a massive search for the winter egg in French vineyards.[78] Planchon, Lichtenstein, and Riley warned against the "winter egg" solution because, unlike *P. quercus*, which overwintered in egg form, Grape Phylloxera hatched during the same season as the egg was laid and wintered as a nymph.[79] Despite their objections, the myth of the "winter egg" as a target for Grape Phylloxera control persisted for decades.[80]

Likewise, when Riley and Planchon independently discovered a predacious

American mite, *Tyroglyphus phylloxerae*, which fed on phylloxera, hopeful vintners besieged Riley with requests for the mite, and Planchon agreed to supervise its introduction in the Midi. Riley cautioned against over-optimism, and in fact the mite proved to be a disappointment. Although Riley expected, on Darwinian grounds, that natural enemies of phylloxera would eventually evolve to check phylloxera in France, he cautioned against hopes that *Tyroglyphus* would provide an immediate control.[81]

Through his investigations and publications, Riley became the authority regarding the scientific name for phylloxera. In 1856, Fitch had named the species *Pemphigus vitifoliae.* In 1866, Fitch's species became the subject of a dispute between Benjamin D. Walsh and Henry Shimer (an Illinois physician-naturalist) regarding its name and classification.[82] Surprisingly, Riley, who wrote his first article on Fitch's *Pemphigus* that year, did not become involved in the dispute. By 1869, however, as Riley learned of the Grape Phylloxera disaster in France, and as he reflected on the Darwinian-evolutionary implications of its presence on both continents, he took charge of the taxonomic issue. In the *American Entomologist and Botanist* (1870) and in his third and fourth annual reports (1871 and 1872), he refereed the intramural scrimmage among American entomologists. Although he held that Fitch, in 1856, knew too little of the insect's "true character" to describe and name it properly, Riley nevertheless recommended combining Fitch's specific name, *vitifoliae*, with the French genus, *Phylloxera*, creating the name *Phylloxera vitifoliae.* Riley reasoned that the French genus was scientifically accurate, but that Fitch's species name, *vitifoliae*, held priority over the French specific name *vastatrix*, plus Fitch's species name was familiar to Americans.[83] He rejected Walsh's classification: the phylloxera was a plant louse (aphid relative), not a scale insect. Riley commended Shimer for recognizing the insect as a louse but scolded him for assigning the species a new name and erecting a new genus when the correct genus had already been established. Riley had little patience for uninformed "species grinding."[84]

Having addressed the confusion on his side of the Atlantic, Riley turned his attention to Europe. Riley rejected Westwood's name, *Peritymbia vitisana*, proposed in 1868, because it did not refer the insect to its proper genus. (Exercising uncharacteristic restraint, Riley diplomatically refrained from calling the Oxford professor a "species grinder.")[85] That same year, Planchon gave the insect the name *Phylloxera vastatrix* ("destroyer"). Once it was known that Fitch's leaf-galling species (1856) and Planchon's root-infesting insect (1868) were the same species, Fitch's *Pemphigus vitifoliae* held priority. At that stage of entomological science, however, entomologists were divided as to whether the rule of priority (first describer) or the rule of general usage should hold sway.[86] At first Riley vacillated. In 1871, he defended Fitch's *Pemphigus vitifoliae*, even though he agreed with Lichtenstein that the name was ungrammatical, *vitis-folii* being preferable.[87] In 1872, he reiterated his adherence to *vitifoliae* although he admitted that most Europeans employed *vastatrix*, and this might be one occasion where "accord" (general usage) should take precedence over priority.[88] By 1873, Riley concluded that Fitch's *vitifoliae* would have to give way because *vastatrix* was being used almost universally and from that time on, he used the name *Phylloxera vastatrix*.[89]

Although Riley's decision was accepted during his lifetime, he did not have the last word. In 1901, after Riley's death, zoologists agreed that the law of priority should prevail.[90] The current approved name is *Daktulosphaira vitifoliae* (Fitch), although other names are commonly used.[91] While modern non-specialists may be excused for not including a detailed history of Grape Phylloxera taxonomy on their bedtime reading lists, Riley's treatment of this subject indicates the paramount place of taxonomy in his economic entomology.

It is noteworthy that Riley's views on the subject, though in flux in the formative stages of taxonomic discussion, were based on scientific rather than personal or nationalistic principles. He did not champion Fitch's name because Fitch was American; in the end, he opted for Planchon's name. It is also noteworthy that Riley, barely thirty years of age, lacking a university degree and indeed any formal training in entomology, challenged recognized authorities twice his age (in 1873, when Riley turned thirty, Signoret was fifty-seven and Westwood sixty-eight). Riley's supporters and detractors alike recognized that his judgments rested on accurate observation, extensive investigation, and—most tellingly in relation to his French contemporaries—his superior command of the relevant literature on both sides of the Atlantic.

A decade elapsed before Riley returned to France in 1884, as chief of the US Division of Entomology, on a trip that concluded his active participation in phylloxera investigations. Meanwhile, in 1878, a new wine pest, Powdery Mildew, had appeared in France. The new pest proved to be yet another American import, adding insult to the injury inflicted by phylloxera.[92]

At the meeting of the Hérault Agricultural Society in June 1884, Riley displayed a spray nozzle that had been developed under his direction in the course of Cotton Worm investigations. Riley also demonstrated three new chemicals designed primarily to control the mildew but with some promise in the control of the leaf-inhabiting phylloxera.[93] Buoyed by the success of Paris Green against the Colorado Potato Beetle and the Cotton Worm, Riley was optimistic about the potential of chemicals in conjunction with the cyclone nozzle. "I consider [the nozzle] the greatest discovery I have ever made," Riley wrote Lichtenstein in 1883.[94] When he appeared in Montpellier, however, Riley apparently feared that expectations for sprayer-chemical control were running too high, because he pointed out that the nozzle was effective against mildew but that the grafting program currently underway was the solution for phylloxera.[95] It is not clear whether Riley was aware of French field tests being conducted at that time of a new fungicide, Bordeaux Mixture, as a control for Powdery Mildew.

Although his active involvement in French phylloxera affairs ceased after his Montpellier appearance in 1884, Riley's reputation among French vintners continued unabated. His Montpellier speech and his phylloxera illustrations were widely disseminated by the Hérault Agricultural Society and by Bazille's translation of the Bush-Meissner Catalogue of American vines.[96] In 1887, Planchon initiated a monthly journal entitled *La Vigne Américaine et la viticulture en Europe* in which Riley's illustrations were featured. The journal was widely distributed until its discontinuation in 1911.

The "Riley nozzle" also enhanced Riley's prominence among French vintners. When he demonstrated the new device in Montpellier in June 1884, the agricultural society delegated Gustave Foëx, professor at the Montpellier Agricultural School, to oversee production of one hundred nozzles for distribution to vintners.[97] Victor Vermorel, owner of a company that manufactured agricultural machinery at Villefranche-sur-Saône, near Lyon, was also present. He soon began producing his version of the nozzle and, having made improvements over the next two years, received a French patent for the "Nozzle-Riley-Vermorel."[98] Vermorel's was the most successful of the half-dozen versions of the Riley cyclone nozzle produced in France during the following decade.[99] In combination with Bordeaux Mixture, the nozzle was so effective in the control of mildew that the Vermorel Company expanded its production force to eight hundred workers. Popularly known as "un Riley," the device in various forms spread throughout the vineyards of southern France and beyond.

In 1887, Pierre Viala, a professor of viticulture at the Montpellier Agricultural School, traveled to the United States in search of American vines suitable as rootstock. His trip was necessary because many of the varieties imported in the early 1870s had proved unsuitable in French calcareous soils. In June 1887, Riley welcomed Viala in Washington and assigned Frank Lamson-Scribner, recently appointed microscopist at the Agricultural Department, to escort the French viticulturist on a six-month trip to New York, Ohio, Missouri, Texas, and California.[100] In Texas, Thomas V. Munson, an expert on vines of the Southwest, led them to the lime-tolerant vines they sought.[101]

In 1889, Riley was placed in charge of the US Department of Agriculture exhibition for the International Exposition in Paris that commemorated the centennial of the French Revolution. In the report of the exhibition, Riley devoted prominent attention to phylloxera.[102] While Riley was in Paris from May to October 1889, various organizations paid tribute to his role in the phylloxera campaign. On May 22, 1889, he was a featured guest at the Entomological Society of France; on June 30, he attended a reception given by the president of the French Republic; in August, he spent a week inspecting the vineyards of Bordeaux.[103] That summer, Riley participated in the grand opening of the Vermorel Viticulture Research Laboratory in Villefranche, where Vermorel personally thanked Riley for his "advice and support."[104] Later that summer, Riley addressed a gathering at a model farm at Sucy en Brie, east of Paris.[105]

Passing through London in September on his return to the United States, Riley was celebrated as the guest of honor at a final banquet. Colman, who as US commissioner of agriculture had appointed Riley to head the Paris exhibit, and M. Leon Grandeau, representing the French government, toasted Riley's contributions to the Paris Exposition and his achievements with regard to phylloxera. Riley, moved by recognition from his two adopted countries, responded first in English, then in French.[106]

Prior to these accolades, Riley received numerous awards for his phylloxera investigations. In 1874, the French minister of agriculture and commerce, together with the Hérault Agricultural Society, awarded Riley a gold medal for his work on phylloxera. The medal arrived in St. Louis just at the time Missouri lawmakers were attempting to abolish Riley's office, and his supporters welcomed French support in their campaign to retain him.[107] In connection with the International Exposition in 1889, the French Foreign Office awarded Riley the Légion d' Honneur, and officials of the Exhibition awarded him the "grand Prix."[108] In 1892, as the restructured French wine industry took shape, Vermorel and the viticulturists from the Saône-et-Loire district presented Riley with a bronze statuette depicting a young couple carrying baskets filled with a bountiful harvest of grapes.[109]

Riley was justifiably proud of his role in the phylloxera campaign. By 1890, 719,000 acres had been replanted, transforming the French viticultural landscape.[110]

FIGURE 6.2 Bronze statuette presented to C. V. Riley by V. Vermorel and the wine growers of the Saône-et-Loire district, 1892. The statuette, which two of Riley's daughters, Cathryn and Thora, donated to the Entomological Society of America in 1962, is now at the society's headquarters in Annapolis, Maryland (courtesy of the Entomological Society of America).

In many respects, Riley's phylloxera campaign constitutes his most enduring achievement. A significant element in his success was the cluster of American scientific institutions with which he was associated: the Smithsonian Institution, the American Entomological Society, and an array of uniquely American agricultural-scientific institutions including state agricultural and horticultural societies, the agricultural press, the US Department of Agriculture, and most directly, his positions as Missouri state entomologist and as chief of the US Division of Entomology. In some respects these resources were superior to those of his European colleagues. In 1873, Planchon declared that Riley, as state entomologist, commanded a "marvelous array of instruments" that allowed him to observe "hour by hour the developmental stages of insects."[111] Of course the comparison worked in both directions. In 1876, Riley credited Balbiani's discoveries on *Phylloxera quercus* and the Grape Phylloxera to his excellent instruments, trained eye, and competent assistants.[112] From whichever side of the Atlantic one viewed phylloxera research, however, it was clear that Riley, supported by American scientific and agricultural institutions, operated as an equal, in some respects superior, partner with his French colleagues.

Riley's phylloxera investigations were vital to his personal and public image. His fame in the Rocky Mountain Locust campaign receded when the locust threat ended abruptly, and his later Vedalia Beetle triumph was clouded by wrangling over credit. Regarding phylloxera, which is a continuing threat throughout the world's wine-growing districts, Riley's contributions are still universally recognized.[113] Of vital importance to Riley, with his high-strung, combative personality, was the respect he enjoyed in France. His adopted homeland offered acceptance, recognition, and a refuge from the political bickering and bureaucratic infighting in St. Louis and Washington. In these ways, the French repaid Riley for his assistance in the phylloxera campaign.

7

Devouring Locusts

[T]hey circle in myriads about you, beating against everything . . . their jaws constantly at work biting and testing all things in seeking what they can devour.

—C. V. Riley, "The Rocky Mountain Locust," *Seventh Annual Report* (1875), 157

[T]he western locust has already ceased to be . . . [an] object of dread . . . and no one intending to migrate west from the Atlantic states or from Europe need be deterred by the fear of such alarming invasions as has occurred in former years.

—USEC, *Second Report* (1880), 322

IN CONTRAST TO the Grape Phylloxera that threatened distant French vineyards, the Rocky Mountain Locust destroyed crops, pasture, and gardens over the vast American plains. Americans awoke to the severity of the locust threat in 1866–69, when locusts repeatedly overwhelmed recently settled districts in Kansas and other western states and territories. Farmers in Illinois feared the grasshoppers devouring their crops were migratory locusts, but Benjamin D. Walsh and Riley pointed out that true locusts had longer wings that enabled them to fly in swarms over long distances, whereas grasshoppers had shorter wings and did not travel far.[1]

As Missouri state entomologist and de facto state entomologist for Kansas, Riley witnessed the swarms that reappeared in 1873 and that in 1874 enveloped the plains from Manitoba to Texas. In central and northern Kansas in 1874, locusts consumed crops, tree foliage, houseplants, curtains, girls' dresses,

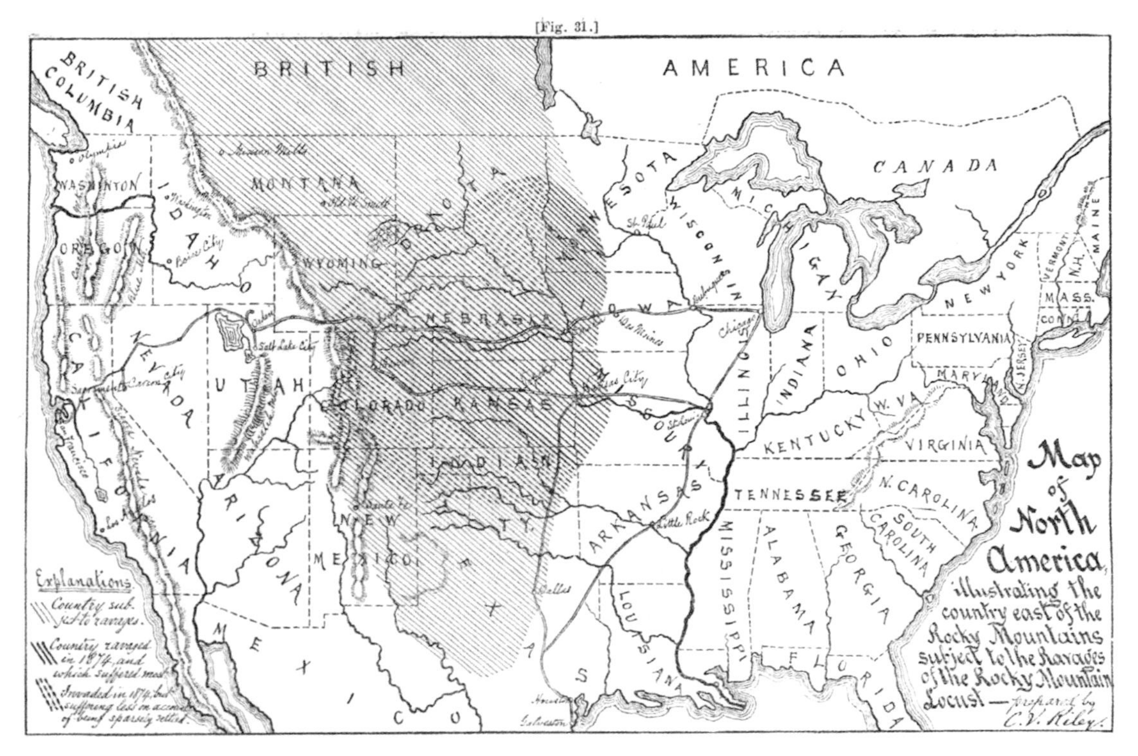

FIGURE 7.1 "Map of North America, illustrating [the range of] the Rocky Mountain Locust," prepared by C. V. Riley, *Seventh Annual Report* (1875), 142 (courtesy of the Albert R. Mann Library, Cornell University).

and leather harnesses. A housewife who frantically covered her garden with carpets watched in horror as locusts devoured the carpets before consuming the remaining vegetables.[2] Kansas governor Thomas A. Osborne convened an emergency session of the legislature to debate relief measures.[3]

The following spring, in 1875, the plague assumed new dimensions when millions of hoppers (nymphs) hatched from eggs deposited in 1874 marched in armies up to a mile wide, devouring grass, grain, and vegetables, leaving only an occasional milkweed stalk. Travelling by buggy through western Missouri and eastern Kansas at the end of May 1875, Riley passed through a landscape as barren as winter.[4] When farmers replanted, Army Worms devoured the second crop. In some areas where Chinch Bugs and Army Worms had destroyed crops during the two previous years, locust hoppers now consumed the third year's crop. Desperate families survived on a diet of bread and water. Rumors circulated of a family of six starving to death in their home. Western Missouri counties sent delegates to St. Louis to plead for aid.[5]

Featured Insect

Rocky Mountain Locust [current name Rocky Mountain grasshopper, *Melanoplus spretus* (Walsh).* (Orthoptera: Acrididae)]. Carte de visite of two adult Rocky Mountain Locusts (courtesy of the Minnesota Historical Society, St. Paul). During Riley's time, this grasshopper was one of the most spectacular and destructive natural phenomenon in the American West, but it had faded into obscurity by the time of his death. Today, it is considered extinct, probably due to alteration of its montane non-outbreak habitats during the late nineteenth century.

* Lockwood, Jeffrey A. *Locust: The Devastating Rise and Mysterious Disappearance of the Insect that shaped the American Frontier* (New York: Basic Books, 2004).

In the summer of 1874, Riley toured the ravaged districts in Kansas and western Missouri, mounted like an entomological scout on the front of a railroad locomotive assessing the strength of the insect enemy.[6] From his command center in St. Louis, he sent questionnaires to newspaper editors, railroad agents, and other correspondents in Kansas, Missouri, Minnesota, Nebraska, Dakota, Wyoming, Colorado, Montana, Indian Territory (future Oklahoma), and Texas requesting information on the size and direction of swarms and estimates of damage.[7] Although migratory locusts were individually smaller than ordinary grasshoppers, they made up for their lack of size in their staggering numbers and voraciousness.[8] In order to place the locust phenomenon in historical perspective, Riley compiled a comprehensive history of locust and grasshopper outbreaks in North America.[9]

Riley portrayed the locust plague variously as a naturalist, artist-poet, scientific prophet, and showman. For *Scientific American*, he sketched a dramatic scene of "Locusts falling upon and devouring a wheat field."[10] In his seventh report, Riley sketched a map depicting the extent of the locust plague and a detailed view of female locusts depositing eggs.[11] Switching from scientific illustrator to drama-playwright, Riley wrote in his eighth report: "The farmer saw his green acres smiling with glorious hope today, and tomorrow [they were] all barren and bleak."[12]

As a scientific prophet, Riley quoted the Old Testament prophet Joel, who twenty-five hundred years earlier described a similar locust species (most likely the desert locust, *Schistocerca gregaria*): "[T]he land is as the garden of Eden before them, and behind them a desolate wilderness."[13]

In the spring of 1875, when Governor Hardin proposed a day of prayer for deliverance from the locust scourge, Riley responded by presenting him with the first copy of his seventh report, in which he predicted that the locusts would begin leaving in early June.[14] The governor, perhaps doubting Riley's ability to predict locust behavior, decided nevertheless to set the day of prayer as June 3, 1875. When William Pope Yeaman, pastor of the Third Baptist Church in St. Louis, learned of the encounter, he publicly reprimanded Riley for insulting the governor and belittling the sincerity of churchgoers who took divine retribution and human penitence seriously. Riley responded by writing that treating the locust invasion as divine retribution implied that western Missouri farmers were more sinful than eastern Missourians, a conclusion he rejected.[15] Locusts, he insisted, behaved according to natural laws, not divine command. Riley reiterated his prediction, based on the locust's natural history, that the fully winged locusts would be gone by the end of June.[16]

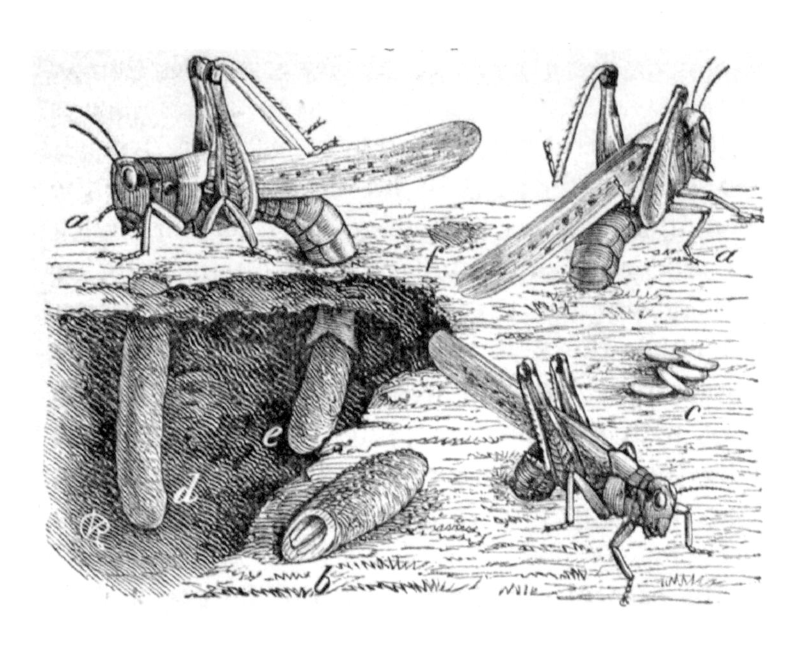

FIGURE 7.2 Rocky Mountain Locust female ovipositing. Illustration by C. V. Riley, *Seventh Annual Report* (1875), 122. Legend: ***a***, female in different positions, ovipositing; ***b***, egg-pod extracted from ground, with the end broken open, showing how the eggs are arranged; ***c***, a few eggs lying loose on the ground; ***d***, ***e***, show the earth partially removed, to illustrate an egg-mass already in place, and one being placed; ***f***, shows where such a mass has been covered up (courtesy of the Albert R. Mann Library, Cornell University).

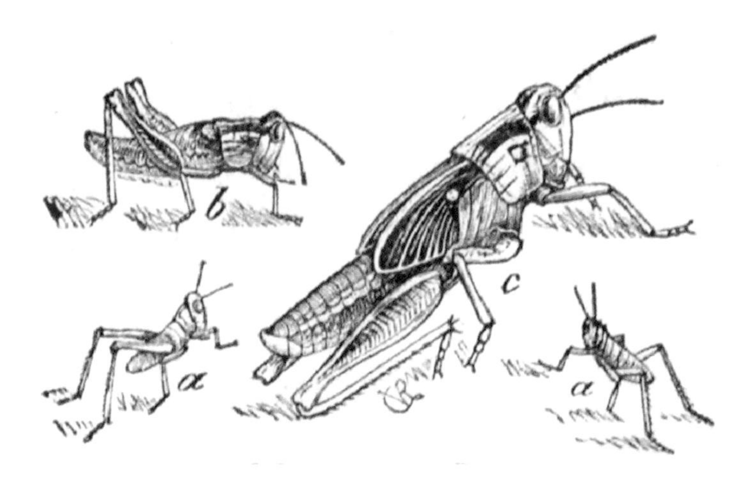

FIGURE 7.3 Rocky Mountain Locust immature stages (now known as nymphs). Illustration by C. V. Riley, *Seventh Annual Report* (1875), 123. Legend: ***a***, newly hatched larvae; ***b***, full grown larva; ***c***, pupa (now known as final nymphal stage) (courtesy of the National Agricultural Library Special Collections, Beltsville, Maryland).

FIGURE 7.4 "A Swarm of Locusts falling upon and devouring a wheat-field," C. V. Riley, *Seventh Annual Report* (1875), 156 (courtesy of the Albert R. Mann Library, Cornell University).

On the appointed day of prayer, June 3, 1875, Missourians assembled in churches across the state. In Osceola, in hard-hit St. Claire County, six hundred penitents overflowed the church sanctuary. That day, Riley was in Warrensburg, Johnson County, to the north.[17] Out for an early morning walk, he urged a local vintner to dig a ditch to protect his vineyard from hoppers, but the vintner explained he had to go to church. A disgusted Riley observed, "[I]nstead of genuflecting on a spade he [went to church], while his beautiful vineyard was literally being gobbled up . . . I respect every man's faith, but there are instances where I would respect his work a good deal more."[18]

Shortly after the day of prayer, Riley predicted again that the locusts would leave by the end of June. He then traveled to England and France, where he spent two months visiting family, attending scientific meetings, and investigating phylloxera. On June 21, 1875, at which time Riley was onboard the *Abyssinia* bound for England, the *St. Louis Post Dispatch* confirmed that locust swarms were departing and were expected to be gone within three days.

In the locust district, farmers replanted and, blessed with heavy rains, produced bountiful crops of corn, grass, and grains.[19] Revisiting the district on his return in September 1875, Riley reminded his listeners that he had risked his reputation by predicting the locusts' exodus, and he referred to the ripening crops and well-fed livestock as verification of his prophecies.[20]

Meanwhile, Riley urged locust victims to turn the tables by eating their tormentors. He pointed out that Native Americans in the southwest and native people in Africa and Asia ate locusts as a regular part of their diet, and that Parisians, who had eaten horses, dogs, and cats during the siege in 1870–71, would have welcomed locusts as nourishment.[21] When a reporter interviewed Riley about eating locusts, Riley offered him a dish of roasted hoppers. At first the reporter declined, but when Riley downed a handful and smacked his lips, the reporter sampled some and liked them.[22]

Announcing that he would subsist on locusts for a twenty-four hour period, Riley experimented with hoppers fried, roasted, fricasseed, and boiled as broth. At the conclusion of the day, he reported feeling full of energy. Having convinced a St. Louis hotel manager to offer locusts on his menu, Riley assumed charge of the cooking and served up locust soup, locust cakes, and baked locusts, followed by a dessert of locusts and honey. When the guests voiced their approval, the regular cook joined Riley in the preparation of more locust delicacies.[23] In Warrensburg in western Missouri, Riley joined Professor Strait, director of the Warrensburg Normal School, in the preparation of locust dishes for the assembled guests. Strait testified that locust broth, seasoned with pepper and salt, tasted as good as beef broth.[24] In Kansas City, John Bonnet, Mississippi Valley's "King of Cuisine," served locust delicacies at his fashionable restaurant.[25] When Riley traveled to Europe in 1875, he carried a box of fried locusts with him, reporting that members of the Entomological Societies in London and Paris found them tasty.[26] Upon his return, Riley addressed the AAAS in Detroit on the topic "Locusts as Food."[27] Whether Riley served locusts to his American scientific colleagues is not recorded.

Riley's energetic response to the invasion of 1874–75 ensured his reputation as the undisputed authority on locusts. In his annual reports, in regional and national newspapers, at agricultural and scientific meetings, Riley reviewed the current state of knowledge of the locust's natural history and offered advice on how to respond to the menace. J. C. Wise, chairman of the Minnesota Locust Commission, wrote to Riley "[O]ur state, and indeed the whole Northwest owe you a debt of gratitude for your investigations."[28] Responding to an Illinois farmer who asked whether in light of the locust invasion he should settle in Kansas, Riley published an article in the *New York Tribune* assuring prospective immigrants that the locust invasion was temporary and that another invasion would not occur for eight to ten years.[29]

Like Charles Dickens, whom he admired, Riley sympathized with the victims of catastrophe, whether of natural or man-made origin, and like Dickens, he applauded voluntary relief efforts to relieve suffering. Riley congratulated local committees, the Patrons of Husbandry, and others who solicited donations of cash, food, and clothing.[30] Unlike Dickens, who distrusted government-sponsored relief, Riley applauded states that reduced or exempted locust victims from taxation and supplied needy settlers with food, clothing, and seed for replanting. He rebuked states, like Kansas, that refused to provide such relief.[31] The contrast between Riley and Dickens regarding government relief reflects in part the difference between the economies of rural Missouri and urban London. Riley predicted, correctly, that a farming economy, given favorable weather conditions, would recover quickly.[32] Alleviating poverty in London, by contrast, required long-term structural changes.[33] Riley's emphasis on government-sponsored relief, even in an agricultural setting, highlights an important aspect of his social philosophy.

Based on his experience in 1875, Riley assured farmers that the hopper menace could be met by coordinated effort at the local and state level. Working together, they could construct barriers of wood, tar, and ditches to stop marching hoppers. He cited, for example, a truck farmer near Kansas City who protected his garden with ditches. He also cited the example of farmers whose land was protected on three sides by an oxbow of the Missouri river who then dug a ditch across the fourth side to prevent hoppers from entering their fields.[34] Riley's experiments in the winter of 1876–77 indicated that locust eggs survived freezing, thawing, and submersion in water but died when exposed to fresh air.[35] Based on these findings, he advised farmers to plow and harrow their fields to destroy the eggs. For egg-infested pasture, he recommended heavy rollers drawn by horses. Furthermore, he advocated state laws offering bounties for the collection of locust eggs in areas where plowing and rolling were not possible.[36]

While he was optimistic about controlling hoppers, Riley warned that swarms of adult locusts could not be controlled locally. They must be prevented from swarming or, if prevention failed, timely warning must be issued to threatened communities. Both prevention and warning depended on mapping the locust's perennial habitat and investigating the conditions that led to swarming. The Agricultural Department was the logical agency to carry out these investigations, but Riley charged that the department under Fredrick Watts was incompetent to carry out the investigations. Riley voiced midwesterners' general dissatisfaction with Watts, a Pennsylvania lawyer-farmer whom President Ulysses S. Grant had appointed against their wishes.[37] Riley also voiced scientists' resentment of Watts for his abrupt firing of Charles Christopher Perry, director of the national herbarium, and for his amateurish pronouncements on scientific issues.[38] Following a heated encounter with Watts, Riley wrote John L. LeConte that Watts had "no ability to discriminate between true and bogus science."[39]

Scientists' resentment of Watts surfaced at the 1873 meeting of the AAAS in Portland, Maine, when LeConte called for the replacement of Watts by a scientific administrator and the reorganization of the Agricultural Department as a scientific agency.[40] Riley, who was in Washington in late 1874, urged congressional leaders to work for Watts's removal, and he wrote articles in the *New York Tribune* toward that end.[41] As he feared, however, Watts remained in place until Grant left office in 1877.[42]

In the course of the locust emergency in 1874, Riley also lost patience with Townend Glover, the Agricultural Department entomologist. "[W]hat has [Glover] accomplished for entomology either from the purely scientific or the economic standpoint?" Riley asked LeConte rhetorically.[43] LeConte was likewise upset with Glover because in his view Glover's insect illustrations that began appearing in 1873 were inaccurate and out of date.[44]

At the AAAS meeting in Detroit in 1875, Riley and LeConte joined forces to propose the establishment of an independent commission to investigate locusts and other injurious insects of national importance.[45] In his annual report, Riley explained: "In cases, as with the Locust, the Chinch Bug, the Cotton Worm, etc., where the evils are of a national character, a [N]ational Commission, appointed for the express purpose of their investigation, and consisting of competent entomologists, botanists and chemists, is necessary, and . . . steps have been taken . . . to memorialize Congress to create such a Commission."[46] Riley charged that Watts and Glover had made no effort to subdue the locust plague.[47] Glover, he said, even confused non-migrating grasshoppers with migrating locusts.[48] Cyrus Thomas, Illinois state entomologist and a naturalist with Ferdinand V. Hayden's survey, agreed. In 1876, when locusts again invaded the West, Thomas wrote to Hayden, "you might as well try to get a prairie dog out of his hole as to get Glover out of his nest."[49] Thomas suggested that Glover be reassigned as curator of the museum at the Agricultural Department, while "an active, working entomologist" be appointed to carry out field investigations.[50]

Scientists at the AAAS supported Riley and LeConte's proposal, but it was circulated too late for state agricultural societies and the National Grange to endorse it, and the proposal foundered in Congress.[51] By 1876, however, Riley had mobilized agricultural organizations and enlisted Kansas senator John J. Ingalls to introduce a bill incorporating their petition in Congress.[52] Riley and LeConte appeared before the House and Senate agricultural committees to urge passage of the bill.[53] Some congressmen opposed the Ingalls bill on the grounds that it duplicated the work of the Agricultural Department. Riley and other supporters argued that a special commission was necessary on the grounds that the locust invasion was a national emergency and that the Hayden Survey, although authorized to conduct scientific investigations in the territories, had no authority to implement controls.[54]

Surprisingly, the bill's most vocal opponent was Illinois senator John A. Logan, a prominent Union army general from southern Illinois, Logan opposed the bill on the advice of his brother-in-law, Cyrus Thomas, who in addition to serving as Illinois state entomologist, served also as naturalist on Hayden's survey. Logan charged that the bill was designed to benefit a certain person (presumably Riley) and he insisted that naturalists with Hayden's survey (presumably Thomas), were competent to deal with the locust problem.[55] Thomas, eighteen years Riley's senior and author of a monograph on the insect family that included the migratory locust, *The Acrididae of the United States* (1873), understandably considered Riley an upstart.[56]

Thomas and Logan trimmed the LeConte-Riley proposal drastically when it reached the Senate Committee on Agriculture. In place of three commissioners appointed for a period of five years each with a salary of $5,000 per year, the revised bill called for only one commissioner appointed for one year at $4,000, and this person was responsible solely for locust investigations, rather than the list of insects earlier proposed by Riley and LeConte.[57] Senator Samuel B. Maxey of Texas moved to reinsert the Cotton Worm, a major pest in his state.[58] Maxey's example inspired Virginia senator Robert E. Withers to include the Tobacco Worm, which in turn prompted California senator Aaron A. Sargent to offer his own tongue-in-cheek amendment:

> Mr. Aaron A. Sargent (California): If it be in order to move to insert another worm - -
>
> President: The Chair will entertain any question that is relevant.
>
> Mr. Sargent: There is one more destructive than another: I move to insert "worm of the still." (Laughter).[59]

The senators then voted to reinsert all the insects in the original bill, whereupon some objected that one entomologist could not possibly investigate all the insects in one year.[60] When someone proposed that the entomological commissioner be placed under the commissioner of agriculture, Senator Justin Morrill objected that paying a subordinate of the commissioner of agriculture more than his boss was absurd. When the amended Ingalls bill passed the Senate, the editor of the *Nation* noted sarcastically: "The Republicans in the Senate . . . have passed a bill to investigate insects injurious to vegetation. . . . [T]he act, should it pass the House . . . will be a new application of the great principle of the division of labor, for in the future the Agricultural [*sic*] Commission will scatter the seed broadcast over the land, while the entomologist will follow closely on his trail and exterminate the various bugs that may attack the ripening grain. We only want now another Commission to harvest the crops, and another to see that they get to deep water, and the husbandman will be entirely relieved from grinding toil."[61] Riley denounced the *Nation*'s mockery of a serious problem, insisting that injurious insects cost the nation's farmers at least $100 million every year. He was so displeased at the mutilation of his bill that he refused to support it.[62] When Morrill asked for reconsideration, the amended bill apparently died without another vote.[63] Riley also opposed a bill introduced by Kansas senator James Harvey on the grounds that it authorized the commissioner of agriculture to appoint a locust commission.[64]

With locust legislation stymied in Congress and the locust threat mounting, Governor John S. Pillsbury of Minnesota called for a conference of governors of states and territories in the locust district to meet in Omaha, Nebraska, on October 25 and 26, 1876. Prior to the conference, Riley and Thomas met and agreed to limit the commission's investigations to migratory locusts, at least for the present. They agreed to have the commission attached to the Hayden Survey (where Thomas served), but Riley convinced Thomas that three commissioners, rather than one, were necessary. The commission would be funded through an addition of fifteen thousand dollars to the Sundry Civil Bill, the catchall legislation passed at the end of each congressional session by which the Hayden Survey was regularly funded. The additional appropriation would pay the salaries and expenses of two entomologists and one ornithologist to investigate "destructive grasshoppers" and to recommend "remedial measures."[65] Thomas insisted on the ornithologist because he believed that a permanent solution to the locust problem depended on the protection of insectivorous birds.[66]

At the Governors' Conference in late October 1876, Riley took center stage. Those who had only heard about Missouri's flamboyant entomologist were eager to meet him. A reporter for the *Omaha Herald* commented, "From the dogmatic way he has of eating his opponents alive in print, from the . . . furor he has kindled up on the grasshopper question, and from [his] numberless prophesies . . . everybody [was surprised] that he is a young man."[67] The reporter, evidently considering himself an expert on grasshoppers, decided to test Riley's familiarity with "western" species:

> Reporter: did you ever see one of those [cannibal grasshoppers] that is [*sic*] sometimes four or five inches long?
>
> Riley: There is no such grasshopper.
>
> Reporter: Yes . . . there are . . . I can show you one . . .
>
> Riley: You don't know what you are talking about. I have an illustration in my last report of the large wingless grasshopper . . . about two inches long.
>
> [At this point Governor Silas Garber of Nebraska joined the exchange.]
>
> Governor Garber: I have seen grasshoppers out on the Republican [River] twice that large.
>
> Riley: There is no other kind of large wingless grasshopper.
>
> Reporter: Prof. Aughey [of the University of Nebraska] has discovered . . . a dozen . . . [previously unclassified plants on the Republican river].
>
> Riley: It's all humbug.
>
> [Ex-Nebraska] Governor [Robert W.] Furnas: I think Prof. Aughey claims that many, if not more.
>
> Riley: Oh, pshaw! It's all humbug.
>
> [At this point, the reporter changed tactics and challenged Riley to support his statement that locusts flew only by day.]
>
> Reporter: How do you know that they don't fly at night?
>
> Riley: It is contrary to their nature . . .
>
> Reporter: I flew a kite at night, covered with tar, and it was covered with locusts.
>
> Riley: That was an interesting experiment. I must take note of it.[68]

The delegates chose Riley, at thirty-four years of age perhaps the youngest participant, as their recording secretary with responsibility for drafting the conference resolutions. Riley essentially reiterated the compromise he had reached with Thomas prior to the conference.[69] The commission would now consist of three entomologists (rather than two entomologists and one ornithologist preferred by Thomas) and would include two "western men." Thomas explained privately that the governors of Minnesota and Nebraska insisted on hiring their favorites. The appropriation was increased from $18,000 to $25,000 to cover these two members.[70] In December 1876, Missouri congressman Robert A. Hatcher introduced a bill incorporating the conference resolutions to the Forty-fourth Congress.[71] With lobbying by Riley, Thomas, and Hayden, the measure passed on March 3, 1877.[72] Perhaps at Riley's urging, the two "western men" were dropped, reducing the requested appropriation to $18,000.[73] Funding the Entomological Commission as an adjunct to the Hayden Survey by means of the sundry bill avoided a repetition of the previous debacle in the Senate Agricultural Committee.

A final change incorporated in the clause in Hayden's budget authorized the secretary of the interior, rather than Hayden, to appoint the entomologists to the commission. Riley likely engineered this change behind the scenes. In December 1876–January 1877, when Rutherford B. Hayes, the incoming president, was selecting his cabinet members, Riley no doubt learned that his fellow Missourian, Carl Schurz, was in line to become secretary of the interior. He may have suggested changing the wording in the Hayden budget appropriation so that Schurz, rather than Hayden, would appoint the Entomological Commission. In any event, Schurz appointed Riley head of the commission. Irrespective of possible cronyism, Riley was best qualified to be the commission's chief. His publications were standard references on locusts, and he had demonstrated leadership and managerial skills at the governor's conference and in his guidance of the proposal through Congress.

Schurz appointed Thomas as "Disbursing Agent" and Alpheus Spring Packard as "Secretary," titles with no practical significance, as the commission members divided responsibilities among themselves.[74] Riley and Thomas were pleased to have Packard join them. Although belonging to the eastern establishment, Packard was a respected economic entomologist and an experienced field investigator with the Hayden Survey.[75] Riley regarded Packard's *Guide to the Study of Insects* (1869) the best single text on economic entomology.

Thomas and Packard's acceptance of Riley's leadership was a tribute to Riley's esteem among his scientific colleagues. Thomas, age fifty-two, was the senior member; Packard, thirty-eight, was midway in his career; and Riley, at thirty-four, was the youngest. Riley and Thomas differed fundamentally in matters of religion and science. Riley shied away from orthodox Christianity, whereas Thomas was an ordained Lutheran minister and a devoted churchman. Riley was a convinced Darwinist, whereas Thomas was for many years a leading anti-Darwinian. Despite these differences, Thomas accepted Riley's leadership and worked with him without reserve. Packard likewise welcomed Riley's leadership, an unusual compliment coming from a university-oriented scholar.

Amid the jockeying for position on the Entomological Commission, Riley conspicuously excluded LeConte, remarking disingenuously to Hayden that the locust commission was "special, and quite different" from their earlier proposal.[76] Thomas reminded Hayden that prior to LeConte's call for reform of the Agricultural Department, he had paid no attention to injurious insects and furthermore that LeConte's specialty was in Coleoptera (beetles), not in Orthoptera, the insect order that included locusts.[77] Having been rebuffed as a possible member of the Entomological Commission, LeConte mounted a campaign to become commissioner of agriculture. When his attempt failed, he settled for a position at the US Mint in Philadelphia.[78]

The Entomological Commission, Riley, Packard, and Thomas, presented a striking contrast to the Agricultural Department, where Watts and Glover were mired in the politics of seed distribution.[79] Primary credit for the establishment of the commission is due to Riley, who conceived the commission and pushed it through. Thomas effectively placed the commission within the Hayden Survey, and Schurz, a champion of civil service reform, ensured the selection of the most qualified personnel.[80] The Entomological Commission also fit well with Hayden's goal of opening the West to settlement.[81]

The commissioners established their headquarters in St. Louis, with a second office at Hayden's quarters in Washington. Meeting in Washington on March 20, 1877, they divided the locust region, consisting of 2 million square miles, among themselves. Riley took the region south of the 40th parallel (the Kansas-Nebraska border) and east of the Rocky Mountains, plus Canada (with Packard); Packard took the northwestern territories of Wyoming, Montana, Utah, and Idaho, as well as the Pacific Slope; and Thomas took the plains region north of Riley's area, including eastern Wyoming, northern Colorado, southeast Dakota Territory, Nebraska, western Iowa, and Minnesota.[82]

Following their meeting, the commissioners took to the field. Riley was impatient to reach the southern theater, as hoppers were already hatching in Texas.[83] In the course of that season, the commissioners, traveling by rail, buggy, horseback, and foot, crisscrossed the vast expanse of prairie, plateau, valley, and mountain terrain. Riley ranged from Texas to Manitoba, and Packard ventured into previously unmapped areas in western Montana, then controlled by Sioux war parties following the annihilation of General George Armstrong Custer's army at the Little Big Horn the previous year.[84]

The commissioners employed a number of special assistants, including Allen Whitman in Minnesota; John Martin in Minnesota, Dakota, and Montana; Samuel Aughey, Lawrence Bruner, and G. M. Dodge in Nebraska; A. N. Godfrey and Francis H. Snow in Kansas; A. J. Chapman in Kansas and Colorado; William Holly in Colorado; and Jacob Boll in Texas.[85] Coordination of mailing, filing, and responding to communications fell to Theodore Pergande and Riley's half-sister, Nina, who joined Riley in St. Louis in April 1877.[86]

Prior to their departure for the field, the commissioners distributed circulars requesting information on the arrival and departure of swarms, egg laying and hatching, soil conditions where eggs were laid, dates when locusts acquired wings, direction of hoppers and winged locusts, estimates of damage to crops, and the success or failure of various control methods.[87] The Santa Fe, Kansas Pacific, and Union Pacific railroads distributed the questionnaires to their agents, who in turn supplied information along the three parallel western rail lines.[88]

Special circumstances influenced the investigations during the first season (1877). First, Riley warned that the enormous number of eggs laid by swarms in 1876 would lead to an explosion of hoppers in 1877. In anticipation, the commissioners issued bulletins with recommendations to erect barriers and to destroy eggs. However, cold, wet weather in late April and early May killed the majority of emerging hoppers. Most of the weakened surviving hoppers fell prey to predators and parasites.[89] In April, Riley assured Kansas governor Anthony that few hoppers survived and farmers could control the remainder.[90]

Another special circumstance involved frequent requests from settlers that the commissioners offer their prognosis of locust danger for that season at public meetings. When addressing local audiences, Riley explained that the swarms in the 1870s only appeared to be more destructive than in the past because westward settlement had spread into locust country.[91] The commissioners' reassurances helped convince settlers to stay, thus reversing the exodus of settlers from Kansas.[92] Although time spent addressing settlers in public meetings diverted them from their scientific investigations, the commissioners considered this an essential service, and they prided themselves on providing accurate and helpful information.[93]

Despite such diversions, the commissioners established in rough outline the extent of the permanent breeding grounds and the invasion area.[94] In the 1860s, Walsh and Riley had established the eastern limit of locust migrations at approximately the Ninety-fourth Meridian, east of the border between Nebraska and Iowa, Kansas and Missouri, and Oklahoma and Arkansas. In 1877, Riley and his fellow commissioners established the approximate northern, western, and southern limits of the permanent and temporary regions. European entomologists had speculated about a breeding ground for migratory locusts that invaded Europe, but, as Thomas pointed out, they drew upon hearsay accounts and neglected to distinguish between three separate species known to invade Europe.[95] In August 1878, Riley announced to his colleagues at the AAAS, "[W]ithin a single year we have been able to map definitely the geographical range of the species and to trace the source of the more injurious swarms. In short, what was a mystery before is mystery no longer."[96]

Riley's statement was calculated in part to dispel lingering doubts among congressmen that the Entomological Commission was necessary. Such doubts surfaced earlier that year when Congress delayed responding to Riley's request to increase the commission's appropriation from $18,000 to $25,000 in order to carry out his original plan to investigate other insects of national importance. In June 1878, near the close of the fiscal year, Congress finally approved $10,000 of the requested $25,000. Riley remained at his post without salary for several months while lobbying for expansion of the commission's mandate. The remainder was approved sometime later, allowing the commissioners to resume their 1878 field investigations.

Despite funding difficulties, the commissioners, over the next several years, completed their mapping of the permanent and temporary regions. In 1878–79, Packard and his assistants established the southern boundary of the breeding grounds in Arizona and New Mexico.[97] In 1880–81, Lawrence Bruner, a Nebraskan farm lad Riley hired as a special assistant for locust matters, established the western border of the breeding grounds in Montana and Idaho. Bruner's mission was risky; Chief Sitting Bull and his Sioux warriors controlled the Judith and Musselshell River Valleys.[98] Bruner reported that friendly Flathead Indians he talked with made a clear distinction between non-migrating grasshoppers and locusts that "fly away."[99] The commissioners concluded that the permanent locust breeding grounds comprised four hundred thousand square miles.[100] More importantly, they concluded that locusts laid their eggs primarily in valley grasslands, avoiding extremely dry, sandy areas, timbered areas, and mountain slopes above timberline. Their second report in 1880 featured a map in six sections, identifying permanent and temporary habitat.[101] Within the breeding grounds, they estimated that about 44 percent (177,000 square miles) was conducive to excessive population buildup.[102]

Two other questions proved more difficult to answer: (1) why did this species alone among hundreds of North American grasshopper species develop migrating swarms? and (2) what governed the distance and direction of their flights? To understand the difficulties the commissioners faced, it is helpful to view their investigations in light of the phase theory proposed by Boris Uvarov, a Russian entomologist, in 1921.[103] Around the Caspian and Black Seas, Uvarov discovered that two species, *Locusta dancia*, a bright green grasshopper that led a solitary existence, and *Locusta migratoria*, a black and orange-red locust that periodically swarmed over farmland from central Asia to central Europe, were in fact two developmental phases of the same species. Uvarov learned that Vassily Plotnikov, an entomologist in Uzbekistan, had induced the change from *L. dancia* to *L. migratoria* by rearing *L. dancia* in crowded cages. Crowding in the nymphal stage became the key to Uvarov's phase theory. Certain species existed in alternate forms that he termed solitary and gregarious phases.[104]

The degree of crowding in turn stemmed from environmental factors such as temperature, humidity, rainfall, availability of food, and mortality due to predation and parasites. When environmental conditions allowed only moderate survival, young hoppers developed like "normal" grasshoppers, responding individually to outside stimuli and avoiding other grasshoppers. When environmental conditions led to overpopulation, crowding in the nymphal stage triggered a hormonal switch that caused the hoppers to develop in the gregarious phase. It is now known that the hormone serotonin is triggered by physical contact, sight, and smell of other locusts in close proximity.[105] Crowded hoppers begin to respond primarily to other hoppers rather than to outside stimuli, imitating each other's movements, congregating in bands, moving in the same direction and, after developing wings, flying in swarms. Gregarious locusts develop longer wings and larger air sacs, enabling them to fly long distances.[106] Particularly perplexing to Riley, Packard, Thomas, and their assistants was the fact that group behavior—crowding and swarming—functioned as a primary stimulus, in the same way that environmental factors like food, heat, and moisture functioned.

The massive locust swarms of the 1870s evoked both scientific curiosity and awe. Samuel Aughey attempted to measure the "countless billions" of locusts that passed over Lincoln, Nebraska, in May and June 1875 by taking a triangulation from buildings at the university campus. He calculated that the swarm was a mile high, a mile wide, and three hundred miles long. Assuming an average of twenty-seven flying locusts per cubic yard, Aughey estimated the swarm comprised a half trillion (500,000,000,000) locusts.[107] Albert Lyman Child, a Signal Service officer in Plattsmouth, Nebraska, reported an even larger swarm passing overhead from June 15 to June 25, 1875. Child estimated the swarm to be one-fourth to one-half a mile high, moving at approximately fifteen miles per hour. He observed the swarm passing overhead for six to seven hours during five consecutive days and he assumed (no doubt correctly) that the flight continued at night. At that rate, he calculated the swarm's length to be eighteen hundred miles. Combining his observations with those of others, he estimated the swarm's width at 110 miles. He calculated that the swarm covered a total area of 198,600 square miles (approximately the combined surface of Kansas, Nebraska, and half of South Dakota). Child himself found the results of his observations "utterly incredible," but as a scientist he was compelled to believe his figures.[108] This swarm represents the all-time worldwide record.[109] Awed spectators compared swarms to thunderstorms, cloudbanks, the Milky Way, and—most frequently—to blizzards with countless snowflakes.[110] Riley waxed poetic over the destructive power of locust swarms:

> Insignificant individually but mighty collectively, locusts fall upon a country like a plague. . . . The farmer plows and plants. He cultivates in hope, watching his growing grain, in graceful wave-like motion wafting to and fro by the warm summer winds. The green begins to golden; the harvest is at hand. Joy lights his labor as the fruit of past toil is about to be realized. The day breaks with a smiling sun that sends her ripening rays through laden orchards and promising fields. Kine [cattle] and stock . . . are sleek with plenty, and all the earth seems glad. The day grows. Suddenly the sun's face is darkened, and clouds obscure the sky. The joy of the morn gives way to the omens of fear. The day closes, and ravenous locust-swarms have fallen upon the land. The morrow comes, and ah! What a change it brings! the fertile land of promise and plenty has become a desolate waste, and old Sol . . . shines sadly through an atmosphere alive with myriads of glittering insects.[111]

While Riley's ode to the destructive power of locusts was tucked away in chapter 8 of the commissioners' first report, the first chapter addressed the scientific classification of the species in sober prose. All three commissioners agreed that the Rocky Mountain Locust was a distinct species, and they agreed on its scientific name, *Caloptenus spretus*, and its common name, the Rocky Mountain Locust. The path leading to this consensus was complicated by diverging views regarding the relevance of Darwinian evolution to locust taxonomy.

About 1860, Thomas sent specimens of locusts he presumably collected in the vicinity of his home at Murphysboro, Illinois, to Philip B. Uhler, the twenty-five-year-old assistant of John G. Morris, lepidopterist and librarian of the Peabody Institute in Baltimore, Maryland, for identification.[112] Among the specimens Uhler returned to Thomas was one he classified as a new species, *Acridium spretis* (*spretis* meaning "despised").[113] In 1862, Thomas described a species under the name of *A. spretis*, but due to wartime exigencies, publication of his description was delayed until 1865. Riley was dissatisfied with Thomas's 1862–65 description because it did not match the thousands of specimens he measured and, furthermore, the range of the migratory locust did not extend to southern Illinois. Both Thomas and Riley recognized the mistake.[114] The ambiguity could not be resolved by reference to the original specimens because these had been destroyed soon after publication of Thomas's paper.

In 1866, Walsh published the first extensive report on the range and natural history of the migratory locust under the name *Caloptenus spretus*, citing Uhler as authority. He did not give a full description but compared the wing length of *C. spretus* with *Caloptenus femur-rubrum*, the (non-migrating) red-legged locust. Two years later, in 1868, Samuel H. Scudder included this species in his *Catalog of the Orthoptera of North America* under the name *Acridium spretum* Uhler. In 1873, Thomas provided a description of *Caloptenus spretus* in his *Synopsis of Acrididae of North America*. The long-awaited description began, "very much like *C. femur-rubrum*."[115] Such is the grist for the mill of taxonomists. Riley, Thomas, Packard, and most entomologists thereafter appended "Thomas" (as the first describer) to the species name, but some, like Snow at the University of Kansas, continued to append "Uhler" to the species.[116] In 1874, one year after Thomas's *Synopsis*, Riley addressed the species issue. Based on measurements of thousands of specimens from the Rocky Mountains to the Atlantic seaboard, Riley identified three closely related species: *Caloptenus spretus*, *Caloptenus femur-rubrum*, and *Caloptenus atlanis*. The latter was a new species that he separated from the other two. Although all three species sometimes became serious pests, only C. *spretus* formed swarms and invaded distant areas.[117]

When the Entomological Commission took up its work, Thomas assumed responsibility for the species name. In the first report, Thomas defended the name "*Caloptenus spretus*," challenging the Swedish entomologist, Carl Stål, who in a recent revision of the Orthoptera renamed the genus *Melanoplus*, thus changing *Caloptenus spretus* to *Melanoplus spretus*.[118] Although Thomas and Riley agreed that the genus *Melanoplus* held priority, they retained the genus name *Caloptenus* because this name had long been in use and they didn't want to confuse the nomenclature with synonyms they considered already too numerous.[119] The priority issue was a central concern of the Entomology Club's nomenclature committee, consisting of Riley, Scudder, and Augustus R. Grote. Riley, Scudder, and Thomas (who was not on the committee) agreed that priority should be assigned to the first describer, but they held that where species names had been in use for some time, "general use" should prevail. Grote argued that priority alone should determine the name, a position naturalists agreed upon a generation later.[120] Riley's name, *Caloptenus spretus*, held during his lifetime, but entomologists later agreed on its current name, *Melanoplus spretus*.[121]

With regard to the common name, Walsh and others initially called the migrating locust the "hateful grasshopper" or the "western grasshopper." Riley scolded Walsh for calling locusts grasshoppers, which he said confused the distinction between non-migratory grasshoppers and migratory locusts. Beginning in 1874, Riley employed the common name Rocky Mountain Locust for *C. spretus*, based on its presumed permanent home, and so the species was known from that time on.[122]

In 1874–75, when Riley separated *C. atlanis* from *C. femur-rubrum* and *C. spretus*, Scudder objected that all three were varieties of a single species. Personal and philosophical factors colored what on the surface seemed to be a purely scientific question. Riley was especially sensitive to criticism from Scudder and other Harvard-Cambridge entomologists associated with the Cambridge Entomological Club. Riley charged that anti-Darwinism lay behind Scudder's objection to his separation of three species: "All discussion . . . as to whether we are dealing with species or varieties, is more or less puerile. Naturalists have no fixed standard as to what constitutes a species, and are fast coming to the conviction that there is no such thing in nature, and that the term is . . . an abstract conception . . . the broad fact [remains] that these . . . forms [*spretus, femur-rubrum, atlanis*]—call them races, varieties, species, or what we will—are separable, and that they each have their own particular habits and destiny."[123]

Riley chided Scudder (and also Thomas and Stål) for erecting species on the basis of a few specimens. Always careful to give the number of specimens upon which his determinations were made, Riley, a "lumper," scorned "splitters." Riley cited a letter to him from Thomas in which Thomas wrote, "Stål's attempt to systemize, if carried out, will give us a genus for nearly every species; and Scudder seems disposed to make a distinct species for each variation in color."[124] For his trump card, Riley again quoted Thomas, who, though sharing Scudder's anti-Darwinism, objected to Scudder's taxonomy: "Although the descriptions of species established by Scudder may be . . . sufficient in other orders; in *Acridii* I have . . . found them quite unsatisfactory. The characters are those most liable to variation, and hence insufficient in describing species. As a . . . consequence, a number of his species are in fact but varieties."[125]

By the early 1870s, both Scudder and Thomas were gradually, reluctantly, accepting evolution. Scudder's conversion began during the debate over mimicry in the Monarch and Viceroy butterflies. Thomas's conversion came about in the course of his investigation of locusts. In 1874, Thomas wrote to Riley regarding *Caloptenus* species: "I am inclined to think *femur-rubrum* the older form and that during the change which produced the desert conditions of the west it was converted in that district into *spretus*. The *Atlanis* form I think is less permanent and more transient, the result probably of suitable climatic conditions continued but a few years, and that as soon as the climate returns to the normal conditions it will revert to the usual form of *femur-rubrum*."[126] Thomas shared Packard's view that climate and other external conditions directly influenced the origin of species.[127] In the commission's second report, he noted a "striking similarity" between the environment of *Caloptenus italicus* (the Italian locust) and that of the Rocky Mountain Locust. He speculated that similar environments had given rise to two migratory locusts species on different continents that, despite their diverse geographic origins, were so closely allied that they belonged to the same genus. This similarity, he said, "cannot be accidental, but results from some law of Acridian life which has not yet been discovered," and he concluded "its solution would carry us . . . into the vexing question of species evolution."[128] In the same report, he speculated further that several local incipient locust species (varieties) in Kansas, Nebraska, Texas, and Minnesota originated when stragglers from Rocky Mountain Locust swarms produced a second brood that, under favorable circumstances, survived and multiplied.[129] Elsewhere, Thomas speculated that local species of *C. atlanis* represented "offshoots" from *C. spretus* and *C. femur-rubrum*, the direction of the variation depending upon the conditions of the locality.[130] By such steps, Thomas gradually accepted a Lamarckian version of evolution.

Riley, the more orthodox Darwinian (in the 1870s), held that the decisive criterion for the classification of species was the ability of a male and female to produce offspring: "It has often occurred to us that the [Rocky Mountain Locust] might . . . through miscegenation [become profoundly modified in the course of two or three generations in the direction of *atlanis*], but the evidence is against this supposition."[131] Contrary to his early expectations, the Rocky Mountain Locust had not interbred with the red-legged locust or other lowland species. He cited this as further evidence that the Rocky Mountain Locust would never become permanently established in the Mississippi Valley.

Underlying the dispute over the scientific and common name was the practical concern that if locusts in the West and East were the same species, farmers in the East faced the prospect of periodic devastation. At the Omaha conference, several governors stressed the fact that, since colonial times, "locusts" constituted a threat in the eastern states and argued that easterners and westerners should remain united in their search for control of the "locust."[132] Although Riley insisted that the Rocky Mountain Locust was a separate species, he refrained from openly correcting the governors, and he retained the clause about the locust threat in the East in the conference resolutions.[133] As chief of the Entomological Commission, Riley continued this balancing act, insisting that the Rocky Mountain Locust was the only true migratory locust in North America while recognizing the "locust" threat in the East. In a not so subtle appeal to eastern districts, the commissioners stressed the fact that other grasshopper species were capable of flights of up to two miles, and they devoted two chapters of the first report to "locusts" in the eastern United States and in other countries.[134]

Philosophical and political considerations aside, the commissioners focused on three more strictly scientific questions with respect to the Rocky Mountain Locust: (1) What caused the species to form swarms? (2) What influenced them to take flight? and (3) What influenced the times, distance, and direction of flights (migrations)?

With respect to the question of why this species formed swarms, Riley and his fellow commissioners distinguished between *remote* causes, for example a series of warm, dry years, and *immediate* causes, such as crowding, hunger, procreative instinct, avoidance of natural enemies, and the search for suitable breeding grounds.[135] They assumed, correctly, that weather conditions led to an excessive locust population similar to seasonal fluctuations in Chinch

Bug and Army Worm populations. Packard and Thomas assumed responsibility for studying the influence of weather, but Riley clearly influenced the investigations and conclusions, particularly in the first report in 1878.

The commissioners were disappointed to learn that weather data from the Signal Service and Agricultural Department indicated no direct relationship between hot, dry years and locust buildup.[136] Although disappointed, the commissioners remained convinced that such conditions favored the buildup of locust populations.[137] Observations by the commissioners and assistants likewise failed to support "immediate" causes of swarming. The search for food failed as a cause because swarms often departed from habitats plentiful in grass, flying over fertile fields before landing in relatively barren areas.[138] The "procreative instinct," the theory that male locusts annoyed females in the process of ovipositing thus causing them to take wing, did not explain why locusts swarmed only in certain years. Likewise, pursuit by natural enemies contradicted the fundamental assumption of the balance of nature, according to which excessive populations resulted, in part, from a shortage of natural enemies.

At the close of the first season, Thomas concluded "[T]he Rocky Mountain Locust is 'constitutionally migratory' and will take to flight when it reaches age."[139] The commissioners were evidently unsatisfied with this formulation because, in the second report, a clearly frustrated Thomas reframed the question: "Are [Rocky Mountain Locusts] normally sedentary in the permanent area, developing under favorable climatic conditions in immense numbers and becoming migratory from some cause connected with this development; or are they essentially migrating in character?"[140] By 1880, he was convinced that the species was migratory regardless of crowding or noncrowding conditions: "It is . . . evident that their flights don't depend upon numbers. . . . [W]hen they attain the proper age [they] will migrate."[141] Thomas admitted, however, that the commissioners' observations were contradictory. Some locusts, or locust "families," were apparently born with an instinct to take to flight whether crowded or not, while others took to flight only when crowded.[142] The commissioners were observing and recording locust behavior accurately but lacked the conceptual framework of the phase theory to interpret their observations.

Riley and his colleagues were ill-equipped to deduce a phase theory because they investigated only locusts that swarmed and destroyed crops, that is, locusts in the gregarious phase. What they did not know or suspect was that swarming locusts were different in physiology and behavior from solitary locusts. Riley's measurements of locusts, for example, were all conducted on adults that (we now know) were of the gregarious phase. It apparently never occurred to him (or he never had the opportunity) to compare locusts in swarming populations with nymphal and adult locusts in their native habitat during non-outbreak years.[143] Even so, the commissioners came tantalizingly close to formulating a phase theory. Thomas noted, for example that Red-legged, Atlantic, and Differential Locusts that took to flight "appear[ed] to be modified in the direction of *C. spretus*."[144] The commissioners also noted that locusts developed enlarged tracheae and air sacs that enabled swarms to fly long distances.[145] They lacked the resources to pursue these leads that might have led them to a phase theory.

Nevertheless, the commissioners made significant steps toward an explanation of locust swarming and migration. When the commissioners took up their work, Riley and Packard believed the swarms traveled the entire distance from breeding grounds in Canada and Montana to Texas in a single season, while Thomas (and also Scudder) believed the swarms from the north flew only as far as Dakota Territory or Nebraska, where they laid eggs that hatched and formed swarms that flew further south and east the following year. In 1877, commission agents tracked the individual flights of swarms that originated in Texas and then flew across Nebraska and Dakota Territory and on to Minnesota and Canada. Their tracking demonstrated that individual swarms flew up to five hundred miles in a single flight and covered thousands of miles in one season. This finding established a fundamental premise with far-reaching implications for locust investigations. European scientists had assumed that locust swarms that invaded Europe required two or three years to travel from their presumed breeding ground in the Crimea to central Europe. The commissioners' findings suggested that these Asian locusts, like Rocky Mountain Locusts, traveled from breeding grounds to areas of invasion in one season.[146]

The commissioners and their assistants confirmed that swarms flew at night and that they sometimes flew so high as to be practically invisible to the naked eye. European entomologists had speculated about both of these phenomena but had not been able to prove or disprove their hypotheses. Night flight and flight at high altitudes explained the sudden appearance of swarms where none had been present hours earlier.[147]

The most difficult question was what determined the direction of swarm flights. In America, the debate revolved around the concept of a "return migration."[148] Most Europeans were skeptical that invading locusts instinctively returned to their ancestral habitat, considering it more likely that locusts that invaded settled regions of Europe and Africa perished where they landed.[149] By contrast, Riley, in 1874–75, postulated a "return migration" for the Rocky Mountain Locust, and most Americans followed his lead. He even claimed that hoppers marched in a northwesterly direction toward the breeding grounds.[150] Based on answers to his questionnaire, Riley concluded that the return instinct was more pronounced among locusts that hatched south of the forty-fourth parallel in southern Minnesota. Locusts that hatched north of this line seemed to march or fly in no particular direction because, Riley speculated, they were closer to the breeding grounds.[151]

Packard, a steadfast believer in environmental influence, maintained that wind alone influenced the direction of swarms.[152] Riley sought to combine internal instinct with the external influence of wind. He theorized that the locusts' instinct directed them to take to the air only when the wind blew in the direction of the breeding grounds to the northwest. Riley may have taken this clue from Fedora Petrovich Köppen, who concluded that locust migrations in southern Russia were directed by a combination of instinct and wind movement.[153]

In the first report, probably in deference to Riley, the commissioners concluded that locust migrations were caused by a combination of wind and the "drive" of locusts toward the northwest.[154] In their second report, Thomas, the main author, wavered, writing at one time that locusts attempted to fly north, even against the wind, and at another time that there were no well-documented cases of swarms flying against the wind.[155] In their third report, based on observations in 1879–81, where one would expect a definitive conclusion on the question of wind influence, the commissioners skirted the matter entirely.[156]

Despite their differences and uncertainties, Riley, Packard, Thomas, and their assistants established the fundamental importance of meteorological conditions to locust management. New light on the question came in the 1950s, when meteorologists discovered the "Great Plains low-level jet," an annual flow of air from south to north across the central plains. In late spring or early summer a two-hundred-mile-wide flow of air, moving at ten to thirty miles per hour, begins flowing from Kansas and Oklahoma northward to Canada. This air stream was most likely the one cited by the Entomological Commission and the one that carried the 1875 world-record swarm of locusts northward. So, was Packard right and Riley wrong? Did wind movements, not instinct, determine the direction of locust swarms? The question remains, how did these wind movements relate to the evolutionary biology of locusts? After all, the Great Plains low-level jet was an environmental factor in the evolution of the Rocky Mountain Locust. Jeffrey K. Lockwood, a University of Wyoming locust expert and historian, notes that migrating birds are known to exploit this airstream in their northward journeys.[157] It seems likely that locusts did the same.

Riley assumed responsibility for investigating invertebrate predators and parasites of locusts. Riley criticized his European colleagues, Carl Eduard Adolph Gerstäcker and Köppen for paying little attention to this topic.[158] By contrast he presented a comprehensive, richly illustrated report of insect predator and parasite controls of locusts, based primarily on his investigations during the years 1875–77. He reported, for example, that tachinid flies, whose larvae feed on internal organs of locusts and emerge prior to their host's death, often decimated locust populations to such an extent that they did not swarm.[159] Riley considered these parasites to be a primary cause of locust mortality in 1877.[160]

In 1876, Riley described the *Anthomyia* egg-parasite, a fly larva (Diptera: Anthomyiidae) that he estimated destroyed 10 percent of the locust eggs in Missouri, Kansas, and Nebraska. The same year, he discovered two locust predators, the Common Flesh-Fly (Diptera: Sarcophagidae), whose larvae sucked the juices from locust eggs, and the carabid beetle *Amara* that fed on locust eggs.[161] In 1877, by rearing hundreds of mites, Riley demonstrated that the six-legged red mite, another locust predator, was the immature stage of the larger eight-legged mite that preyed on locust eggs.[162] Until then, they had been assigned to two different genera. In the autumn and winter of 1877–78, Riley kept a Cylindrical Tiger Beetle in a vivarium, feeding it two locusts per day. In the larval stage, the insect was large, slow, and practically sightless, and therefore a poor hunter, but in the adult stage it was an efficient predator. Collectors prized the adult beetle because of its large size, ferocious appearance, and rarity.[163]

How Riley found time, while traveling from Washington, DC, to Texas, Colorado, and Manitoba, meanwhile coordinating the Entomological Commission, to hand-feed a Tiger Beetle and to breed hundreds of parasitic mites remains a mystery. His discoveries continued after 1877, when his attention was focused primarily on the Cotton Worm. In 1877–80, he discovered that blister beetles (Coleoptera: Meloidae), whose larvae parasitized locust eggs, sometimes overwintered in the larval stage to emerge the second or third year. Riley pointed out that, among parasites whose existence depended on the irregular appearance of hosts like the migratory locust, staggering the emergence from the larval stage over several years increased their chances for survival.[164] In 1877–79, aided by Dodge in Nebraska, J. G. Lemmon near Sacramento, California, and Samuel Wendell Williston, a Kansas State University entomologist recently moved to New Haven, Connecticut, Riley demonstrated that bee fly larvae (Diptera: Bombyliidae) parasitized locust eggs. Next to the *Anthomyia* egg parasite, Riley considered bee flies to be the most effective enemies of locusts.[165]

Citing these and other examples of invertebrate predators and parasites, Riley assured farmers that natural controls would eventually reestablish nature's balance. This sometimes occurred in unexpected ways. Under the subtitle "No Evil without Some Compensating Good," Riley argued that ravenous hoppers in the spring of 1875 left practically no nourishment for the Chinch Bugs that, during previous seasons, had destroyed grain crops. Riley therefore predicted little damage from Chinch Bugs in 1876, and his prediction was correct.[166]

In addition to his investigation of invertebrate predators and parasites, Riley also investigated the survival rate of locusts hatched from eggs laid in the lowlands. From Texas, Kansas, Nebraska, and Iowa, he sent Pergande locust eggs with instructions to hatch them and to test them under various conditions. In St. Louis and in Carbondale, Illinois, the hoppers were confined in twelve-foot square enclosures with zinc-covered walls. Most hoppers died in the second and third molts. Of the few that developed into winged adults, none laid eggs.[167]

In Nebraska, Aughey performed similar experiments, comparing the strength and vitality of locusts hatched on the plains with locusts captured from swarms from the northwest. Attaching the legs of both broods to a spring by means of a silk thread, he found that those originating in the breeding grounds were 25 to 50 percent stronger than those hatched in Nebraska. When confined without food, locusts from invading swarms lived longer than their lowland cousins. Citing his and Aughey's experiments, along with testimony from questionnaires, Riley assured farmers that locusts that hatched in the Mississippi Valley were so weakened by disease and subject to parasites and predators that they were incapable of sustaining a permanent population there.[168]

Thomas's account of vertebrate predators included few new discoveries, vertebrate species being fewer in number and better known than invertebrates. Thomas engaged Aughey to investigate the feeding habits of insectivorous birds.[169] Having dissected some 630 specimens representing ninety species of birds, Aughey concluded that birds did not compete with humans for grain. On the contrary, blackbirds and other species devoured insects that fed on grain.[170] He railed against the misguided massacre of beneficial species and the slaughter of birds shot for sport at state and county fairs: "[I]s it not the sport of a barbarian and . . . savage? . . . [F]uture ages will look on the wanton killing of birds . . . with the same . . . disgust that we feel in reading the stories of the animal contests of the Roman arena."[171]

The response to the commissioners' question, "[T]o what extent have birds . . . been useful in destroying [locusts]?" indicated a growing public awareness of the beneficial role played by vertebrate predators, especially birds.[172] Though unable to quantify the precise effect avian predation had on locust and other injurious insect populations, the commissioners demonstrated that insectivorous birds were beneficial and worthy of protection.[173] Riley admitted that, prior to Aughey's investigations, he had not fully appreciated the importance of birds in the control of insects. Although he presumably was aware of the report of California Gulls staving off the destruction of Mormon settlers' crops by "Mormon Crickets" in the 1840s, he may have considered this an undocumented report.[174] The commission's findings prompted several states to revise state laws in favor of protecting insectivorous birds. Unfortunately, these laws were often not enforced, and effective protection of birds was delayed for at least a generation.[175]

Over the objections of his fellow commissioners, Thomas insisted that the commission import foreign insectivorous birds.[176] Aughey pointed out that the English sparrow, imported in the 1850s, had become a serious pest in North America. Undeterred, the commission, in 1878, imported twenty-four English rooks. Perhaps fortunately, the birds were severely weakened during the voyage and during their detention at New York customs. A year later, only one rook was still living.[177]

Among other controls proposed by the commissioners was the hopper dozer, a horse-drawn machine that gathered hoppers, then crushed or burned them.[178] Riley, the inveterate tinkerer, developed a hopper dozer at Manhattan, Kansas, that was featured in *Scientific American*, in the commission's report, and later in the report of the commissioner of agriculture.[179] The commission reiterated Riley's recommendations for state laws offering bounties for eggs and hoppers and for the recruitment of manpower to dig ditches and collect and destroy hoppers. Such laws were passed in Kansas, Nebraska, and Minnesota; however, they were ineffective because farmers resisted the notion

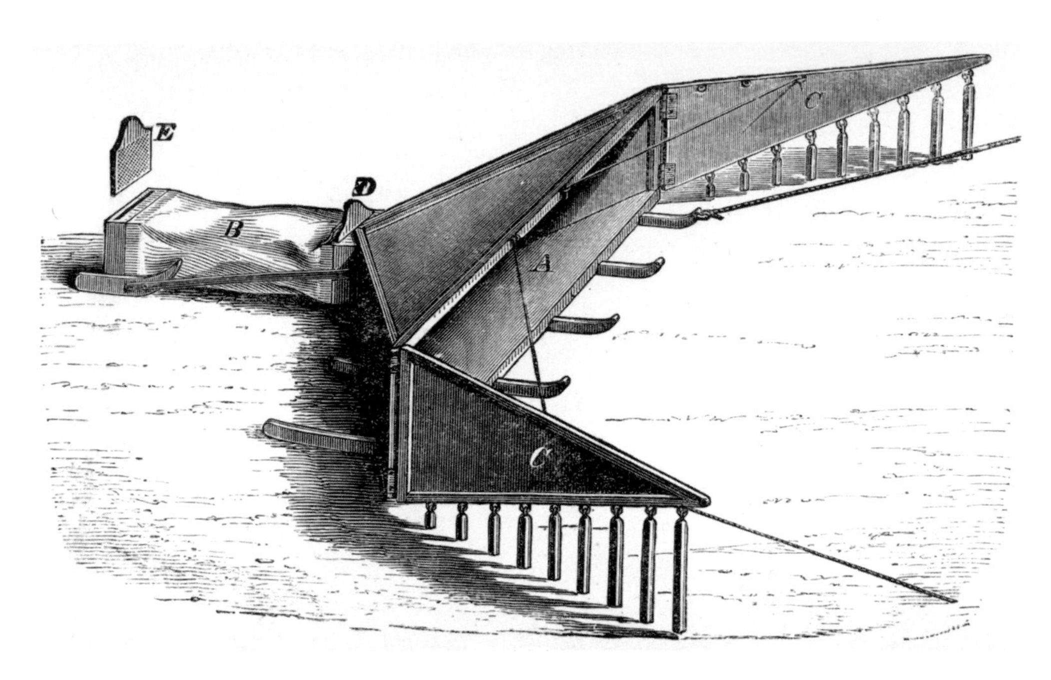

FIGURE 7.5 “The Riley Locust Catcher,” *ARCA* (1883), plate IX (Image courtesy of Albert R. Mann Library, Cornell University).

of conscripted labor. The recruitment of a “grasshopper army” to combat locusts never materialized.[180]

Regarding control of locust swarms, all three commissioners agreed: “[M]an is powerless before the mighty host.”[181] They elaborated Riley’s proposals to prevent the formation of swarms, or failing this, to warn of approaching danger. The commission proposed: (1) organizing teams of US Army and Northwest Mounted Police to burn hopper infested prairie in the breeding grounds; (2) equipping the Signal Corps, US military, and Indian agents to survey the breeding grounds and to issue warnings of approaching swarms or the likelihood of locust invasions for the coming year; and (3) facilitating settlement of the West, especially the breeding grounds.

Thomas, the primary author of the second and third locust reports (in 1880 and 1883), at first focused on burning hopper-infested prairie. The commissioners were not unanimous in support of burning, though they refrained from airing their differences in public.[182] The maps accompanying the second report highlighted known areas of intense locust reproduction with the optimistic notation, “Grass-covered. Easily burned.”[183] Thomas conceded that destroying eggs over twenty-five thousand to thirty thousand square miles of locust breeding grounds was impossible. He contended, however, that burning selected prairie in the spring would destroy enough hoppers to prevent the buildup of swarms. Following Riley's earlier proposal, Thomas proposed the extinguishment of prairie fires in these areas in the fall and burning the prairie when hoppers emerged in the spring.[184] He admitted that the plan was risky. Locusts hatched at different times and fires burned unevenly, the flames often skipping pockets of congregated hoppers. He also admitted that burning hoppers in Minnesota had not been effective.[185]

From his reading of *Man and Nature* (1864) by the pioneer ecologist George Perkins Marsh, and based on his own observation of western settlement, Thomas concluded that settlement of a farming population in the permanent breeding grounds would solve the locust problem: "[O]ur plan . . . is to place an agricultural population in the very home of the [locusts], which from necessity would be compelled to wage a constant warfare against them."[186] As incentive to colonists, Thomas proposed that the federal government sponsor irrigation, forestation, and the construction of a rail and wagon road network. Like others of his generation, Thomas believed that irrigation and forestation would bring about an increase in rainfall and transform the arid region into fertile cropland.[187] He applauded the commissioner of agriculture's proposal to cover the northern plains with trees, and he assured Governor Pillsbury that Minnesota would soon be comparatively free from locusts because Dakota Territory was being rapidly settled and timbered, thus reducing locust breeding grounds.[188]

Supplementing irrigation, forestation, and road building, Thomas proposed that the federal government issue land grants to private companies that agreed to promote settlement in key locust breeding grounds, preferably by Russian peasants who had experience fighting locusts.[189] In some respects, Thomas's plan clashed with the proposals of John Wesley Powell, who as director of the US Geological Survey, advocated increasing the size of homesteads in the arid west from 160 acres to approximately 2,000 acres to facilitate ranching, and aligning political boundaries along watersheds, rather than on a rectangular grid, to facilitate the organization of irrigation districts.[190] In contrast to Powell's emphasis on ranching, Thomas favored irrigation farming on traditional quarter-section homesteads. Convinced that Powell had underestimated the amount of irrigable land, he called on Henry Gannett, a Harvard educated geologist, to produce a revised estimate. Gannet complied by doubling Powell's estimate of irrigable land, an area Thomas considered sufficient to support his envisioned farming and locust-fighting population.[191] Riley did not comment on Thomas's proposals for western settlement; however, he likely favored irrigation farming over ranching.

In addition to Thomas's proposals of burning and agricultural settlement, the commissioners reiterated Riley's call for annual surveys of locust breeding grounds, including warnings of migrating swarms, similar to weather forecasts.[192] Riley's proposal for locust surveys in Canada apparently never went beyond the planning stage, but it was incorporated in Lawrence Bruner's annual locust forecasts that, beginning in 1880, were published by the Agricultural Department and circulated in regional newspapers.[193]

The success of Paris Green as a control for the Colorado Potato Beetle aroused hopes that chemicals might be useful against locusts. Riley was skeptical because of the magnitude of locust swarms and because of possible danger to human health from applying poison on a large scale.[194] Despite his skepticism, the commission engaged R. L. Packard (no relation to A. S. Packard), a chemist at the Patent Office, who conducted field tests with various chemicals in Colorado. Packard concluded that contact insecticides were of limited effectiveness against locusts.[195] In 1885, farmers in the San Joaquin Valley, California, developed a grasshopper poison bait consisting of bran, white arsenic, and sugar. This mixture, with some refinements, was widely used in the control of grasshoppers and other insects until replaced by synthetic organic insecticides (carbaryl, malathion, DDT, and others) after World War II.[196]

For the third report, the commission engaged A. H. Swinton, a British specialist in insect migration, to investigate the relationship between locust outbreaks and sunspots.[197] Comparing data from the commission, locust records from other continents, and records of sunspots and earthquakes, he predicted that the next outbreak of the Rocky Mountain Locusts would occur within the next four to ten years. This was clearly not the kind of answer sought by the commissioners, who prided themselves on providing annual predictions that would be of practical use to farmers. Swinton's report was printed, accompanied by the footnote, "[W]e hereby tender our acknowledgments without endorsing all of the author's views."[198]

Meanwhile, nature stepped in, fulfilling the commissioners' prediction that the locust invasion of the 1870s was a temporary phenomenon. The year 1877 was in fact the last year American farmers suffered significant damage from the Rocky Mountain Locust. In Canada, the last serious outbreak occurred in 1875.[199] By the 1880s, the commission's maps showed only isolated locust pockets in Montana, Idaho, and Texas, with virtually no damage to agriculture. Rocky Mountain Locusts were so scarce that Packard, who was preparing a paper on the embryological development of locusts, had to substitute eggs of Atlantic locusts (*C. atlanis*) because he could not obtain Rocky Mountain Locust eggs.[200]

By the early 1880s, the locust threat having receded, the commissioners and the general public were turning their attention to other matters. Riley now focused his attention on the Cotton Worm. Thomas turned his attention from entomology to ethnology and joined Powell's Bureau of Ethnology in 1882.[201] What the commissioners did not know was that the cycle of locust invasions had been permanently broken. Except for limited outbreaks in Montana, the Dakotas, Minnesota, and Canada in the 1880s and 1890s, the Rocky Mountain Locust was already history. The last specimen was collected in Canada in 1904, and the species is now considered extinct.[202]

In his book, *Locust*, Lockwood notes that during the first half of the twentieth century, entomologists doubted that the Rocky Mountain Locust had really disappeared. Many entomologists surmised that *Melanoplus mexicanus*, a near relative of the Rocky Mountain Locust, was the solitary phase. However, when forced into crowded conditions that should have triggered its transformation to the gregarious phase, *M. mexicanus* failed to transform. Also, more rigorous taxonomic criteria indicated that *M. mexicanus* and *M. spretus* were two separate species. Other theories regarding extinction of the Rocky Mountain Locust involved climate change, cessation of burning by Indians, alfalfa cultivation, and the removal of the bison.

Conner Sorensen in *Brethren of the Net* (1995) favored the bison theory proposed by Canadian entomologist Paul Riegert. According to Riegert, buffalo had periodically overgrazed the range, reducing available food and pulverizing the soil, thus providing ideal conditions for egg laying. Abundant hatch and scarcity of forage led to crowding and swarming. When the buffalo disappeared, the locusts ceased to multiply in outbreak numbers.[203]

At the time Sorensen's book appeared (1995), Lockwood was retrieving century-old Rocky Mountain Locust carcasses from receding glaciers in the vicinity of Yellowstone National Park. Dissatisfied with the bison theory and other explanations, Lockwood proposed that intensive agriculture in the form of plowing, irrigating, over-grazing, and flooding in the mountain valleys in the 1870s destroyed the breeding grounds of the Rocky Mountain Locust.[204] The removal of this ecological bottleneck spelled extinction for this once omnipresent species. Lockwood and his associates found that the area for optimal egg-laying conditions was even more limited than the commissioners estimated, comprising perhaps as little as three thousand square miles (as compared to thirty thousand square miles). Although still vast in extent, these mountain valley breeding grounds were rapidly transformed by farmers, ranchers, and livestock. In the 1870s, two million pioneers settled in the mountain states and territories, and in the 1880s even more came. Many of these pioneers established commercial farms to feed the mining population. By 1884, these settlers grazed some forty million cattle and probably as many sheep in mountain valleys. Livestock on this magnitude compared in numbers to the bison that were removed from the plains.[205]

Lockwood's findings, like those of Uvarov, confirmed the central conclusions reached by Riley, Packard, and Thomas. For example, Uvarov discovered air sacs in *Locusta migratoria* similar to those found by American entomologists in the Rocky Mountain Locust.[206] Given time and resources, the commissioners might have discovered the ecological bottleneck Lockwood proposed over a century later. Lockwood and his students found that the preferred soils comprise only one percent surface of the state of Wyoming.[207] Riley at one time narrowed the primary breeding grounds to the three forks of the Missouri River and along the Yellowstone.[208] Riley and the commissioners also found that water was deadly to locusts, especially in the early nymphal stages. In one of his last publications on the locust, Riley recommended irrigation to suppress the Rocky Mountain Locust. He was apparently unaware that 70 percent of the farms in the mountain valleys located in the permanent zone were already irrigated by the 1890s.[209] Riley's finding that locust eggs perished when exposed to fresh air takes on new significance when viewed in light of Lockwood's finding that, by the mid-1870s, farmers in the mountain states and territories were equipped to plow 5 percent of the permanent breeding grounds in a single day.[210] Twentieth-century researchers also found that alfalfa and other legumes were distasteful, even poisonous to locusts and grasshoppers, a finding Riley anticipated when he noted that ravenous locusts ate almost all plants but avoided legumes.[211] Even Riley's seemingly quixotic proposal to utilize locusts as food turned out to have a solid basis. Nutritional analysis has shown that locusts are rich in protein and calories.[212] Edible insects are now considered a significant factor in human food security in many regions of the world.[213]

Although the menace of the Rocky Mountain Locust receded, locust and grasshopper outbreaks continued to plague farming sectors in North America, South America, Africa, Asia, and Australia. Beginning in the 1880s, the Division of Entomology under Riley assisted locust-plagued countries like Cyprus, Algeria, and Argentina.[214] Today, grasshopper and locust experts, in the tradition of Riley, Packard, and Thomas, work on behalf of governments in many countries to monitor grasshopper and locust populations and to apply controls.[215]

Riley's Entomological Commission, which soon morphed into the Division of Entomology, reoriented agricultural science from a passive, discipline-orientated practice to a "problem centered" approach, with scientists from various disciplines pooling their resources to solve specific problems.[216] The flamboyant Riley, the pious Thomas, and the erudite Packard were not only pioneer locust investigators; they were also master communicators. Their prediction that the locust cycle of the 1870s was a temporary phenomenon served to reinforce the general impression that the Entomological Commission helped end the plague. In 1880, an article in the *American Naturalist*, probably by Packard, stated, "It is believed that this locust will never be so destructive as in the past, and due credit is given [to the] U. S. Entomological Commission [and to] Congress in ordering the investigation."[217]

Riley's locust investigations and his leadership of the Entomological Commission propelled him into the nation's capital, where he spent the remainder of his career. Riley relished the excitement and camaraderie there, but he perhaps underestimated the cost to his mental and physical health during what turned out to be the remaining seventeen years of his life.

8

Washington Gadfly, 1879–1881

An army of oxen led by a lion is better than an army of lions led by an ox. The farmers are being led by an ox.

—The *Prairie Farmer*, referring to Commissioner of Agriculture William G. LeDuc, Riley's primary opponent in the Hayes administration, 1877–1881

Riley's management of the locust crisis facilitated his advancement from Missouri state entomologist to chief of the US Entomological Commission in March 1877. His advancement was slightly tarnished by the refusal of Missouri lawmakers to continue the office of state entomologist and their refusal to pay Riley for six months of service prior to his move to Washington.[1]

Although experienced in politics, Riley found that politics in Washington required special skills. The spoils system of political patronage that operated in the nation's capital mirrored the unrestrained capitalism that dominated the nation's financial centers. Captains of industry (or robber barons, depending on one's orientation), not presidents and legislators, set the dominant tone of the Gilded Age. Ambitious individuals in that era turned to finance, railroads, lumber, oil, meatpacking, or flour milling, and left politics to their more cautious fellows. Politicians maintained order and stability but did not try to initiate change, the assumption being that major ideological issues, like unification versus secession and freedom versus slavery, had been settled by the Civil War.

Political leadership resided primarily in the US Senate, the balance having tipped from the president to the Senate with the impeachment of President Andrew Johnson. In the 1870s, two Republicans, "King" Roscoe Conkling, boss of the New York state machine, and "Prince" James Blaine of

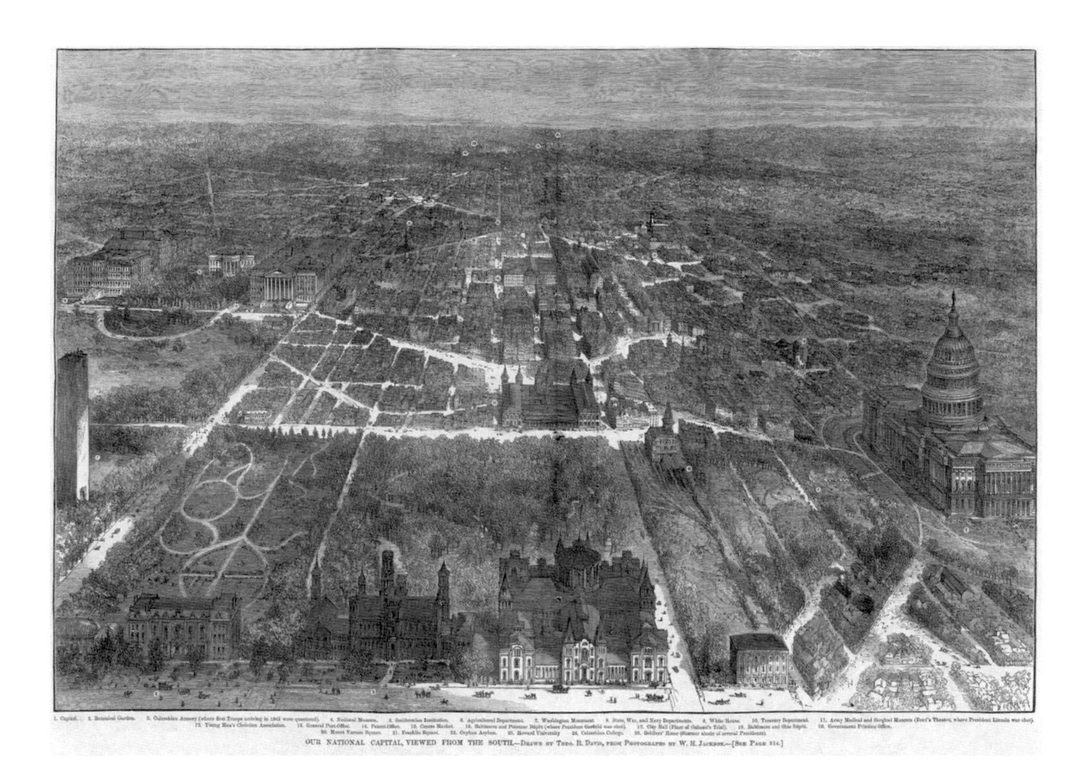

FIGURE 8.1 "Our National Capital," drawn by Theo. R. Davis, from photographs by W. H. Jackson, *Harper's Weekly* (May 20, 1882), 312. In the foreground, from left to right, are 7. Washington Monument (unfinished), 6. Agricultural Department Building, 5. Smithsonian Institution "Castle", and 4. US National Museum. In the center, from left to right, are 9. Executive Mansion (White House), the Mall, 15. Central Market, 14. Patent Office, 18. Baltimore & Potomac Railroad Station, and 1. Capitol Building. Riley's first residence is located left of center, further back. (Image supplied by HarpWeek.)

Maine, vied for leadership. Although the president was the primary appointing agent, in practice Conkling and Blaine often overruled him. Until 1900, presidents (with the exception of Grant) served only one term, while senators often remained in office for decades by reelection every six years. Furthermore, powerful senators influenced the selection of presidential candidates and cabinet officials. Senators were elected by state legislatures, rather than by popular vote, and political machines associated with powerful senators controlled state legislatures. According to "senatorial courtesy," US senators could appoint or influence the appointment of thousands of federal and state officials. The two most coveted cabinet positions after the secretary of state, who still enjoyed considerable prestige, were the postmaster general, who appointed local postmasters comprising nearly half the federal bureaucracy, and

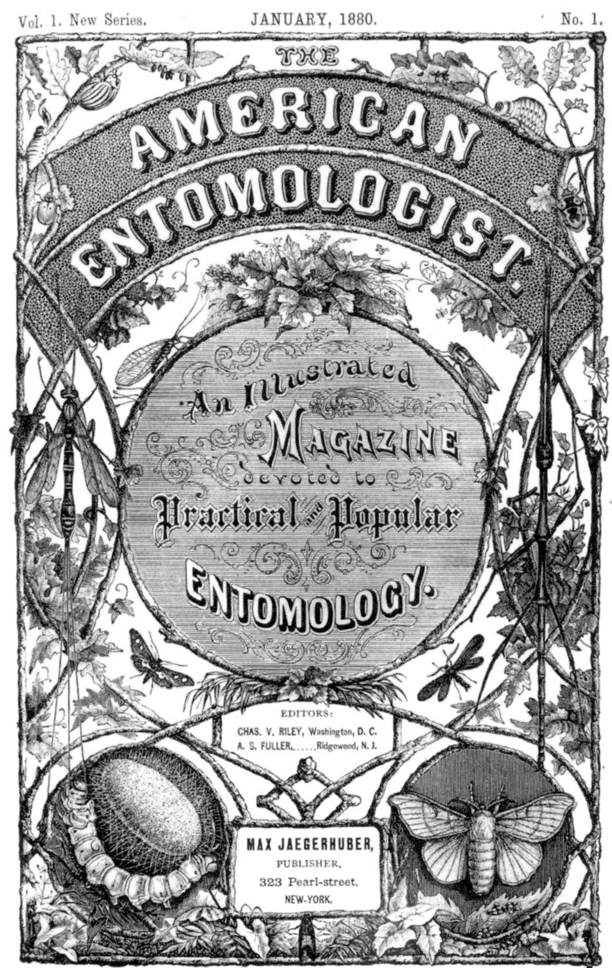

FIGURE 8.2 *American Entomologist*, front cover of second series, vol. 1 (January 1880) (Riley Family Records, National Agricultural Library Special Collections, Beltsville, Maryland).

the secretary of the Treasury, who appointed the customs collectors and employees in port cities like New York, Boston, and Baltimore. Customs officials had enormous influence because import duties and penalties for infraction of these duties collected by treasury agents accounted for the lion's share of federal revenues, the New York Customs House alone accounting for one-third of the federal budget.[2] Senators (and to a lesser extent representatives) "assessed" annual contributions from their appointees to finance their next campaign in what was designed as a perpetual hold on power. The least coveted position was commissioner of agriculture, who, unlike full cabinet members, could not enter the House or Senate chambers without an explicit invitation, and whose budget request went through two committees, first, the Committee on

Agriculture and then the Committee on Appropriations, thus increasing the chances for amendments and budget cuts. The commissioner of agriculture's annual salary of $3,000 was the lowest of any cabinet-level position.[3]

In June 1877, four months after Riley established the Entomological Commission's Washington headquarters, President Hayes named William G. LeDuc to replace Fredrick Watts as commissioner of agriculture. Hayes appointed LeDuc upon the recommendation of his private secretary, favoring him over equally or better qualified contenders, like Jacob R. Dodge, the departmental statistician, and John L. LeConte, whom Riley supported, both of whom had substantial support from farming and scientific constituencies.[4] LeDuc, a Minnesota lawyer, farmer, land speculator, and former brigadier general who had marched with General William Tecumseh Sherman through Georgia, was an autocrat who countenanced no disagreement on matters of policy and who sought no advice from his subordinates. His goal as commissioner was national self-sufficiency, which he pursued by promoting sorghum as a domestic source of sugar, silk from domestic silk growers, and an experimental tea farm in South Carolina.[5]

Legislators might agree with LeDuc's goal of national self-sufficiency, but they found the free distribution of seeds by the Agricultural Department a more effective means of ingratiating themselves with voters. In the two and a half decades during which agriculture was under the commissionership, seed distribution averaged 25 percent of the departmental budget.[6] By the time LeDuc took the helm, Congress had taken charge of the distribution of seeds, the packets being prepared in the department and then delivered to congressmen and senators for distribution.[7] When seed merchants cried unfair competition, Congress responded by sending free seeds to merchants for resale to the public. Riley and other critics complained of the wastefulness of seed distribution, charging that seed packets often contained weed seed. LeDuc himself admitted that congressional seed distribution seemed designed to do "the greatest political rather than agricultural good."[8]

Riley's ambition to expand the scope of federal entomology depended on his ability to secure influential congressional support while maintaining the good will of the agricultural commissioner. During the congressional session of 1877–78, Riley developed a southern strategy, buttonholing senators and congressmen with his plan to investigate the Cotton Worm, a project that appealed not only to southern legislators but also to northern and western legislators as a means to promote the economic development of the South. Riley's lobbying for Cotton Worm control at that time speaks volumes about his buoyant optimism regarding economic entomology. Never mind that Congress delayed his request for increased funding for the Entomological Commission; never mind that the commission was associated in most people's minds with locusts only; and never mind that the commission's first report was still in preparation. Riley sought new challenges.

Riley had had his eye on the Cotton Worm for many years, having encountered this pest during his Army tour of duty in 1864. Although relatively few Missouri farmers raised cotton, Riley featured the Cotton Worm in his Missouri Reports.[9] At the height of the locust crisis of 1874–76, Riley ranked the Cotton Worm as the third most injurious insect nationwide, surpassed only by the Rocky Mountain Locust and the Chinch Bug.[10]

The Cotton Worm had been a serious pest since the late eighteenth century, in some years destroying almost the entire cotton crop. From Thomas Affleck in Brenham, Texas, the best authority on the Cotton Worm, Riley learned that female Cotton Worm moths laid four hundred to six hundred eggs that matured in the "incredibly short space" of fifteen to twenty days.[11] When, in June 1872, the National Agricultural Congress was organized in St. Louis, Riley announced to the delegates that his experiments using Paris Green against the Fall Army Worm indicated that this insecticide would prove effective against the Cotton Worm as well.[12] At the second meeting of the National Agricultural Congress at Indianapolis in May 1873, Riley confirmed the effectiveness of Paris Green, and he cited testimony from cotton planters who had obtained favorable results. His address was published widely in the southern papers.[13]

In the spring of 1878, Riley secured a strategic ally in Mississippi representative Van H. Manning. With the help of fellow southern congressmen and senators, Manning delivered an appropriation of $5,000 to investigate the Cotton Worm and other insects affecting cotton.[14] Riley's southern strategy meshed with current national politics. The removal of federal troops from the three remaining southern states in 1877 marked the resumption of southern home rule. Southern leaders were planning an agricultural and industrial exposition in Atlanta in 1881 to stimulate the southern economy after more than a decade of war and reconstruction. Northern legislators willingly funded projects that promised to restore the southern economy. With estimated losses from Cotton Worm depredations running at $15 million per year, Riley's Cotton Worm investigation had wide appeal.[15]

While promoting expansion of the Entomological Commission's mandate, Riley simultaneously maneuvered to replace Townend Glover, whose health was failing, as entomologist in the Agricultural Department. Testimonials urging Riley's appointment reached LeDuc's desk as early as February 1878.[16] By March, farm and scientific journals were announcing that LeDuc was about to name Riley to replace Glover.[17] In May 1878, Glover resigned, and LeDuc named Riley as his successor. Having secured the appointment, Riley now urged legislators to fund the Cotton Worm investigations under the Agricultural Department, at the same time continuing to lobby for funding of locust and other insect investigations by the Entomological Commission.

With his appointment as federal entomologist, Riley began recruiting a staff that became the core personnel of the Division of Entomology. Riley had difficulty finding entomologists with practical skills.[18] American entomologists in the 1870s, most of whom were associated with the American Entomological Society, were primarily concerned with systematics. Entomologists typically had expertise in classification but not in researching insect life histories, in field observation, insect rearing, and control methods.[19]

Riley first hired Theodore Pergande, his trusted assistant in St. Louis.[20] Riley next chose Eugene Amandus Schwarz. One year younger than Riley, Schwarz had studied entomology at the University of Breslau and the University of Leipzig before immigrating to the United States in 1872 and joining Hermann Hagen at the MCZ.[21] At Harvard, he struck up a lifelong friendship with Henry Guernsey Hubbard. In 1873, Schwarz left the MCZ and joined Hubbard at his home in Detroit, Michigan. Hubbard's father, Bela Hubbard, geologist, attorney, real estate dealer, and founding member of the AAAS, funded the two entomologists on collecting expeditions to Florida and to Lake Superior.[22] When the AAAS convened in Detroit in August 1875, Schwarz met Riley, LeConte, and other scientists, many of whom were guests at the Bela Hubbard estate. LeConte was so impressed with the Coleoptera Schwarz and Hubbard had collected that he incorporated their findings in his new revision. In 1876, LeConte, Hubbard, and Schwarz authored "The Coleoptera of Michigan," and in the spring of 1878, LeConte financed Schwarz to collect insects in Colorado. Upon Riley's appointment as federal entomologist in May 1878, he telegraphed Schwarz in Colorado with an offer of employment. Schwarz accepted, and Riley dispatched him to survey Cotton Worm conditions in Texas, the Gulf States, and the Bahamas.[23]

Shortly thereafter, Riley recruited Hubbard as a special agent for the cotton insect investigation. Hubbard's fragile health, combined with the death of his two brothers in a sailing accident in 1879, precluded his full-time employment. That year, Hubbard moved to Florida to manage the family orange plantation. In 1880, Riley assigned Hubbard to investigate scale insects in Florida orange groves. His report, *Insects Affecting the Orange* (1885), was considered one of the best reports produced by the division.[24]

In July 1878, Riley appointed Augustus Radcliffe Grote as special agent.[25] Born in England in 1841 to a German father and a Welsh mother, Grote immigrated with his family to a farm on Staten Island in 1848. Under the guidance of John Ackhurst, New York's leading lepidopterist, Grote developed a specialty in moths, including the noctuids, to which the Cotton Worm belongs. In 1864, Grote proposed the name *Anomis xylina* for the Cotton Worm, to correct the synonomy he discovered between Thomas Say's *Noctua xylina* and Achille Guenée's *Anomis bipunctina*.[26] In 1874, Grote discovered that the insect he referred to as *Anomis xylina* (Cotton Worm) had been described earlier by Jacob Hübner as *Aletia argillacea*. Since Grote's compromise name (*Anomis xylina*) was too recent to be considered "well established," he replaced his recent name with Hübner's. Riley, along with other entomologists, accepted Grote's taxonomy. Thus, the Cotton Worm, against which Riley was assembling his forces, was now officially *Aletia argillacea*.[27]

In 1870, Grote married and moved to Demopolis, Alabama, where he began investigating the life history of the Cotton Worm that was damaging local cotton fields. Between 1871 and 1875, he published five articles on the Cotton Worm and tried, without success, to gain congressional funding for a Cotton Worm investigation under his direction. Following his wife's death in 1873, Grote moved to Buffalo, New York, as director of the Buffalo Society of Natural History. There, he initiated the *North American Entomologist*, a short-lived rival of Riley's *American Entomologist*. When the Entomological Commission was established in 1877, Grote hoped to be appointed, but Riley, preferring Thomas and Packard, effectively blocked Grote's appointment.

Although Riley accepted Grote's name, *Aletia argillacea*, for the Cotton Worm, he disagreed with Grote's theory of its migration. Grote held that all Cotton Worms in the United States perished each year and that these moths were replaced by a new population migrating from subtropical lands the following year. Riley rejected the migration theory, arguing from the analogy of related moth species that the Cotton Worm overwintered in the United States.[28] At the Hartford, Connecticut, AAAS meeting (1874) and the Saratoga Springs meeting (1879), Grote and Riley crossed swords on this issue. Lacking direct observations of hibernating insects (larvae, pupae, or adults), both Riley and Grote cited indirect evidence, like the "sudden appearance" of Cotton Worms in the northern reaches of the cotton belt, which might indicate Cotton Worms emerging from hibernation; the existence or non-existence of parasites, which might indicate whether or not the Cotton Worm was indigenous (hibernating) in North America; and the hibernation or non-hibernation of related species of moths.[29] Riley emphasized the importance of this issue to the development of control measures. If the Cotton Worm hibernated as egg, pupae, or adult in the continental United States, as he maintained, this might represent a weak point in its life cycle where controls might be effective, whereas if the moths migrated from outside the United States, as Grote contended, control would be more difficult.[30]

Riley's assignment of Grote to investigate the evidence supporting migration or hibernation of the Cotton Worm was equivalent to a prima donna inviting her prime rival to share the stage. Grote's contorted conclusion, at the end of his report, that "the Cotton worm passes the winter, when it survives at all, as a moth," and that the full history of the worm "awaited further investigation," evidently constituted his bid to be retained as a team member while maintaining views contrary to the team leader.[31] It therefore came as no surprise when Riley dropped Grote from the Cotton Worm investigation for the 1879 season.[32] In the final report on the Cotton Worm, Riley pointedly omitted Grote's name from the acknowledgments.[33] Riley and Grote continued to express their differences in print. Those aware of their rivalry interpreted Grote's description of Thaddeus William Harris as the "founder of American economic entomology" as a backhanded slap at Riley. In 1884, Grote moved to Bremen and later to Hildesheim, Germany, so that Riley and his erstwhile critic were separated for the remainder of their careers.[34]

Riley's next three appointees, John H. Comstock, Leland O. Howard, and William Stebbins Barnard, were all associated with Cornell University at Ithaca, New York. In late June 1878, Riley recruited Comstock, a twenty-nine-year-old invertebrate zoology instructor, to investigate the Cotton Worm in Arkansas, Tennessee, Mississippi, and Alabama.[35] Comstock, who was engaged to marry Anna Botsford and needed money to pay for a house, eagerly accepted Riley's offer of $100 per month. From July to October 1878, Comstock investigated Cotton Worm depredations in the South, and then returned to his teaching post at Cornell.[36] Neither he nor Riley suspected that seven months later he would replace Riley as entomologist at the Agricultural Department.

Riley would also have been surprised at the ramifications of his next appointment. On September 4, 1878, when meeting with Comstock in Nashville, Tennessee, Riley asked Comstock to suggest someone to serve as his office assistant. Comstock recommended Howard, who had just finished a year of postgraduate study in preparation to enter medical school. By November, Howard was at work in the Agricultural Department office. At first Howard performed clerical tasks, but when Riley learned of Howard's abilities, he assigned him to entomological projects.[37]

Riley recruited Barnard in 1880, by which time he had transferred the Cotton Worm investigations from the Agricultural Department to the Entomological Commission (see below). Born and raised on an Illinois farm, Barnard entered the University of Michigan in 1867 and, a year later, transferred to Cornell University, graduating in 1871. He then spent two years at universities in Germany, Italy, and France, receiving his doctorate under Ernst Haeckel, "Darwin's German Bulldog," at Jena. In his thesis, he compared the muscular structure of the orangutan to that of humans.[38] Returning from Germany in 1874, Barnard lectured on zoology at Cornell, followed by teaching appointments in Mississippi, Illinois, Wisconsin, and Iowa. Riley met Barnard at the Detroit AAAS meeting in 1875, where Barnard delivered three papers: his orangutan thesis and papers on the opossum and on protozoa but none on entomology.[39] The next recorded meeting between the two was sometime in the spring or summer of 1880, when Riley hired Barnard to assist in the Entomological Commission's Cotton Worm investigations.[40]

Why did Riley hire Barnard? Up to 1880, Barnard had published only two articles on entomology, only one of which was on economic entomology, hardly enough to impress the author of nine Missouri Reports.[41] The spark between the seasoned entomologist and the talented youngster was apparently their mutual interest in contriving mechanical devices to solve problems in applied science. Riley had designed and built insect cabinets, hopper dozers, insect rearing cages, and insect breeding tents. He found a kindred spirit in Barnard, who had contrived a system of portable fences, a machine for cutting, gathering, and shocking corn, and an ingenious system for taking notes and sorting them. In 1877, he applied for a patent for specialized bookends, and he is credited with devising the "Harvard book rack."[42] One may imagine the animated discussion in Detroit between the two midwestern farmer-scientist-tinkerers as they exchanged ideas about insecticides and application devices to control the Cotton Worm. In fact, Riley recalled later that he chose Barnard because of his expertise in natural history and his ability to contrive mechanical devices.[43]

In the summer of 1878, Riley hired George Marx, a pharmacist and entomologist-arachnologist in Philadelphia, as draftsman and illustrator.

Riley also recruited Lillian Sullivan, who supplied illustrations for the final Cotton Worm report.[44] With the addition of the two illustrators, Riley's core personnel was complete.

At the close of the fiscal year, June 31, 1878, Congress delivered only $10,000 of Riley's request of $25,000 for the Entomological Commission.[45] Nevertheless, Riley launched his Cotton Worm project by drawing from the $10,000 authorized for the Entomological Commission plus $5,000 that Manning and his southern colleagues had secured for Cotton Worm investigations by the Agricultural Department. In typical Riley fashion, he expended funds from both agencies for the Cotton Worm investigations.

Without skipping a beat, Riley prepared for his June wedding, moving household items from St. Louis to the new brick home he and his bride-to-be had purchased at the corner of Thirteenth and R streets, about a mile and a half north of the Agricultural Department.[46] On June 20, 1878, Charles and Emilie were married at the bride's father's house in St. Louis.[47] Surprisingly, there was no public announcement of their marriage. Although St. Louis newspapers normally reported marriages of prominent personalities in meticulous detail, no public announcement appeared when the daughter of one of St. Louis's prominent businessmen was married to the Missouri state entomologist. The lack of fanfare was no doubt Emilie's wish. The couple enjoyed a short honeymoon in eastern cities, then returned to St. Louis for more packing and the final move to Washington.[48]

Domestic affairs now settled, with funding in hand and supported by a capable staff, Riley turned his full attention to the Cotton Worm campaign. In July 1878, Riley circulated a Cotton Worm questionnaire requesting dates and extent of outbreaks, observations bearing on the question of hibernation, and the effectiveness of controls.[49] He and his assistants, Schwarz, Comstock, and Grote, reconnoitered regions of severe Cotton Worm outbreak, while local agents consisting of planters, doctors, judges, and professors reported on local conditions throughout the cotton belt.[50]

In Selma, Alabama, Comstock witnessed the planters' resignation as they watched the season's crop being devoured.[51] Harriet Beecher Stowe, author of *Uncle Tom's Cabin*, who together with her husband and partners had purchased a Florida plantation, watched helplessly as Cotton Worms stripped the cotton field in two days.[52] Although damage in June and July was substantial, heavy rains in August delayed the further proliferation of worms, and that fall, freezing temperatures soon ended their depredations for that season. In August 1878, yellow fever broke out in Mississippi. Panic over yellow fever quickly overshadowed concerns with the Cotton Worm as planters fled north, leaving field hands to cultivate and harvest the remaining cotton. Riley traveled through Tennessee, Georgia, the Carolinas, and Virginia but dared not venture into Mississippi, the center of yellow fever contagion.[53]

Notwithstanding yellow fever and limited opportunities to observe the full destruction inflicted by Cotton Worms, Riley considered the first season's investigations successful. One key discovery was the food of the adult moth. Although naturalists had long known that the moth was attracted to sweets, they had not identified its food plant. In early September 1878, near Baconton, Georgia, Riley, Comstock, and special agent J. E. Willett observed moths feeding on a liquid secreted from cotton plant leaves and stems.[54] Riley immediately published their observation in the *Atlanta Constitution*. Subsequent investigations indicated that the moths also fed on secretions of cowpeas and grasses.[55] The discovery that the moth fed on plant secretions encouraged some growers to resume "sugaring" with poison as a means of control. Riley opposed sugaring because it destroyed beneficial as well as harmful insects. Riley also reported that he and his assistants had identified nine parasites. This assertion later became a bone of contention between him and Comstock.[56]

As the year 1879 opened, Riley's prospects looked good. He and his wife lived comfortably in their new house. His locust and Cotton Worm projects ushered him into the elite circle of Washington scientists. He was widely regarded as the nation's leading economic entomologist. From all indications, he enjoyed a good relationship with LeDuc, which was surprising in light of his criticism of the department's inadequate response to the locust emergence and his campaigning for rival candidates for the office of agricultural commissioner. LeDuc, in his annual report for 1878, urged renewal of the $5,000 appropriation for the Cotton Worm investigation and suggested additional compensation for the department's new entomologist.[57]

By March and April 1879, however, this harmonious relationship had soured into acrimony that resulted in Riley's firing/resignation. While accounts differ as to the details, the overall issue seems clear. In the congressional session of 1879, Riley continued lobbying for funds for the Entomological Commission and for the Cotton Worm investigations of the Agricultural Department. LeDuc considered congressional liaison his prerogative and was becoming increasingly impatient with his headstrong entomologist who also headed a rival agency. Riley's direct approach to congressmen and senators and his promotion of the Entomological Commission smacked of insubordination. The former Union general expected his staff, especially former privates, to follow orders. It was a case of an irresistible force colliding with an immovable object. Something had to give. Sometime in mid-April 1879, Riley and LeDuc had a showdown; LeDuc demanded Riley's resignation, and Riley complied. After less than one year in office, Riley was suddenly outside the department.[58] Depending on the source, Riley was fired, resigned in a fit of anger, resigned because the commissioner insulted him, or resigned for health considerations. A member of LeDuc's staff sniped that Riley then sought out sympathetic newspaper reporters to publicize his "scroll of malicious slanders."[59] To LeDuc's supporters, Riley was known as the "crushed bugologist."[60] Riley hoped that Thomas would succeed him, but LeDuc distrusted any close colleague of Riley's.[61] LeDuc asked Andrew D. White, president of Cornell University, to suggest a candidate. White recommended Comstock, who accepted immediately, took leave from Cornell until the anticipated change of US presidential administrations in 1881, and assumed his official duties May 1, 1879.

Riley promptly transferred the Entomological Commission from the Agricultural Department to his home, taking along portions of his insect collection, notes, and other data relating to the Cotton Worm project.[62] Riley's staff faced a choice: go with Riley or remain in the department under Comstock. Howard, overjoyed at Comstock's appointment, remained. Pergande, Marx, and Sullivan also remained.[63] Comstock added his wife to his staff as clerical assistant and illustrator.[64] Schwarz, a steadfast Riley confidant, went with Riley and the Entomological Commission. Riley hired W. H. Patton, an entomologist about whom little is known, and Riley's half-sister, Nina, continued as clerk.

Following the showdown with LeDuc, Riley persuaded Congress to transfer full responsibility for the Cotton Worm investigation from the Agricultural Department to the Entomological Commission.[65] LeDuc and Comstock opposed Riley's perceived hijacking of the Cotton Worm program. In late 1879, they circulated an unsigned document entitled "Digest of Appropriations 1879," to members of Congress, in which they charged that the Entomological Commission was superfluous and that the Agricultural Department was best equipped to carry on the Cotton Worm investigations.[66]

Faced with two requests for funding of Cotton Worm investigations, Congress, in 1879, produced a compromise that reflected the standoff between Riley's Entomological Commission and the Agricultural Department. Congress appropriated an initial $10,000 for the Entomological Commission, to be used for the locust and Cotton Worm investigations, and $5,000 for the

Agricultural Department (its usual annual appropriation) for ongoing operations including the completion of Comstock's Cotton Worm report.[67] In other words, Congress charged both agencies to investigate the Cotton Worm, pitting them against each other in a race to investigate, compile data, propose control measures, and produce reports on the same subject.

In preparation, Riley circulated a Cotton Worm questionnaire similar to the one he had sent from the Agricultural Department the year before.[68] Comstock issued his own questionnaire. Some correspondents, understandably puzzled to receive questionnaires on the same subject from Riley and Comstock, good-naturedly responded to both.[69] In July 1879, the competing teams departed Washington for the cotton belt. Schwarz and Patton headed first to Texas, then to the cane breaks of Alabama, where Riley joined them. Later, Schwarz traveled over much of the cotton belt, as far as the Bahamas.[70] Comstock was eager to head south but was hampered by insufficient funds, the department's travel budget, he claimed, having been exhausted by Riley. He managed to rustle up enough money to send William Trelease, his Cornell assistant, to the field headquarters at Selma, Alabama.[71]

Both teams addressed the perennial question of whether the Cotton Worm hibernated or migrated. In May 1879, shortly after his resignation, in an address before the National Academy of Sciences, Riley again disputed Grote's migration theory and defended his hibernation theory.[72] With Grote's exit from the staff, Riley's hibernation theory was now the official position of the Entomological Commission.[73] Surprisingly, on this issue, Comstock sided with Riley rather than Grote. He agreed that a few moths hibernated in the southern edge of the cotton belt and started or augmented the population the next year.[74]

While the dispute over hibernation-migration represented a central theoretical issue, the more immediate question was which agency would conduct the investigations. Riley and Comstock engaged in a race to generate attractive, authoritative reports that would serve as ammunition in the battle over appropriations. Riley scored first with the commission's *Bulletin 3*, "The Cotton Worm," dated January 28, 1880 (preface dated November 3, 1879). This report, Riley announced, was "the most complete treatise published up to that date," and he later referred to it as the first edition of the final Cotton Worm report that was published in 1885.[75]

Four months later, on May 18, 1880, Comstock responded by depositing pre-publication copies of his Cotton Worm report on the desks of congressmen.[76] To the casual reader, the two reports might appear to be alternate treatises on an insect pest of regional importance. To insiders, the subtext of both reports was clear: my/our report is authoritative, theirs is riddled with scientific errors; I/we have struggled heroically to produce the best report in a timely manner, theirs is either too late or too little; I/we are best positioned to carry on the work relating to control of injurious insects, they have wasted large appropriations of the taxpayers' money and are clearly incapable of carrying out this investigation. The conclusion was clear: only my/our agency can be trusted with appropriations to conduct the investigations.

Beneath gentlemanly nods to the opposition lay barbed remarks. Riley praised his "successor" (Comstock) as well qualified to prepare the materials he (Riley) had collected but added that "practical questions" would take longer to resolve, i.e., stay tuned for the Entomological Commission's final report.[77] Comstock's report described the scene near Selma in September 1878 where Comstock, Riley, and Willits discovered the food of the adult moth. While acknowledging that Riley had suggested that the moth fed on secretions from the cotton plant, Comstock emphasized that he (Comstock) actually discovered the moth sipping nectar.[78] Comstock's subtext was clear: I made the key discovery that resolved the question of the adult moth's food source. Riley, in his final report, insisted that Comstock's report "really consists of the first year's work [1878] conducted by us [Riley et al.], as . . . Entomologist of the Department."[79]

In their discussion of Cotton Worm parasites, the protagonists took off their gloves. Riley, in *Bulletin 3*, asserted that in 1878, when he began investigating the Cotton Worm, no parasites had been scientifically described.[80] Comstock, a few months later, listed six parasites that Glover and others had identified and given "tolerable popular descriptions." With thinly veiled sarcasm, he continued, "the undoubted ability of [Riley and Grote] renders their statements all the more singular."[81] He went on to describe thirteen parasites he and others in the department had reared.[82] Riley, in his final report, again asserted that no parasites had been "scientifically" described prior to his entry in the field, and he produced a scorecard of "scientifically recognized" parasites published up to 1883: Riley six, Comstock two, [Thomas] Say two, Howard one, Fallén one, and Br. [?] one. The winner according to Riley's scorecard was clearly Riley.[83]

The race to accumulate parasite points soon spilled over into non-government publications. Here, once again, Riley had the advantage: his colleague Alpheus Spring Packard edited the *American Naturalist* and his friend Charles Bethune edited the *Canadian Entomologist*.[84] In 1880, Riley revived

FIGURE 8.3 Calling card for Prof. C. V. Riley, United States Entomologist, with initial mark (Riley Family Records, National Agricultural Library Special Collections, Beltsville, Maryland).

the *American Entomologist*, providing an additional forum for his views. The wrangling over parasites sputtered on for decades. In their respective reports, Riley and Comstock disagreed as to whether *Phora aletiae* parasitized the Cotton Worm.[85] Riley called on testimony from Schwarz, Hubbard, and Baron Carl Robert Osten Sacken, the world's authority on American Diptera, then residing in Heidelberg, Germany. Osten Sacken agreed with Riley that it was probably not a parasite but rather a scavenger that entered the host after it died. Howard, fifty years later, claimed that he wrote the section on *P. aletiae* in both Comstock's and Riley's reports, agreeing with Comstock in 1879 that *Phora* was a parasite but concluding with Riley in 1883 that it was a scavenger.[86]

During his years of exile (1879–81), Riley's public prominence overshadowed that of his rival, Comstock. In 1879, Riley was thirty-six years old, compared to Comstock's thirty, but the contrast in their careers reflected more than a six-year difference. Having been appointed Missouri state entomologist at age twenty-five, Riley produced nine internationally acclaimed reports, helped to uncover the evolutionary origins of mimicry in butterflies and the fertilization of yucca, worked in the forefront of the phylloxera battle, and initiated the US Entomological Commission in response to the locust crisis. Furthermore, Riley's athletic figure, prominent mustache, wavy black hair, and fastidious dress easily overshadowed the prosaic professor from Cornell. Comstock's public performance suffered because, as Howard recalled, he stuttered. Riley, as chief of the Entomological Commission, drew a salary of $3,000, augmented by substantial income from his business ventures and the assets of his wife's family. By contrast, Comstock relied on a modest teaching salary from Cornell and, as federal entomologist, drew a salary of $1,900 per year.[87] Howard recalled that, during Comstock's tenure, he (Howard) often pawned his watch toward the end of the month in order to purchase theater tickets for himself, the two Comstocks, and Trelease.[88]

At AAAS meetings, Riley operated as an insider, presenting papers at virtually every meeting beginning in 1872 and serving on committees of the Entomological Club and the general AAAS. In August, following his removal from the Agricultural Department, Riley was elected secretary of the AAAS Natural History Section. From his desk at the head of the chamber, the handsome, immaculately dressed secretary announced the names of new members.[89]

When the Entomological Club convened at the Saratoga Springs meeting of the AAAS in 1879, Joseph A. Lintner, the New York state entomologist, opened his presidential address by announcing the transfer of the Cotton Worm investigation from the Agricultural Department to the Entomological Commission. He made no mention of Comstock's Cotton Worm investigations, although Comstock was present at the meeting. Lintner's promotion of Riley indicates the pecking order in the entomological fraternity. In the following sessions, participants apparently avoided the Cotton Worm as a topic of discussion.[90]

At the Entomological Club meeting of the AAAS in Boston the following year, Scudder (a Comstock supporter), in his presidential address, mentioned neither the Cotton Worm investigations nor the competing reports by Riley and Comstock. Comstock was occupied with citrus insect matters

FIGURE 8.4 Agricultural Building, Washington, DC, viewed from the west, with the Smithsonian Institution "Castle" in background (courtesy of the National Agricultural Library Special Collections, Beltsville, Maryland).

in California and did not attend. Riley gave two papers, one on the Cotton Worm and the other on insecticides for the protection of cotton.[91]

Outside the halls of (presumably) neutral science, Riley spoke his mind freely. In a keynote address at the Mobile Cotton Exchange on July 9, 1879, two months after his removal as federal entomologist, Riley urged additional appropriations for the Entomological Commission.[92] At the National Agricultural Association in New York City in December 1879, Riley ridiculed the Agricultural Department as a "laughing stock."[93] Throughout 1879 and 1880, Riley furnished the agricultural and southern press with frequent updates on the Cotton Worm and other investigations being carried out by the Entomological Commission. Southern papers praised his work and rated the Entomological Commission reports as superior to those of the Agricultural Department.[94]

In the Washington scientific and social scene, Riley clearly outshone Comstock. On October 12, 1878, five months after his appointment as federal entomologist, Riley was elected a member of the Philosophical Society of

Washington.[95] Joseph Henry, secretary of the Smithsonian and founder and president of the Philosophical Society, strove to lift the proceedings of the society above the sordid atmosphere of business and politics. Formal meetings were held at the Smithsonian, where members dressed in uncomfortable evening attire listened to dry, technical papers. Until his death in early 1878, Henry served as permanent president, other officers being drawn from a rotating list of the National Academy of Sciences. Riley presented numerous papers at the Philosophical Society and was elected to the nine-member General Committee.[96]

At the time of Henry's death, many Washington scientists were looking for an alternative to the Philosophical Society. John Wesley Powell, the dashing explorer of the Colorado River and soon-to-be director of the US Geological Survey, asked why Washington should not have a club like the Scientific Club of London or the Century Club of New York, where scientists combined social and scientific intercourse. In November 1878, Powell, with friends and colleagues, formed the Cosmos Club, where they could share their professional and scientific experience over billiards, cards, and chess in a comfortable ambience of leather upholstered furniture, cigars, wine, and brandy. Riley was one of sixty founding members and the sole entomologist. Having modified his views on alcoholic beverages since his Templar days in Chicago, he mixed well with this elite assembly.[97] The Cosmos Club was an instant success, moving from rented quarters in the Corcoran Building to lavish rooms of its own on Lafayette Square in 1886.[98]

Riley's entry into Washington's two leading scientific societies whetted his appetite for more. On November 19, 1880, Riley; George Brown Goode from the National Museum; Richard Rathbun, also of the Museum; Theodore Gill, zoologist at Columbia College and curator at the National Museum; and Lester F. Ward, at that time librarian at the Treasury Department, but also a recognized botanist and student of evolution, founded the Biological Society of Washington. Spencer F. Baird was named as an honorary member of the society. Corresponding members outside the city included Packard and Asa Gray. The Biological Society was also an immediate success, its membership increasing to 161 by 1885.[99] Biologists were united by their interest in living organisms and their interpretation of living forms in the context of evolutionary theory.[100] The first issue of the society's proceedings was dedicated to Darwin, on the occasion of his death in 1882. There, Riley reviewed Darwin's impact on entomology.[101] Four years after the founding of the Biological Society, on February 29, 1884, Riley, Schwarz, and Howard invited Washington-area entomologists to Riley's residence, where they formed the Entomological Society of Washington. The new society also grew quickly from 26 members in the first year to 114 by 1902.[102]

Riley's only disappointment with regard to Washington's scientific establishment was his failure to gain membership in the National Academy of Sciences. As noted above, Riley addressed the academy on Cotton Worm hibernation shortly after his resignation.[103] A reporter at the session described "Professor Riley," as "tall and slender . . . the handsomest member [*sic*] of the Academy except for Dr. Elliott Coues, and the one to whom Tom Edison's criticism of the academician's scorn of the tailor does not apply."[104] Despite Baird's backing, academy members rejected Riley's bid for membership. They kept no records of discussions regarding individual candidates, but they most likely considered economic entomology too utilitarian. With their academic orientation, they tended to rate what they considered "pure" or "philosophical" sciences like math, physics, and chemistry at the high end of an informal scale with geology somewhere in the middle and applied science, like economic entomology, at the low end. Riley's colleague Eugene W. Hilgard, for example, was elected a fellow in 1888 not because of biological investigations but because of his geological studies.[105]

In contrast to Riley, Comstock projected a low profile in Washington scientific circles. Howard recalls that Comstock attended meetings of the Biological Society in 1879–80, but he apparently did not make presentations there.[106] Whether Riley actively worked to exclude his rival from Washington's scientific clubs or whether Comstock (and his assistants, Howard, Trelease, and Pergande) lacked the initiative to push for involvement is not clear.[107] What is clear, however, is that in Washington, as on the national scene, Riley far outshone Comstock.

From his home-office, Riley issued the Cotton Worm bulletin and announced plans for a series of bulletins on the Army Worm, the Hessian Fly, the Chinch Bug, and other destructive insects. At the AAAS, at the National Agricultural Congress, and in the farm press, he took center stage, leaving LeDuc, Comstock, and the Agricultural Department in the sidelines. Behind the scenes, however, Riley steadily lost support in Congress and within the administration. In April 1880, when the House Committee on Agriculture considered funding for investigations of injurious insects, the National Cotton Exchange called for $50,000 for the Entomological Commission to continue investigating the Cotton Worm and other cotton insects, while LeDuc and Comstock called for zero appropriations for the commission.[108]

The committee majority recommended an appropriation of $25,000 for the Entomological Commission to complete its investigation of the Rocky Mountain Locust, the Cotton Worm and other cotton insects but with the proviso that the commission was not to investigate additional insects and that it report to the commissioner of agriculture at the end of the fiscal year (June 30, 1881). Because cotton planters continued to support Riley, the majority held that the success of the Cotton Worm investigations "greatly depends on the special fitness and ability of those charged with the work [i.e., Riley]" and recommended against transferring the investigations to the Agricultural Department.[109]

The committee minority, comprised of four members who sided with LeDuc and Comstock, charged that the Entomological Commission, originally mandated to investigate locusts, had been granted multiple appropriations to "complete" the locust report but had expanded its mandate to include the Cotton Worm and other insects, and had then issued only "pamphlets" on the Cotton Worm and the Chinch Bug. They claimed that, by contrast, the Agricultural Department, with a meager budget of $5,000, had produced a comprehensive report on the Cotton Worm (Comstock's report). Warning that Riley was creating a permanent agency in competition with the Agricultural Department, they cited Interior Secretary Schurz's recommendation that the Entomological Commission be transferred from his department to the Agricultural Department.[110]

The growing strength of LeDuc and Comstock's position in Congress, together with Schurz's about face, spelled serious trouble for Riley and the Entomological Commission. With every season of calm on the locust front, the rationale in support of the Entomological Commission lost force. As Riley admitted to Packard, "[O]ur admission that our work on locusts was essentially finished knocks that influence [of the West] from under us."[111] Increasingly pessimistic about his prospects in Washington, Riley responded hopefully to news in March 1880 from Charles E. Bessey, professor and proponent of agricultural-oriented botany at Iowa Agricultural College, of a possible opening for a state entomologist in Iowa.[112]

More promising, however, was the prospect of reinstatement as federal entomologist following the presidential election of 1880. President Hayes having announced he was not running for another term, the appointment of a new commissioner of agriculture and a new federal entomologist was almost certain. When the Republicans convened in Chicago in June 1880, the "Half-Breed" and "Stalwart" wings of the party fought each other to a standstill, finally settling on a dark horse, James A. Garfield.

In the election that resulted in Garfield's victory on November 4, 1880, personal ambitions, party organization, fundraising, and bribery, rather than principles, determined the outcome. From his election victory in November 1880 to his inauguration on March 4, 1881, Garfield spent most of his time fielding requests from hopeful office seekers.[113] With hopefuls far outnumbering vacancies, Garfield inevitably made more promises than he could possibly fulfill. As the *New York Times* quipped, "[I]f all reports are true, President Garfield's cabinet will contain about one hundred and twenty-five persons."[114] Impatient and anxious, Riley traveled to Garfield's home in Mentor, Ohio, on February 19, 1881, to "pay his respects" to the incoming president.[115] The outcome of this meeting was not recorded, but Riley clearly felt that Garfield would look out for his best interests.

Selection of the commissioner of agriculture was delayed repeatedly. Despite LeDuc's efforts to rally support for his own retention in office, given the workings of the patronage system, the question was not whether a new commissioner would be appointed, but how to select a politically viable candidate to replace the incumbent.[116] After considering various candidates in terms of party standing, support of special interest groups, and regional balance, Garfield's advisers concluded that George B. Loring, physician, retiring congressman from Massachusetts, gentleman farmer, and leader in agricultural organizations, met these requirements. Power plays over various issues unrelated to the agricultural commissioner blocked a decision until early June 1881, when Loring was finally appointed.[117] The incoming commissioner, described by one historian as "a rather pompous physician, devotee, and patron of the new agriculture, and politician," had angled for the position for a decade.[118] Riley had known him since 1872, when they shared the platform at the National Agricultural Congress. Loring and Riley agreed at that time that reform of the Agricultural Department must start with the replacement of Watts.[119] Riley supported Loring early on, writing to Packard, who resided in Loring's home state, to "Work earnestly for Loring."[120] In the months leading up to Loring's appointment, Riley met with him several times and hosted him for dinner at his home at least once.[121] Riley credited his own influence in the West and his standing among southerners as critical factors leading to Loring's appointment.[122]

While focusing his primary attention on Garfield's selection of a commissioner, Riley cajoled congressmen in yet another round of budget hearings. When Manning proposed $10,000 for continuation of the Entomological Commission's Cotton Worm investigation, the discussion degenerated into sectional wrangling reminiscent of the debate on the locust commission in 1876. Representative James Proctor Knott of Kentucky allowed that $10,000 would be better spent on purchasing turkeys to catch and eat Tobacco Worms (tobacco being more important to his state than cotton), whereupon Agricultural Committee chairman Edward K. Valentine of Nebraska proposed a $5,000 rider to the Cotton Worm appropriation to fund the Signal Service to report on movements of the Rocky Mountain Locust.[123]

In March 1881, in the midst of the deadlock blocking Loring's appointment, Riley became embroiled in a new controversy with LeDuc, who was now most certainly a lame duck. The outgoing commissioner charged Riley with removing valuable Cotton Worm insect specimens from the department. The issue dated back to the summer of 1879, when Riley engineered the transfer of the Cotton Worm investigations to the Entomological Commission and took what he considered to be the relevant materials and specimens to his home. Now charged with insect larceny, Riley returned the disputed insects to the department with the request that those not needed be sent back to him. Apparently Comstock sent about fifty specimens back to Riley, which LeDuc interpreted as further evidence of theft by Riley. LeDuc appealed to the district attorney of the District of Columbia, who declined to charge Riley.[124]

Upon announcement of Loring's appointment in June 1881, Riley declared his approval of Garfield's choice.[125] Comstock, having learned some tricks of Washington politics, circulated a letter among scientists and friends requesting that they send Loring letters supporting his own retention. Riley, past master at orchestrating testimonials, mailed out his own circular to friends across the country.[126] The response in favor of Riley overwhelmed Loring. Following a feeble effort to conciliate both sides, first with a proposal that both Comstock and Riley work in the department, then with a proposal that Comstock direct a branch of the Division of Entomology at Ithaca, Loring reappointed Riley as entomologist in the Agricultural Department. Loring announced Riley's appointment on July 9, 1881, one week after he became commissioner.[127] Comstock submitted his resignation on July 5, effective August 1, 1881. To his wife he wrote, "[M]y case . . . should make the heart of every lover of our glorious country ache with shame."[128]

Just as Riley made plans to reclaim his former quarters in the Agricultural Department, as Comstock prepared to return to his professorship at Cornell, and as Americans across the country were getting familiar with their new postmasters, an event in Washington threatened to set the political carousel spinning anew. On July 2, 1881, one day after Loring officially replaced LeDuc as commissioner, three days before Comstock's announced resignation, and one week prior to public announcement that Riley would replace Comstock, Charles Guiteau shot President Garfield in the back as he was preparing to board a train at the Baltimore and Potomac Railroad Station in Washington.[129] The mortally wounded president lingered through the summer and died on September 19, 1881. The next day Vice President Chester Alan Arthur was sworn in as president.[130]

Chester who? The question posed by average Americans in the twenty-first century is the same question their nineteenth-century counterparts asked. One political insider commented, "Chet Arthur? President of the United States? Good God!"[131] For government employees, however, the prominence or obscurity of the new president was of secondary importance to the question how the shakeup in patronage would affect their positions. That Arthur, as president, would reshuffle the cabinet was a foregone conclusion. After all, he represented the Stalwart Republican faction, and James Blaine (whom Garfield had named secretary of state) and the Half-Breeds had outmaneuvered Arthur's Stalwarts in the selection of Garfield's cabinet. Both Riley and Comstock believed that Garfield favored them, and both believed that, should Garfield survive, they would be confirmed as federal entomologist.[132] Riley was so convinced of Garfield's good will that he named his first son, born in November 1881, William Garfield Riley. As Garfield's condition deteriorated in the summer of 1881, the primary question for Riley was whether Loring would be replaced. In the end, Arthur retained Loring, which assured Riley's retention as entomologist in the Agricultural Department.

Throughout the political wrangling during his exile from the department, Riley remained convinced that he was destined to implement applied entomology on the national scene. Following two years of political turmoil, he now reclaimed his position as federal entomologist and proceeded to expand the government's entomological activities. Now that the locust menace had receded, he focused on the Cotton Worm. He outlined his agenda for expanded federal entomology in his keynote address at the Cotton Exposition, scheduled to open in October 1881 in Atlanta, Georgia.[133]

9

Assisting Nature's Balance

Then followed 13 years of slow but steady growth and accomplishment [at the Division of Entomology].

—Leland O. Howard, referring to the years 1881–1894, *History of Applied Entomology* (1930), 90.

Following his reinstatement as federal entomologist, Riley's life assumed a somewhat normal routine. Economic expansion in the 1880s nourished the "steady growth" of the division described above by Leland O. Howard. In that decade more land was brought under cultivation than in the two and a half centuries since the founding of the first colonies. During that same decade, the gross national product of the nation almost tripled.[1] Construction crews laid track for feeder lines to the transcontinental railroads that totaled one and half times the track laid up to 1880.[2] With Europe's expanding population calling for American agricultural products, American farmers supplied a surplus over domestic demand that produced an extraordinarily favorable balance of trade.[3] Tariffs produced surpluses in the federal treasury that translated into steady increases in Riley's budget. In 1879, when Riley was forced out of the department, the appropriation for entomology was $10,000. When Riley returned, he negotiated a $20,000 appropriation, and by 1889 the division's budget peaked at $50,508.[4]

In contrast to his years of strife with Watts and LeDuc, Riley now enjoyed the support of two successive commissioners of agriculture. He had cordial relations with George Loring, who served from 1881 to 1885.[5] When

Grover Cleveland and the Democrats won in 1884, Norman J. Colman, Riley's friend and colleague from Missouri, replaced Loring. A delighted Riley confided to Joseph A. Lintner, "[F]rom now on I shall . . . have my own way."[6] Riley's division was also blessed with fresh, young graduates from the land grant universities. For example, Charles Marlatt, a graduate of Kansas State University who joined the division in 1889, eventually became its third chief.

Under such favorable conditions in the 1880s, Riley oversaw a fundamental reorientation in pest control. With newly developed pesticides and delivery systems, entomologists placed increasing emphasis on chemical controls. While Riley frequently discussed insect control in traditional balance of nature terms, he preferred to address specific insect problems rather than general theories of control. For example, when he revived the *American Entomologist* in 1880 and was travelling almost constantly in the course of the Cotton Worm campaign, he faithfully answered readers' questions about specific pests, somehow keeping up his correspondence while crisscrossing the country by train.[7]

On a few occasions, however, Riley took time to explain his philosophy of pest control. At a meeting of the Georgia State Board of Agriculture, in Savannah, on February 12, 1884, Riley spoke on "General Truths in Applied Entomology." As usual pressed for time, Riley confessed that his address comprised only "hasty notes" made en route from Washington.[8] In fact, Riley had delivered a similar address at the Philosophical Society of Washington a few days previously. The themes he presented in Savannah and Washington also appeared in articles he authored for *Stoddert's Encyclopaedia Americana* (1883) and for the *Encyclopaedia Britannica, American Supplement* (1889).[9] In these articles and addresses Riley discussed two broad topics: (1) the function of insects in the economy of nature; and (2) the recent development of new insecticides and spraying devices for controlling insect pests.

In his Georgia presentation and in similar talks and articles, Riley first reviewed the role of insects as predators, parasites, scavengers, pollinators, and as food for other organisms in the economy of nature. He then explained that European-style agriculture had altered the primitive balance between insects and plants in North America: "In the primitive condition . . . as the White man found it, insects . . . took their proper place in nature's economy, and rarely preponderated . . . to the injury of the wild plants . . . [b]ut civilized man violated this primitive economy. . . . [T]he . . . cultivation in large tracts . . . of one species of plant . . . gave exceptional facilities for the multiplication of such insects as naturally fed on such plants. In addition . . . many other [injurious species] have been unwillingly imported. [T]he worst . . . insect pests . . . are importations from Europe."[10] Several assumptions underlying

Featured Insect

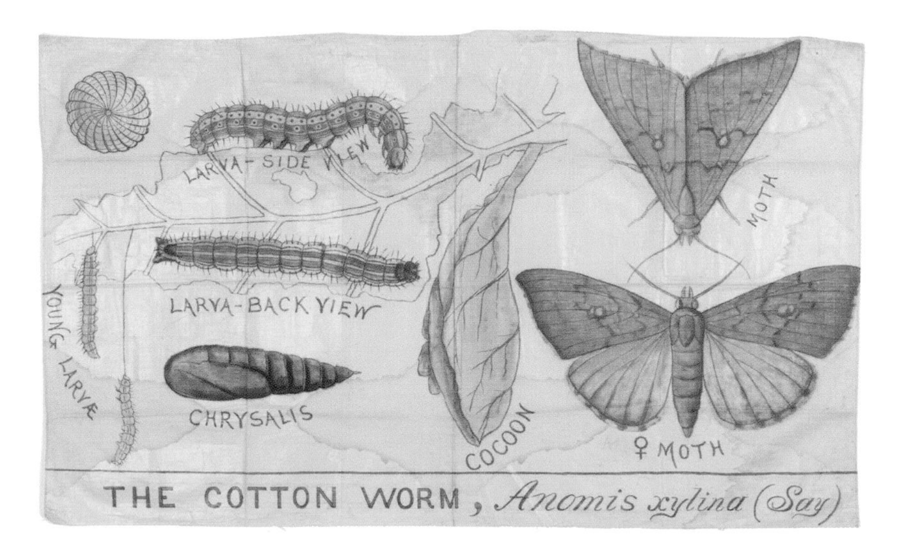

Cotton Worm [cotton leafworm, *Alabama argillacea* (Hübner)(Lepidoptera: Noctuidae)]. Riley Insect Wall Chart (Courtesy of Morse Department of Special Collections, Kansas State University Libraries). Five caterpillars are considered significant pests of cotton in the US, but Cotton Leafworm is no longer among them. Known as the most serious cotton pest during Riley's time, the Cotton Leafworm "has had a dramatic reversal of fortune," with the last recorded collection in the US in 1992.* Wagner believes a combination of insecticide treatments in North, Central, and South America, as well as the changing geography of cotton plantings and the lack of dormancy in this insect (which Riley mistakenly claimed it possessed), have resulted in its disappearance from the US.

*David L. Wagner, "Ode to *Alabama*: The Meteoric Fall of a Once Extraordinarily Abundant Moth," *American Entomologist* 55 (2009): 170.

Riley's statement bear examination. Like many of his contemporaries, he was acutely aware of ecological changes resulting from monoculture, where extensive fields of corn, wheat, and other crops replaced a patchwork of subsistence farmland, forest, and grassland. He seems, however, to have assumed a rather simplified, romanticized, primitive harmony in North America. Riley never mentioned Indian subsistence agriculture, although clearings for Indian corn, squash, and bean fields facilitated the rapid settlement and cultivation of land by European-American settlers.[11] Neither did he comment on the replacement of the bison and the Plains Indian culture by farming and ranching, although he observed this transition firsthand in the American West.[12] Furthermore, he did not consider that farming, by replacing forests with a mosaic of farmland, woodland, and grassland, fostered an overall increase in insect (and other) fauna, including beneficial and harmless species. The spread of milkweed in the rural America, for example, favored an increase in the Monarch Butterfly population.[13] In short, his simplified account of a presumed primitive harmony represented a rather superficial analysis of how insects fit into the massive ecological transition in North America.

Like many British immigrants, Riley associated a presumed harmony in primitive North America with memories of his native England. The historian Steven Stoll identifies such British nostalgia as a primary inspiration for agricultural improvers along the Atlantic seaboard who promoted cattle manure as a means of regenerating exhausted soils.[14] Along with this distinctly British perception, Riley shared the American assumption that cultivated farms and orchards would ultimately dominate the North American landscape. "Culture," for Riley and his contemporaries, meant the selection and growing, or "cultivation," of plants that were valued for their nutrition, beauty, or otherwise desirable qualities. The term "culture" as employed by Riley and his horticultural colleagues was quite different from the term "culture" as employed by anthropologists around the end of the century and later, when they referred to the aggregate social behavior of humans. Nineteenth-century horticulturists and agriculturists associated "culture" with human activity but in a fundamentally different way. They believed that those who actively "cultured" or improved plants and crops also improved, or "cultivated," themselves. The act of cultivating or improving fields and orchards tended to improve the cultivators. As a result, active cultivators were better qualified to pursue and admire literature, art, and music, activities they also referred to as "culture."[15]

Riley's romantic notions of natural harmony and culture, drawing on memories of well-tended fields and gardens along the Thames, around Dieppe, and along the Rhine, informed his view of agriculture and horticulture in America. For Riley, the Book of Nature was ideally represented in productive and aesthetically pleasing fields, gardens, orchards, and pastures. His vision of a domesticated North America practically excluded wilderness, nomadic Indians, and uncultivated land. While his fellow midwesterner, John Muir, glorified the wild Rockies, the High Sierra, and remote Alaskan wilds, and his fellow English immigrant, John Wesley Powell, found inspiration in the unmapped canyons of the Colorado River country and the vanishing customs and language of its Indian inhabitants, Riley crisscrossed the American West without commenting on its natural wonders or aboriginal inhabitants except to depict the destruction of crops by locusts. For him, the significant points of reference were cultivated fields and orchards in Kansas, irrigated farms around Denver, and citrus groves in Los Angeles.

Riley, like other "cultivators," considered injurious insects a necessary evil. Liberty Hyde Bailey, a prominent agricultural spokesman at the turn of the century, noted that some sort of primitive harmony might have existed in North America prior to European settlement, but added that it was a harmony that lacked the production of vital foodstuffs and other necessities of modern society.[16] Marshall P. Wilder, founder and president of the American Pomological Society, compared harmful insect species to moral evil in an imperfect world: "Every plant imported from abroad brings with it a new insect or disease, and the dissemination of new plants and varieties without which there can be no progress in horticulture inevitably disseminates their insect enemies . . . but as long as moral evil exists in the world, so long may we expect there will be evil in the natural world, and he who is not willing to contend against both, is not worthy of the name either cultivator or Christian."[17] Riley advised cultivators to accept some losses as a necessary evil associated with agricultural or horticultural improvement and increased production. His goal as economic entomologist was not to eliminate all losses due to insects but to "prevent as much of the loss as possible and at the . . . least expense."[18]

At the AAAS meeting in 1880, Riley explained that "new" destructive species made their appearance in three ways: (1) some insects were recently introduced species from outside North America (usually Europe) that had no natural enemies; (2) some were native species, previously unobserved or unrecorded, that found a domestic crop or vegetable to their liking; and (3) some native species, which were well known to entomologists but not previously recorded as injurious, changed their eating habits, becoming pests on plants they previously avoided.

In explanation of the third category, Riley held that "[insect] characters [could] originate within periods that are very brief compared to those [of] the higher animals."[19] In 1880, he reported that the Streaked Cottonwood Leaf Beetle that had previously fed exclusively on willows had recently begun feeding on cottonwoods. Riley compared the beetle's altered behavior to the apple maggot that had recently begun attacking cultivated apples.[20] In 1882, Riley reported that two beetle species had suddenly begun feeding on eggplants. The causes of such "remarkable and sudden change[s] in the food-habit of a . . . common . . . species," he said, remained unclear.[21]

In 1880, Eugene W. Hilgard reported that phylloxera in California's Sonoma Valley was confined to about two square miles, whereas it had ravaged all the vineyards in nearby Napa Valley. He speculated that that a wingless race, variety, or species had developed that could not spread through the dispersion of winged females.[22] Both Hilgard and Riley were disappointed when, later that summer, winged phylloxera females were reported in the Sonoma Valley.[23] Hilgard still held to the slim hope that these females were "sterile" (how else to account for its slow spread in Sonoma Valley?), but this last hope was swept away when Riley determined that the winged specimens sent to him were indeed fertile.[24]

As the nation's chief entomologist, Riley sought to mitigate insect damage regardless of whether insect pests were introduced, newly acclimated to a food plant, or had recently evolved harmful habits. On occasion, he advocated exclusion of insect pest species, but he considered the eradication of established insect pest species impossible. Despite disruption of the primitive balance, he was optimistic that proper management would ensure plentiful crops and productive orchards and gardens.

The second theme in his Savannah address and related statements was the recent development of new insecticides: arsenic compounds (primarily Paris Green and London Purple); kerosene emulsion; and pyrethrum, plus new spraying technologies, chiefly the cyclone nozzle.[25] Riley and his Cotton Worm team played a central role in each of these developments.

As noted above, Riley's promotion of Paris Green to control the Colorado Potato Beetle accelerated its widespread use. In 1878, prompted by the popularity of Paris Green, a British chemical firm sent several barrels of an arsenic-based waste product of aniline dye production, to Charles Edwin Bessey at Iowa Agricultural College. Bessey gave it the name "London Purple."[26] Bessey's tests indicated that London Purple was effective in the control of the Colorado Potato Beetle. In the winter of 1878–79, Riley persuaded the British manufacturer Hemingway & Co. to send quantities of London Purple to the Agricultural Department, where the departmental chemist also found it had potential as an insecticide. Riley then sent samples to his agents in Georgia, Alabama, and Texas, with instructions to compare its effectiveness with Paris Green as a control of the Cotton Worm. Based on reports from Schwarz and other agents, Riley concluded that London Purple was superior to Paris Green in the control of chewing insects.[27] Unlike Paris Green, London Purple was purely a waste product, and it was thus cheaper, costing three to five cents per acre compared to fifteen cents to one dollar for Paris Green. Riley downplayed reports from William Trelease (John Comstock's protégé) that London Purple damaged plants, remarking that Trelease had applied excessive dosages.[28]

Due in large part to Riley's promotion, London Purple seemed destined to replace Paris Green.[29] It was cheaper, more concentrated, and more economical to ship in quantity, adhered better to plant foliage, and its bright color alerted users and others to the presence of poison. Among its disadvantages, Riley noted that it acted more slowly than Paris Green, requiring two to three days rather than a few hours to kill its victims, and rain seemed to wash London Purple away more readily than Paris Green. Also, London Purple was more difficult to maintain in water suspension than Paris Green, making it more difficult for users to maintain a consistent spray of correct proportions.[30] Within a few years, however, gardeners, farmers, and entomologists concluded that London Purple damaged the plants it was supposed to protect, and by the 1890s it practically disappeared as an insecticide. At about the same time, lead arsenate, introduced against the Gypsy Moth in 1893, replaced both Paris Green and London Purple, and remained the insecticide of choice until after World War II, when DDT replaced it.[31]

Beginning in the 1870s and well into the twentieth century, entomologists, medical doctors, and others debated possible human and animal health hazards resulting from contact with arsenical compounds. Townend Glover, Albert J. Cook, Lintner, and others frequently voiced opinions mostly for but sometimes against the application of Paris Green, London Purple, and other arsenicals. For example, Cook, who generally scoffed at the supposed dangers of arsenicals, admitted that bees (his specialty) that collected pollen from sprayed blossoms suffered injury.[32]

Riley vacillated between caution and enthusiasm regarding arsenical compounds. When, in 1875, John LeConte proposed a ban on Paris Green and other toxic insecticides in an address before the National Academy of Sciences, Riley characterized his warnings as "alarmist." Acknowledging his own initial suspicions, Riley asserted that further experience with Paris Green had convinced him that the compound, when diluted and applied with caution, was harmless.[33] Responding to concerns that London Purple was even more toxic than Paris Green, Riley cited two bizarre incidents in support of his view that London Purple posed no life-threatening danger to humans. Two African Americans who mixed London Purple with bran when preparing pancakes became quite sick, but neither one died. A ship's cook on a Mississippi steamer also used the scrapings of London Purple dust that had leaked out on the deck to color ice cream. Again, no one died.[34] On the other hand, when Cook recommended Paris Green to control the Strawberry Leaf Beetle and the Apple Worm, Riley warned against unnecessary spraying of fruit and berries, because this posed a health hazard to consumers.[35] Overall, Riley appears to have been more cautious than most of his entomological colleagues regarding the danger of arsenic poisoning.[36]

James Whorton, an historian of pesticide use, concludes that the consistent flaw underlying scientists' and physicians' blindness to the danger of arsenical insecticides was their failure to recognize that arsenic poison accumulates in the organism. Physicians and entomologists typically tested the immediate effect of one, or a few, applications. In the twentieth century, scientists and physicians slowly recognized that arsenic accumulates over time, eventually reaching toxic levels.[37]

During the summer of 1880, Riley and his agents sought to exploit the Cotton Worm's habit of congregating on the underside of cotton leaves as a potential weak point in its life cycle. Riley instructed William Barnard to design a device for spraying the underside of cotton leaves.[38] In September, Riley joined Barnard in Selma. They concluded that they must develop a new kind of sprayer because most spray nozzles up to that time were designed to apply water, not poisonous insecticides, and as Cornell horticulturist E. G. Lodeman later pointed out, "[T]he best spray nozzle, so far as efficiency, simplicity, and cheapness are concerned, is the end of a hose and a man's thumb."[39] Barnard proposed to force the liquid tangentially into a chamber, causing it to whirl and to eject the "atomized" liquid as a fine spray. Using transparent watch crystals, he designed a chamber where they could observe the "atomizing" of the liquid. By late September 1880, Barnard had a working prototype that he and Riley called the "cyclone nozzle" because of the whirling action of the liquid.[40] In November 1881, Riley and Barnard presented a working model of the nozzle at the Atlanta Cotton Exposition. A host of visitors, including cotton growers, manufacturers of agricultural implements, and financiers attended the Atlanta Exposition, an event that heralded the emergence of the "New South" following the Civil War and Reconstruction.

On November 4, 1881, Riley mounted the exposition platform. Like a southern evangelist addressing his congregation, Riley reminded cotton growers of the enormous losses they suffered, and he praised southern legislators for funding Cotton Worm investigations. In an emotional aside, he related how,

FIGURE 9.1 “The Cotton Worm,” USEC, *Bulletin* 3 (1880): plate 1. (Courtesy of the National Agricultural Library Special Collections, Beltsville, Maryland).

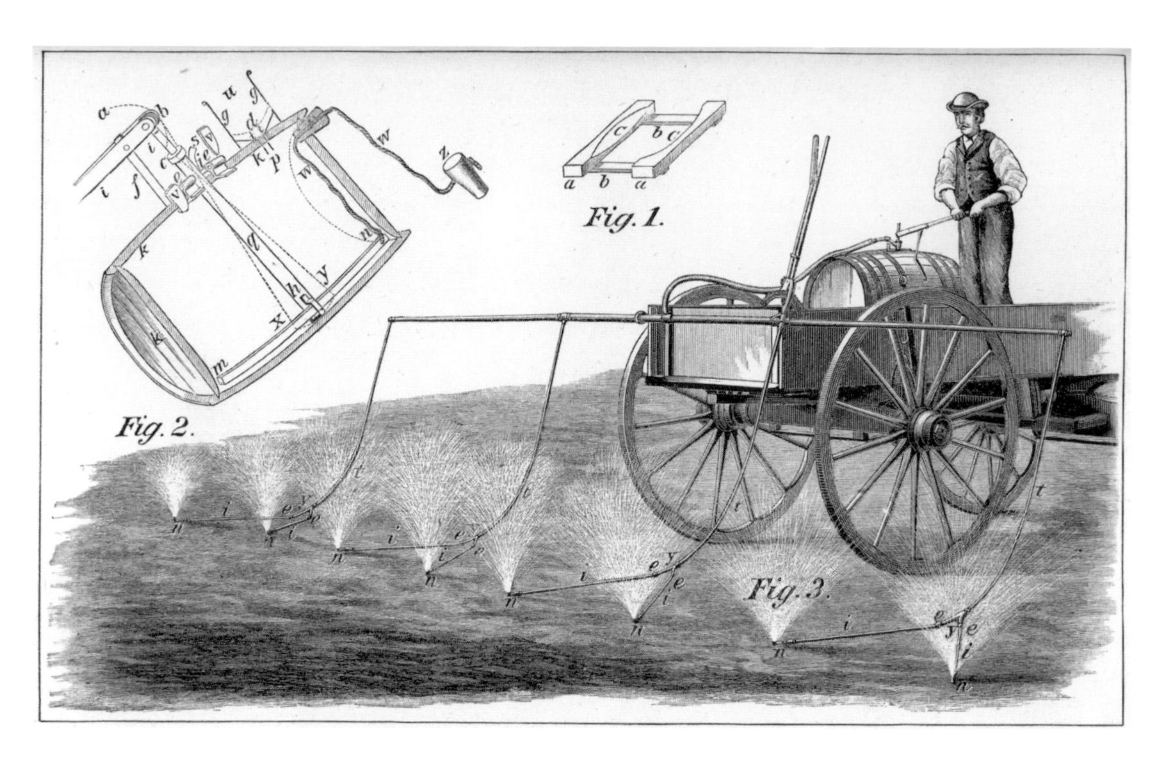

FIGURE 9.2 Cotton Worm Sprayer, *ARCA* (1881–1882), plate IX, following p. 159. (Image courtesy of Albert R. Mann Library, Cornell University).

the previous fall, Barnard had nearly died of yellow fever during the investigation. Then, with a dramatic flourish, Riley unveiled a bigger-than-life illustration of the Cotton Worm, saying, "Here is the culprit familiar to you all."[41] He recalled that when he and his co-workers began their investigation in 1878, many planters still believed the worm originated by spontaneous generation. The Cotton Worm, Riley explained, was so prolific that each female moth was theoretically capable of producing twenty billion progeny in one season. The Cotton Worm's extraordinary fecundity explained why natural checks by predators failed to control its proliferation.

Then, with another flourish, Riley unveiled the cyclone nozzle. With its capacity to deliver insecticide spray from below, it was "the best yet invented."[42] Noting that the trial cotton patch on the exposition grounds had been totally defoliated by Cotton Worms, Riley claimed that, with their new sprayer, he and his team could have saved the cotton plants in one hour using less than one dollar's worth of poison. The *Atlanta Constitution* praised Riley's speech, especially his illustrations and spontaneous remarks: "[H]e has a way of treating an otherwise technical subject so as to interest and instruct all."[43] Following Riley's address, listeners crowded around the model spraying device. It featured a gravity flow system with the insecticide fluid flowing from a barrel elevated high over the wagon, eliminating the need for a pump to supply pressure to the sprayer. This machine, Riley claimed, could spray from twelve to twenty rows of cotton in one swath. It cost ten to fifteen dollars to construct, and with it one man with a team of horses could apply poison to 150 acres of cotton in a day.[44] Shortly thereafter, however, field tests of the sprayer revealed numerous problems. When operating on even ground with uniform rows, a single operator could indeed spray eighteen rows of cotton twenty yards wide; however, when operating under normal field conditions, with uneven ground and uneven rows, the operator could spray only four rows. Stiff pipes that didn't adjust to uneven rows, leakage, varying size of cotton plants, and above all unevenness of rows and uneven ground required changes in the initial design and operation of the sprayer.[45]

Over the next several years, Barnard supervised the development of many variations of the spraying device, variously referred to as the cyclone nozzle, atomizer, or eddy chamber. At the World's Industrial and Cotton Centennial Exposition in New Orleans, in 1884–85, Riley, Barnard, and other staff members prepared a display of economic entomology accompanied by a catalog with twenty pages devoted to various nozzles designed by Barnard and his coworkers at the Division of Entomology, including the Eddy-Chamber Rose, the Slot-Rim Eddy Chamber Nozzle, the Centrifugal Spray Nozzle, and the Thick-Lip Eddy Chamber Nozzle.[46] When Riley was in Europe two years later, he reported that most of the sprayers produced overseas were reproductions of machines designed and developed in the division.[47] As noted above, sprayers came into widespread use in France in the late 1880s and 1890s. In the 1890s, American manufacturers began mass-producing similar devices. Beverly T. Galloway, chief of the Division of Plant Pathology, published detailed plans for the construction of knapsack pumps. Such pumps for the application of pesticides, familiar to professional and hobby gardeners ever since, came into widespread use.[48]

Yet another key innovation developed by Riley's Cotton Worm team was kerosene emulsion insecticide. Horticulturists and gardeners had often experimented with kerosene, widely used for lighting in lamps, but had found that, unadulterated, it harmed or killed plants. When Riley, in 1880, directed his agents to develop a kerosene solution or emulsion that would kill insect pests but leave the plant undamaged, Barnard proposed whipping the kerosene with milk into a kind of "butter," an emulsion that could be diluted with water to

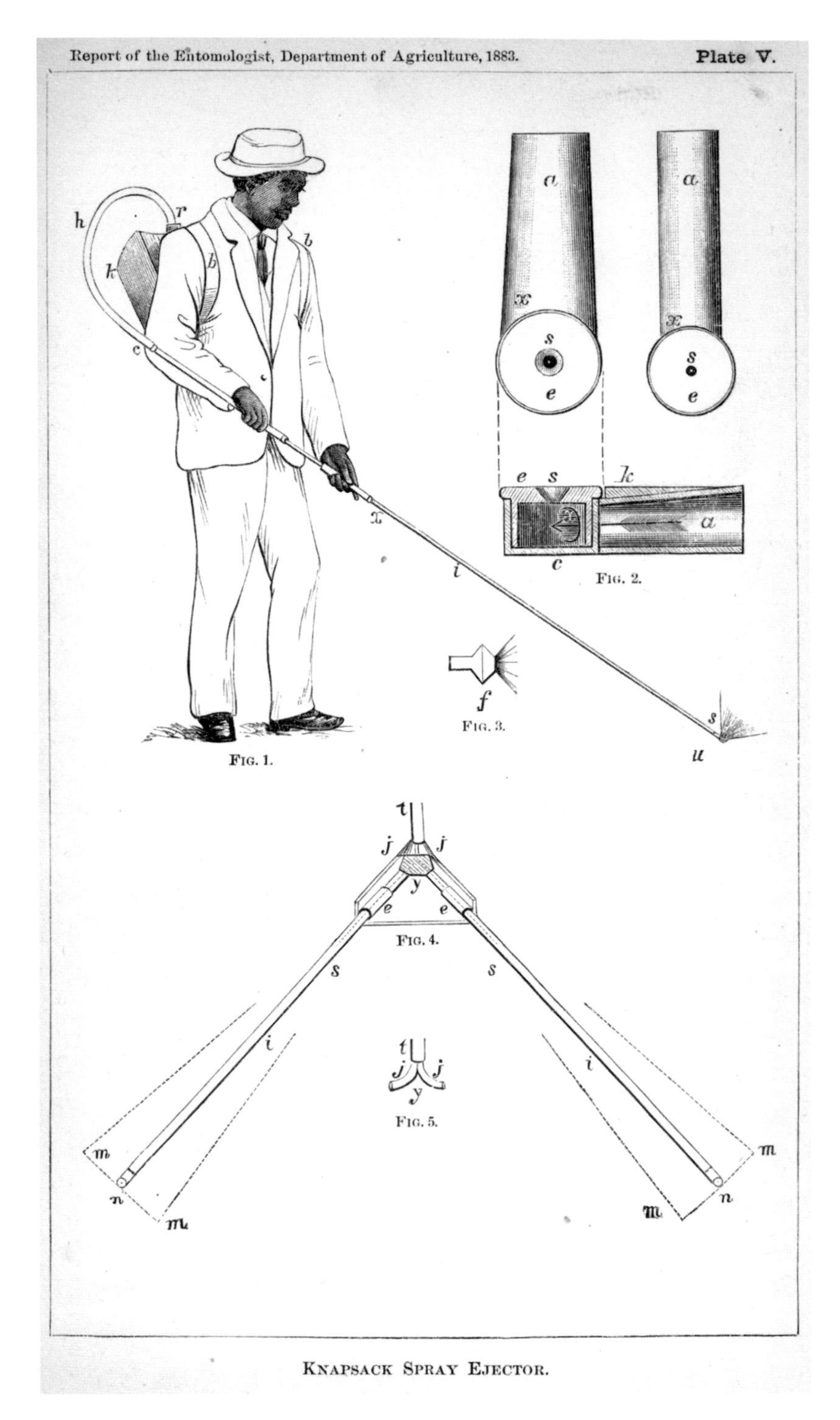

FIGURE 9.3 "Knapsack Spray Ejector," *ARCA* (1883), plate V. Note spray ejected from below (courtesy of Albert R. Mann Library, Cornell University).

the desired proportions for use as an insecticide. Unfortunately, kerosene-milk emulsion damaged cotton plants even in thinly diluted form.[49]

In 1881, the development of kerosene emulsion took a new direction. Riley, now reinstated as federal entomologist, hired Henry Hubbard, who now lived in Crescent City, Florida, to investigate insect pests of citrus groves. Hubbard discovered that Barnard's kerosene-milk emulsion was an effective control of scale insects that infested orange trees. Unsatisfied with Barnard's kerosene-milk formula because it required reheating and mixing at regular intervals, Hubbard developed an alternate kerosene-soap emulsion that retained its consistency for months without losing its potency. Diluted at one part emulsion to twelve parts water, it killed scale insects on orange trees but left the trees uninjured. The Hubbard formula caught on quickly. In 1882, Riley recommended it over kerosene-milk emulsion. By 1883–84, it was being used not only against scale insects but also against a host of plant-sucking insects in orchards, greenhouses, and gardens.[50] Riley promoted kerosene emulsion for the control of plant-sucking insects but favored arsenates for plant-chewing insects on the grounds that these targeted only harmful species that fed on plants.[51]

The commercial success of cyclone nozzle sprayers, kerosene emulsion, and arsenic insecticides, all developed in full or in part by agents or employees of the Agricultural Department, raised the question of who was entitled to patent devices developed within the federal government? Riley held that innovations by federal employees automatically became public property and could not be patented. In 1885, in response to an article in the *Rural New Yorker* that credited Riley with the invention of the cyclone nozzle, he responded that neither he nor any one individual had invented the nozzle but rather a team under his direction was responsible for the invention. Riley noted further that some "private parties" had tried to patent its essential principles.[52]

By "private parties," Riley meant Barnard. The sequence of events leading to his dispute with Barnard is recorded in the decision of the commissioner of patents in 1892.[53] When Riley replaced Comstock in June 1881, Comstock returned to his professorship, and Barnard (who filled Comstock's position at Cornell during the academic year 1880–81) received a call as professor of chemistry and biology at Drake University. Barnard postponed assuming this professorship for five years, until 1886, in order to continue as Riley's special assistant in the further development of the cyclone nozzle and kerosene emulsion.[54]

In 1882, Barnard (unknown to Riley) applied for patents on kerosene milk emulsion (May 31) and the cyclone nozzle (August 4).[55] It is not clear exactly when Riley learned of Barnard's patent applications, but in October 1885,

Riley together with Colman filed a protest with the commissioner of patents.[56] They contended that (1) Barnard was a member of a team that developed the kerosene emulsion and the cyclone nozzle, and although he made the initial suggestions of mixing milk with kerosene and forcing rotating fluid at an angle through a nozzle, both "inventions" represented team efforts in which Riley and Hubbard deserved as much credit as Barnard; (2) that Barnard had agreed in writing that any discoveries or inventions he made while a government agent would be public property and could not be patented by him; and (3) that experiments and development carried out by Barnard and other agents were funded by the government and were therefore pubic property.[57]

The second point referred to Riley's discovery, or suspicion, in May 1885, that Barnard had filed for patents. He and Colman then required Barnard to sign a statement relinquishing to the government the patent rights to anything he worked on while a government agent.[58] By October 1885, when Riley and Colman filed their protest, Barnard was already *persona non grata* in the Division of Entomology.[59] It was apparently too late that year for Barnard to assume the professorship at Drake; he therefore waited until the fall of 1886 to take the position. Barnard died a year later, on November 13, 1887, apparently from a combination of overexertion and sickness, having served only one year at Drake.[60]

Barnard's application proceeded through official channels for seven years until, on June 2, 1892, the commissioner of patents ruled in his favor.[61] The commissioner ruled that, because Barnard originally suggested using milk, he was therefore the "inventor" of kerosene emulsion and entitled to the patent.[62] The commissioner also ruled that, because Barnard had made the original suggestion of utilizing a rotating motion to force the spray upward, he was therefore entitled to the cyclone nozzle patent. Riley claimed that the idea of rotation had come in the course of conversations between the two of them, and therefore belonged to both of them. Regarding Riley's more fundamental contention that procedures or devices resulting from the research and development of government agents automatically belonged to the public, the commissioner sidestepped the issue. The commissioner cited the Supreme Court case, *Solomons v. United States* (1890) in which Spencer M. Clark, chief of the Bureau of Engraving and Printing, had patented a die that he had designed on government time at the request of his superiors. The Supreme Court ruled in favor of Clark on the grounds that Clark had signed an explicit agreement that allowed the government to use the invention at no cost but implicitly reserved the right to patent the invention himself.[63]

According to Katherine L. Fisk, an historian of patent process, when the justices decided in favor of Clark in the *Solomons* case, they attempted to combine the principles of a democratic-egalitarian tradition emphasizing the rights of individual inventors regardless of employment status with the hierarchical tradition of the master-servant relationship. From the beginning of the patent system, courts had favored individual inventors regardless of employment status, but in the 1880s the courts began favoring employers, arguing (like Riley) that corporate enterprises (whether public or private) furnished the environment and the means for invention.[64] The *Solomons* decision revealed a Supreme Court of two minds. The court decided in favor of Clark, the employee, but based its decision on reasoning that upheld the rights of the employer, in that case the government.[65] The patent commissioner and Riley cited different and contradictory sections of the decision. The commissioner argued that Barnard, like Clark, was entitled to patents for inventions he made on government time. Riley's position, that corporate support enabled employees to make inventions, and patents to products or processes therefore belonged to the employer, was more in tune with the future, however, because by the early twentieth century, the courts ruled almost unanimously in favor of corporate enterprises over individuals.[66] Had Riley entered his protest ten years later, he would most certainly have won. In 1892, however, the commissioner upheld Barnard, citing the fact that he had not assigned rights to use his invention to the government.[67] Furthermore, the commissioner held that his mandate was to issue patents to inventors, whereas decisions as to who had the right to apply for or receive patents resided with the courts.[68] In overruling Riley's protest, the commissioner indulged in a flourish of legal jargon, calculated to further inflame Riley's hostility to the Patent Office and to frustrate any subsequent attempt to fathom his meaning: "The discussion of the questions as to Barnard's right to the practical development of that idea, of his right to take a patent for an invention made by him while he was an employee of the Government, and of the Government's rights in the premises because of expenses it may have defrayed in the same connection, had with reference to Barnard's application for patent filed May 31, 1882, decision whereupon is of even date herewith, is equally to this case. [!?]"[69]

In 1897, the two patents were issued to Barnard's widow.[70] Apparently neither she nor other family members attempted to develop or market products based on the patents. They probably considered any attempt to compete with the multitude of sprayers and insecticides on the market to be futile. Barnard's family, however, clearly felt that he had been slighted. Barnard's son later complained that Barnard received neither credit nor monetary recompense for his invention of the cyclone nozzle.[71] E. G. Lodeman, a professor at Cornell and an associate of Riley's rival Comstock, asserted in 1896 that Barnard deserved primary credit for invention of the cyclone nozzle and should have enjoyed the monetary benefits accruing from his patent right to it.[72]

In his Savannah talk and elsewhere, Riley listed pyrethrum, along with Paris Green, London Purple, and petroleum emulsions, as one of the "most common and most effective" insecticides. Riley's advocacy of pyrethrum dates from his search for non-toxic substitutes for arsenic-based insecticides. In 1879–80, Riley instructed his team to identify and test substances that would be effective against the Cotton Worm but non-poisonous to humans and domestic animals.[73] Pyrethrum, which was commonly used against pests on houseplants, seemed a likely candidate. With typical thoroughness, Riley corresponded with naturalists in the Caucasus region to learn more about this plant.[74]

Pyrethrum had been produced for centuries by growers in the Caucasus Mountains from two species in the genus *Chrysanthemum* (Asteraceae, the Composite or Daisy family). Growers there maintained strict secrecy regarding the plant's cultivation and processing, thus maintaining a monopoly on pyrethrum powder with its insecticidal properties. In the 1820s, an Armenian merchant named Jumtikoff who had business connections in the Caucasus discovered the origin of the powder, and in 1828 his son commenced commercial production of pyrethrum. By the 1850s, imported pyrethrum powder had become a household commodity in the United States.[75]

Pyrethrum's great advantage was that it was nearly harmless to humans, vertebrate animals (although it is somewhat toxic to fish), and plants. Its disadvantages were that it lost its potency after a few hours exposure to sunlight and that it was expensive.[76] An unnamed Frenchman, in 1857, and William Saunders, in 1879, applied pyrethrum to field crops, the latter with some success. In 1879, Riley requested that Hilgard at the University of California analyze the insecticidal properties of pyrethrum. Hilgard reported that the insecticidal substance could be extracted as a volatile oil from all parts of the plant (not only the flower, the traditional source of pyrethrum powder). Encouraged by Hilgard's findings, in 1880 Riley supervised the cultivation of pyrethrum plants on the Agricultural Department grounds. Testing by department chemists indicated that the potency of pyrethrum extracted from these plants was equal to imported pyrethrum. One pound of extract dissolved in two hundred gallons of water proved to be effective against Cotton Worms. Reporting these findings at the Entomological Club in August 1880, Riley predicted that pyrethrum, when cultivated and processed in bulk, promised to be a major breakthrough in the control of the Cotton Worm, and indeed in the control of a wide range of injurious insects.[77]

Having demonstrated that pyrethrum could be raised and processed in America, Riley sought to promote its large-scale cultivation, production, and distribution in order to make it available as an inexpensive field insecticide. In the winter of 1880–81, he procured a large quantity of pyrethrum seed from abroad, which he distributed free of charge to cotton growers in the South and in California.[78] Riley was confident that once pyrethrum was produced in bulk, its price would drop to levels comparable to Paris Green.[79]

Despite distribution of free seed, growers were reluctant to devote resources to the cultivation of an unfamiliar crop with uncertain rewards. Growers who did experiment reported difficulties in planting, cultivating, harvesting, and processing pyrethrum. By 1884, only one grower-processor, G. N. Milco, of Dalmatian heritage, who operated a farm near Stockton, California, offered quantities of domestically produced pyrethrum, which he sold under the trade name of "Bulach."[80] The price of Bulach, although somewhat lower than imported pyrethrum, remained too high for pyrethrum to compete with Paris Green or London Purple. Riley, however, remained optimistic. In his 1884 address before the Georgia State Board of Agriculture, he noted, "[Pyrethrum's] action on many larvae is marvelous."[81] Riley predicted a wide range of uses for pyrethrum extract, even beyond its use as an insecticide. He suggested that it be tried against various microorganisms that Louis Pasteur and Robert Koch had demonstrated to be the cause of many diseases.[82]

Pyrethrum was never used widely as a field insecticide. In the catalogue accompanying the Division of Entomology's exhibit in New Orleans (1884), pyrethrum was recommended for the control of the Colorado Potato Beetle (also arsenic mixtures), the Harlequin Cabbage Bug, the Cabbage Plutella (also kerosene emulsion), the Imported Cabbage Worm, the Grapevine Aphis (also arsenic mixtures), and the Rose Chafer (also hand picking).[83] In 1889, Riley still listed pyrethrum along with arsenicals and kerosene emulsion as one of the three main insecticides in use, but his 1889 list of insects for which pyrethrum was recommended included only the Imported Cabbage Worm, the Southern Cabbage Butterfly, the Cabbage Plusia, and the Harlequin Cabbage Bug. Conspicuously absent in both lists was the Cotton Worm, the insect Riley had in mind when he first suggested pyrethrum as a field insecticide. By the 1890s, the Department of Agriculture recommended pyrethrum only for the control of insects on household and greenhouse plants.[84]

Pyrethrum was one of many plant extracts Riley's agents tested in the search for alternatives to arsenic based insecticides. In all, his agents at Selma rendered over five tons of various plants into extracts. One promising candidate was Ox Eye Daisy, an indigenous American plant that Riley's Canadian colleague William Saunders suggested had insecticide properties similar to pyrethrum. When Riley asked Barnard in Ithaca (prior to his joining Riley's team in Selma later that summer) to test the plant, Barnard reported that Ox Eye Daisy held no promise as an insecticide.[85] Experiments with other plant extracts carried out that summer by Eugene Schwarz and James Roane at Selma, R. W. Jones at Oxford, Mississippi, Barnard at Ithaca, and J. P. Stelle in Texas produced no new source of insecticide with the same effectiveness as pyrethrum and no viable alternative to arsenic-based insecticides.[86]

Bordeaux Mixture, the final control agent that figured prominently in the new economic entomology, had nothing directly to do with insects; it was a fungicide to control fungal plant pathogens. Although Riley had little or nothing to do with its discovery, Bordeaux Mixture had significant implications for his division.

Until the 1870s, scientists had no explanation, let alone a cure, for various maladies known as "rust," "blight," "smut," "mildew," etc., that weakened or destroyed crops and orchards. In that decade, Pasteur and Koch discovered that microorganisms (bacteria and fungi) caused such diseases.[87] From 1878–82, Downy Mildew spread rapidly in French vineyards, threatening to deliver a second blow to the French wine growers only just recovering from the phylloxera scourge. No known remedy seemed to work. French vintners noticed, however, that strips of vines next to walkways that had been sprayed with verdigris and salt of copper to discourage pilfering of grapes remained unaffected. When botanists Edouard Prillieux and Pierre-Marie-Alexis Millardet investigated, they identified a copper compound as the active agent. In August 1883, Millardet successfully controlled Downy Mildew using copper compounds. By 1885, French scientists had developed a standard formula, known as Bordeaux Mixture, that effectively controlled many fungus diseases like White Rot, Bitter Rot of Grapes, Strawberry Leaf Blight, and Apple Scab.[88] Seemingly out of nowhere, an entirely new discipline of plant pathology sprang up. It was a hybrid mix of botany and bacteriology, complete with ready-made control measures.

In 1885, Congress approved the establishment of the Division of Mycology, soon renamed the Division of Vegetable Pathology, in the Agricultural Department. The exponential growth of the new division was due in large part to Beverly T. Galloway, a University of Missouri professor who was hired as an assistant in 1887, and who became its head in 1888.[89] Slight in stature, with artistic hands suggestive of a surgeon or concert pianist, Galloway proved to be a skillful administrator who assembled a young and highly motivated staff. In rapid succession, his team controlled a multitude of plant ailments that had baffled gardeners and horticulturists: Downy Mildew of Potatoes, Tomato Black Rot, Brown Rot and Powdery Mildew of Cherries, Leaf Blight and Cracking of Pear, Rose Leaf Spot, Plum Pockets, Apple Rust, Leaf Spots on Maples, Cottonwood Leaf Rust, Peach Yellows, and Celery Leaf Blight.[90]

In the late 1880s and early 1890s, plant pathologists and growers, experimenting with various sprays, discovered that Bordeaux Mixture had some insecticidal value and that Paris Green, the standard insecticide, was effective against some fungi. By combining the two, horticulturists hoped for an all-purpose spray capable of controlling anything that hindered the healthy growth of plants. Orchardists soon began employing a blanket spray consisting primarily of Paris Green and Bordeaux Mixture.[91] The twin disciplines of economic entomology and plant pathology seemed to have converged to assure growers a steady increase in the production of fruits and vegetables. In 1886, Riley published the formula for Bordeaux Mixture in a division bulletin. Hilgard in California, Colman in Missouri, and Frank Lamson-Scribner at the USDA Division of Mycology published similar recommendations.

The effectiveness of the new insecticides and fungicides enhanced the prestige of agricultural science. With the disciplines of plant pathology and entomology growing exponentially, Riley and Galloway, once viewed as mere practitioners, now enjoyed a status as scientists on a level with Spencer F. Baird and Powell. In formal and informal settings (e.g., at the Cosmos Club), the fraternity of federal scientist-administrators developed a shared vision of fulfilling America's potential through the application of science.[92] Within this fraternity, Galloway represented a formidable rival to Riley, long the prime mover and shaker in and out of the Department of Agriculture. A showdown seemed inevitable. As it turned out, Riley left the department in 1894 before the tinder ignited. Howard continued the rivalry with Galloway, in most instances losing out to him.[93]

Riley's increased emphasis on insecticides is evident in his 1885 final report on the Cotton Worm.[94] As in the past, Riley ridiculed "humbug" remedies like sprinkling cotton plants with holy water and, as in the past, he began by discussing natural (biological) controls.[95] When discussing insect parasites, however, Riley took a new approach. In his review of Stephen A.

Forbes's 1881 publication "On Some Interactions of Organisms," Riley had signaled his agreement with Forbes that, from an evolutionary point of view, parasites depended on the survival of the host species. Natural selection tended to bring about an adjustment between a host species and its parasites that in the end promoted their common interests.[96] Based on this reinterpretation of the parasite-host relationship, Forbes and Riley concluded that parasites did not function primarily as "enemies" of their hosts but as partners and, therefore, they were ultimately ineffective as controls of the host population. Whereas in the 1870s Riley had held that parasites were primary control agents of locusts, by the 1880s he considered parasites to be ineffective in the control of the Cotton Worm.[97] With its tremendous fecundity and extraordinarily rapid development through multiple generations, the Cotton Worm outstripped natural enemies that often came too late to save the cotton crop.

Whereas, by the mid-1880s, Riley accepted Forbes's theory of parasitism, he resisted Forbes's parallel findings that insect disease held promise as an effective control. In the early 1880s, farmers in Illinois and the upper Midwest experienced the most widespread Chinch Bug infestation up to that time. Forbes began experimenting with bacterial and fungal diseases (pathogens) as an alternate to traditional controls. He was following the lead of Pasteur, who in 1865–70 had isolated the bacteria that caused pébrine (blackspot) in silkworms, a disease that had decimated the French silk industry. Pasteur had also developed sterilization techniques that cured or prevented silkworm disease. Forbes, now equipped with improved microscopes with which bacteria could be clearly observed, isolated and tested bacterial diseases (pathogens) affecting the Chinch Bug. At the AAAS meeting in August 1883, he announced his initial success with contagious disease as a control strategy.[98]

In response to Forbes's presentation, Riley recalled that LeConte, in 1874, had proposed using fungal disease as control, but that subsequent investigation had not supported this proposition. While recognizing the scientific merit of Forbes's investigations, Riley expressed his doubts that insect disease would prove effective against the Cotton Worm and other pest species.[99] Riley's doubts signaled a change from his views in the 1870s, when he had been intrigued with the concept of insect disease as a control measure. In his *Fourth Annual Report* (1872), he drew a parallel between predators and parasites that operated as checks at the species level and disease that operated as a check at the microorganismal level. Riley explained how, in his view, the microorganisms identified by Pasteur in silkworm disease operated as controls:

> These seeds of disease . . . play a most important part in the economy of Nature. They are omnipresent guards wisely ordained to keep order and harmony in her Domain . . . to right the wrong which man's ignorance begets . . . fulfilling God's will . . . in prompting us to . . . higher effort . . .
>
> "All nature is but art, unknown to thee,
> All chance, direction which thou canst not see;
> All discord, harmony not understood;
> All partial evil, universal good."[100]

Why Riley changed his mind about insect disease as a control is unclear. For whatever reasons, by the mid-1880s, he doubted that insect parasites and insect pathogens offered effective controls.

Riley continued to recommend farm cultural controls when applied in combination with insecticides and biological controls. For the Cotton Worm, he recommended planting cotton varieties that matured prior to July and August, when worm broods typically overwhelmed biological controls; planting strips of jute and other plants that were obnoxious to Cotton Worms; leaving stumps in the cotton fields as protection for ants (one of the main predators of Cotton Worms); and the retention of shade trees on the edge of fields (Riley had no explanation for the apparent effectiveness of this practice).[101] Riley emphasized the necessity of coordinated action and the application of controls early in the season, before the worms reached outbreak numbers.[102] As with the Rocky Mountain Locust, Riley envisioned a Cotton Worm warning system, which, given the density of settlement in the South, he argued would be more effective than locust warnings in the sparsely settled West.[103] Ultimately, however, Riley held that chemical controls were the key to control of the Cotton Worm. Furthermore, insecticides and delivery systems developed for the Cotton Worm could be applied to other injurious species.[104]

And what became of the Cotton Worm, the species that prompted Riley's organization of the Division of Entomology, the development of the cyclone nozzle, the development of arsenic-based insecticides, petroleum emulsion, and pyrethrum, and that prompted Riley's reevaluation of natural controls? Up to the mid-twentieth century, the Cotton Worm continued to devastate southern cotton fields, while growers and entomologists responded with lead arsenate, which reached a record amount applied to cotton plants in 1911, and ever-greater quantities of new insecticides available after World War II.[105] Today, the Cotton Worm no longer exists in the United States and is rare or unknown throughout its former permanent range in the subtropics. The last specimens in the United States were collected in the 1990s.[106]

According to David Wagner, entomologist and historian, the demise of the Cotton Worm is due to several interrelated factors. First, beginning in the 1940s, the application of DDT, a pesticide that was more effective than arsenic, delayed the population buildup in south Texas so that the northward migration was less destructive. Second, farmers in northern Mexico discontinued growing cotton, removing a major stage in the annual advance of the Cotton Worm from its permanent home south of latitude twenty. Third, growers in the "Cotton South" shifted from cotton to sugar cane and other crops. Fourth, agribusiness firms developed cotton varieties with resistance to caterpillars; these varieties now account for 70 percent of cotton planted in Texas. Fifth, cotton growers now clear cotton fields after the harvest, removing the host plant upon which late generations of Cotton Worms previously multiplied.[107]

As a result, the Cotton Worm, like the Rocky Mountain Locust, disappeared as a significant pest in American agriculture. Thus, two of the four major species about which Riley engaged in a struggle of wits and means disappeared or receded into insignificance. Riley could not have foreseen the eventual decline of the Cotton Worm, but he did recognize its obstruction of economic development in the South. In the course of his Cotton Worm campaign, Riley organized the Division of Entomology and modified his approach to insect control, incorporating insecticides and application technologies developed in the 1880s. His agency and the technologies he developed and promoted set the pattern of economic entomology for the next century.[108]

10

Years of Fulfillment

I should consider the present the happiest period of my life.

—Riley to J. A. Lintner, April 7, 1885, Lintner Corr., NYSM

For three-quarters of a century, the nation's capital as envisioned by Pierre Charles L'Enfant, George Washington, and Thomas Jefferson remained a sleepy southern village. During congressional sessions, senators and congressmen took up quarters in Washington's inns, filled their pork barrels, and hightailed it back to their home districts. With the outbreak of the Civil War, Washington was jolted into action. While Union armies guarded southern and western approaches to the capitol, a makeshift federal bureaucracy struggled to recruit, provision, and transport armed forces on a continental scale. Federal employees tripled in number, setting in motion an expansion only temporarily stalled by the Panic of 1873. By the 1880s, Washington's population had reached about two hundred thousand, making it the country's eleventh largest city. Like other Washington residents, Riley expected Washington's boom to continue into the indefinite future, a reasonable expectation borne out by the addition of another one hundred thousand residents during his lifetime.[1]

In the postwar years, Washington eclipsed Boston-Cambridge as the nation's natural history center. Spencer Baird at the Smithsonian Institution established the US Fish Commission, the United States National Museum, a marine laboratory at Woods Hole, Massachusetts (home base for the cutter *Albatross*), and secured steadily increased appropriations for the Smithsonian. By 1880, more naturalists lived and worked in Washington than any other

American city, and young, energetic naturalists strove to emulate Baird's success.[2] These scientist-administrators craved power, not primarily the power of wealth or political influence, but the power of ideas and knowledge, power based on investigation and research and utilized to improve society and to strengthen the nation.[3] John Wesley Powell's vision centered on rational management of land and water resources in the West and investigating and caring for its aboriginal population. Riley's vision centered on controlling insect pests in order to expand America's agricultural and horticultural resources.

Riley's pending appointment as entomologist at the Agricultural Department in early 1878 very likely prompted his decision to move to Washington.[4] His appointment may also have motivated him to change his citizenship. Although he had served in the Union army and had been appointed chief of the US Entomological Commission, Riley remained a British citizen until December 1878, when he became a citizen of the United States.[5]

Charles and Emilie purchased a red brick town house at 1700 R Street NW, at the corner of Thirteenth and R. Three stories high, positioned elbow-to-elbow with adjoining houses, it featured elaborate stone cornices at the roof, spacious windows on the street side, wrought iron fence in front, and servants' quarters in the basement. Guests were entertained in the front parlor with an oak-mantled fireplace and an upright piano. Upstairs were the home office and bedrooms.[6]

The Riley neighborhood, along with Washington in general, was in the midst of transition fueled by racism and real estate speculation. In the 1880s, Washington's African American population comprised one-third of the total, and was the most prosperous African American community in America.[7] Below the surface, however, the city was deeply divided along racial lines. In 1878, when the Reconstruction District Government foundered into bankruptcy, the white majority urged Congress to assume control. The District of Columbia Organic Act of 1878 assured the district's solvency but also curtailed local (meaning African American) participation in politics.[8] In the 1880s, Riley's neighborhood in Northwest Washington was still integrated, but the uneasy detente was eroding. By the 1890s, the Jim Crow system arrived in full force, and Washington's black population was forced into segregated neighborhoods.[9]

During the decade and a half following the family's move to Washington, Emilie gave birth to six children. Alice was born on March 12, 1879, less than a year after the move.[10] Two and a half years later, on November 26, 1881, William Garfield was born. "Willy" immediately assumed the position of

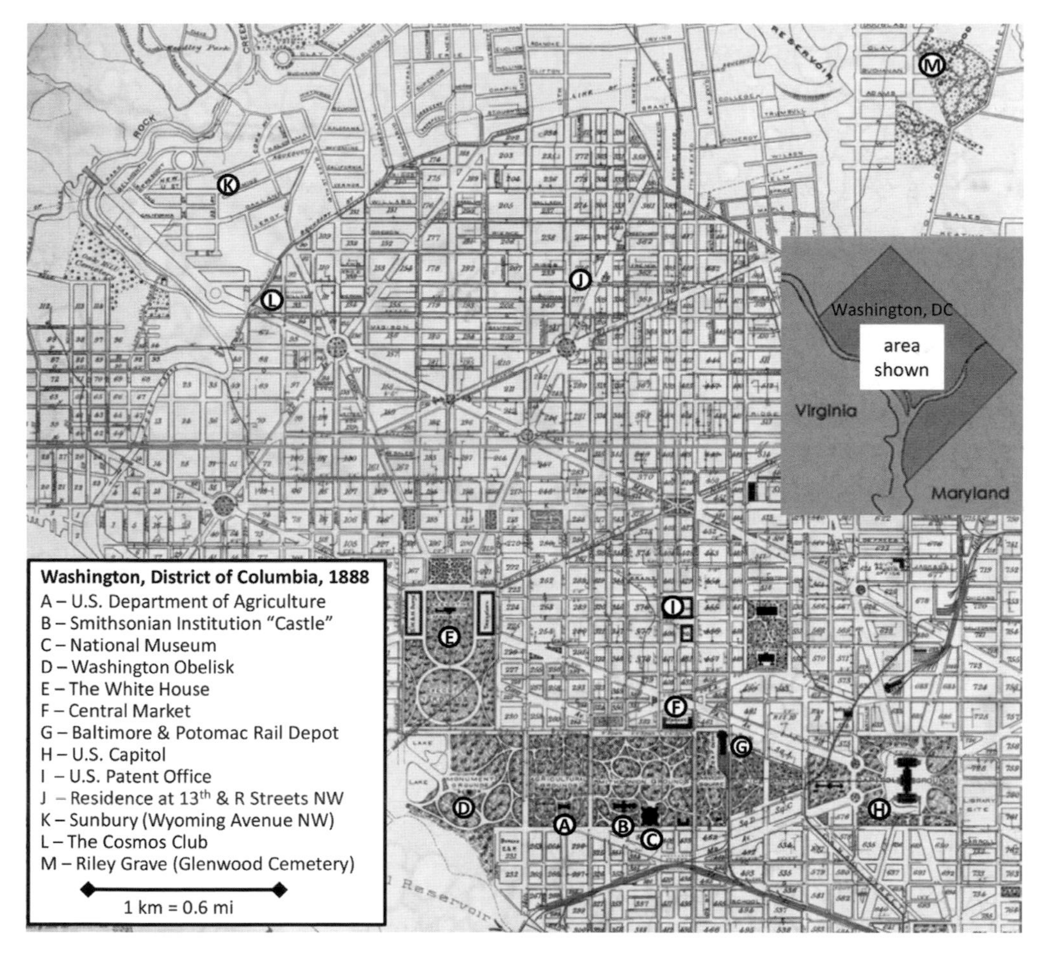

FIGURE 10.1 Washington, DC, 1880s, showing locations of Riley's activities (map prepared by Don Weber from historical sources, 2016).

"favorite son." A year and a half later, April 22, 1883, a second daughter, Mary Gene, was born. In what appeared to be a pattern of girl/boy births in cycles of eighteen months, Harold Gottlieb followed on October 21, 1884. The pattern was broken, however, when in the winter of 1884–85, all four children became ill with whooping cough and scarlet fever, and Harold Gottlieb died on March 13, 1885. Riley's spirits rose again with the birth of Helen Petrea, born on Christmas Day 1885. Two more girls followed, now at longer intervals. Thora Miranda was born September 27, 1888, and Cathryn Vedalia on January 30, 1891 (the latter named after the Vedalia Beetle).

Emilie gladly accepted the responsibility for her extended family of nine. With dark hair parted in the middle, high-button blouse fitting snugly around

FIGURE 10.2 Riley Family Portrait (1891). From left to right: Mary, Helen, Cathryn (held by Emilie), Thora (seated on Charles V.'s lap), William, and Alice (Riley Family Records, National Agricultural Library Special Collections, Beltsville, Maryland).

the neck, and plain dress extending to the ankles, Emilie fit the image of her Danish and German forebears.[11] Devoted to her husband, whom she revered for his exceptional talents, she managed her home with competence, as she had with the household of her father and two brothers.

Riley's half-sister, Nina, who moved with Charles and Emilie from St. Louis, was five years younger than Charles. Born of the unstable union of Mary Cannon and Antonio Hipolito Lafarge, Nina likely felt unwanted as a child. (Mary had two miscarriages prior to Nina's birth, and her doctor had advised her not to become pregnant again.) Nina, Charles, and George knew each other from childhood, and the two boys may have lived with their half-sister, mother, and stepfather before going to school in Dieppe.[12] According to Riley's Aunt Mira, Nina's parents lived separately, though they remained legally married.[13] Following Hipolito Lafarge's death, probably in 1871, her Aunt Mira intensified her care for Nina.[14] In 1876, when Nina prepared to leave England for St. Louis, Riley composed a poem dedicated to her:

> "One more April 27 shall pass, I hope, between the time when Nina sails,
> May smiles from Heaven play round her on the Canard Line."[15]

Charles was the protective big brother, and Nina was the adoring little sister. Shortly after her arrival in St. Louis in April 1877, Nina wrote to Mira regarding Charles: "I spell universe B.R.O.T.H.E.R."[16] Their bond was no doubt strengthened when, one month after Nina's arrival, their mother died in London.[17] Nina performed clerical duties for the Entomological Commission, assisted Emilie with the children and the household, and accompanied the family on vacation.[18] In 1887, Nina apparently accompanied Riley on a visit to England.[19] She returned to live in London in 1894 or 1895, where she was in close touch with Mira and Tim but lived at a separate address. Nina and Charles remained in close contact until his death in September 1895. Two years after his death, Nina wrote to Emilie, "When Charlie came over from Paris [in 1894 or 1895] he was delighted to find I had a fire. It was so good to see how he enjoyed it. There never was anyone it so paid to give pleasure to."[20]

As noted above, Riley's brother George was committed to the St. Louis Poor House, where he died in 1888. With the death of Riley's mother and brother and Nina's joining him in America, Riley's sole remaining family ties in England were his Aunt Mira and Uncle Tim. Grandpa Conzelman joined the family for Christmas and other holidays until his death in March 1885.[21]

Sometime in the 1880s, Riley's daughter Helen, who was fascinated by her father's English ancestry, was rummaging through the attic with a friend when they discovered a cache of family letters. There the adolescent girls discovered much to their surprise and delight that Helen's father was born illegitimately, news they immediately shared with Helen's sisters. Emilie, shocked at the revelation of family secrets, scolded the girls for prying into private matters and commanded them to keep their discovery secret. Emilie may then have destroyed potentially revealing correspondence, which would explain why only one or two letters between Charles and Emilie are extant, despite their frequent correspondence when Riley was away from home.[22]

Riley, with his "tall, athletic figure, drooping mustache, [and] broad-brimmed light felt hat," was one of the most striking personalities in Washington.[23] Many of Riley's scientist-colleagues were also his neighbors. Otis Mason, ethnologist, lived across the street from Riley; the geologists Karl Grove Gilbert and William John McGee lived a block and a half to the west; and Frederick W. True, mammalogist, lived two blocks to the south. Baird lived on Thomas Circle, six blocks in the direction of the White House; the malacologist and paleontologist William H. Dall and the geologist Powell lived a few blocks east of Thomas Circle.[24]

Riley's workday began with a walk or bicycle ride to the Agricultural

FIGURE 10.3 Sketch of Charles V. Riley from *St. Louis Globe-Democrat* (November 12, 1893), Charles Valentine Riley Papers, 1866–1895, Record Unit 7076, Box 23, Scrapbook 95 (courtesy of Smithsonian Institution Archives, Washington, DC).

Department. On rainy days or during Washington's infrequent snowfalls, the Fourteenth Street horse-drawn streetcar provided alternative transportation.[25] Along his route was the massive Treasury Building on the right, with the Executive Mansion (White House) just behind. A block to the left was the National Theater, where he sometimes attended performances or lectures.[26] Ahead were the partially completed Washington Monument, the Smithsonian Castle, the National Museum, and the Agriculture Building. The US National Museum that housed the insect collection was Riley's favorite workstation, but most of his time was spent in the Agriculture Building. Crossing B Street (now Constitution Avenue), Riley passed through flower gardens and an arboretum designed by William Saunders, the department's superintendent of grounds. Arnold Cluss, architect of the central market, designed the Agricultural Department building, completed in 1868. The three-story brick structure situated next to the Smithsonian Castle projected an air of dignity and permanence.[27]

Riley had the option of entering the main building through the massive oak front door and passing by the commissioner's office, but he normally entered by the back entrance, ascending the stairs that led directly to the entomological room.[28] Across the hall in the southeast corner of the east wing was the "Cabinet of Entomology," a small exhibit of insects, featuring examples of insect-damaged plants and insect architecture with insect drawings by Townend Glover decorating the walls.[29] The design of the public collection bore Riley's stamp. Riley's personal collection, prior to its transfer to the National Museum in 1885, was maintained in the entomological room, away from the public.

The center section of the second floor was occupied by the Museum of Agriculture. The entomologist and his staff occupied the east wing and the statistician with his staff the west wing.[30] The Museum of Agriculture, designed by Glover, featured the agricultural and manufacturing productions of the United States. Fruits and vegetables, modeled in plaster of Paris, dated back to Patent Office days. Riley might agree with featuring the nation's agricultural productions, but he may have questioned devoting most of the second floor to the entertainment of the public when problems like the Cotton Worm and the Chinch Bug cried for manpower, a working collection, laboratories, and office space. He likely also resented the allocation of space on the third floor to a taxidermist and a "modeler" who prepared birds and fruit for the museum's exhibits.[31]

Riley's visits to the commissioner's office during William LeDuc's tenure often ended in strife, whereas visits during the Loring-Colman years were harmonious. If Riley ever visited the seed division in the basement, he was likely appalled at the sight of employees sorting and packaging seed packets for distribution by senators and representatives. In 1882, the seed division moved to a separate building, where two hundred employees prepared 4.5 million packages for mailing.[32]

Always an immaculate dresser, Riley knew how to present his family and his home to good advantage, and he socialized easily after hours. When Riley transferred the Entomological Commission headquarters to his home in 1879, he remodeled three rooms on the second floor of their house as office space.[33] As noted above, the organizing meetings of the Biological Society of Washington and Entomological Society of Washington took place in his home.

The children attended Mrs. Louise Pollock's pioneering kindergarten seven blocks to the south. Mrs. Pollock followed the philosophy of Friedrich Froebel, a nineteenth-century German educational reformer who strove to balance children's natural self-expression with a judicious degree of discipline.[34] In 1895, when Alice and Mary were respectively sixteen and twelve years old, their parents sent them to a finishing school in Paris.[35] Family recreation included visits to the Zoological Park and the National Museum, concerts, weekend outings to the countryside, and special Washington events like the annual Easter egg rolling on the White House lawn. On Thursday—market day in Washington—Emilie took the streetcar to the central market, the largest and most modern food market in the country.[36]

Emilie accompanied her husband to selected social events. On November 9, 1888, the "Six O'clock Club," comprised of forty-seven prominent Washington men with their wives, celebrated its anniversary (it's not clear which one). At the Willard Hotel, the center of Washington's social scene, at the corner of Fourteenth and Pennsylvania Avenue, the assembled guests enjoyed a fine dinner, during which various members spoke on the topic, "What have you been after this summer, and did you get it?" Among the speakers were Adolphus W. Greely, chief of the Army Signal Service, Lester Ward, head of the Division of Fossils at the Geological Survey, William T. Hornaday, taxidermist and curator of living animals at the National Zoo, and Riley.[37]

Riley's finances were based on his salaries as chief of the Entomological Commission and the Division of Entomology, Emilie's dowry and inheritance, profits from real estate sales and rentals, and honoraria derived from his illustrations, articles, and lectures. In 1877–78, his first year in Washington, Riley's allotment of $5,000 as chief of the Entomological Commission, on paper, exceeded that of the commissioner of agriculture, who drew a salary of $3,000.[38] The $5,000 allotment, however, covered expenses for assistants, equipment, office rental, travel, etc., which reduced Riley's actual salary to roughly the level of the commissioner. From about May 1878 to May 1879, Riley served as both chief of the Entomological Commission and as federal entomologist. That year he drew salaries for both offices, which, after deducting expenses for the Entomological Commission, amounted to considerably more than the commissioner. Such double appointments, though unusual, were legal and accepted. Powell, for example, held concurrent positions, with salaries, as director of the Bureau of Ethnography and director of the Geological Survey.[39] When LeDuc and John Comstock challenged Riley's right to salaries for two positions, the first comptroller ruled in Riley's favor, pointing out that there was no statute prohibiting dual appointments.[40] Following the first two fat years in Washington came two lean years, when funding for the Entomological Commission shrank and Riley had to subsidize some commission assistants from his own pocket.[41] Following his reinstatement in 1881, Riley's salary returned to a comfortable plateau of somewhere around $2,500. Considering that salaries in the Agricultural Department, from the commissioner down, lagged behind those of other departments, Riley had reason to be satisfied.[42]

Riley had even more reason to be satisfied with the windfall of Emilie's inheritance. Father Conzelman's largesse allowed the newlyweds to pay cash for their house. Other government officials, like Powell, had to finance up to half of their house purchases with mortgages.[43]

Anticipating profits from Washington's real estate boom, the Rileys drew on Conzelman wealth in 1882 to purchase lots and to construct a row of six townhouses on Thirteenth Street at a total cost of $25,000.[44] Although some rental units remained unoccupied for a time, the investment apparently paid well. Two years later, in 1884, they constructed three additional houses on R Street, between Thirteenth and Fourteenth Streets, all designed by the up-and-coming architectural firm of Hornblower and Marshall. In 1885, the Rileys constructed two additional residences on Thirteenth Street, designed by the same firm.[45] In 1883, one of the houses on R Street was valued at $4,800.[46] Five years later, in 1889, one of the houses on R Street was offered for $8,800, double the value in 1884.[47] The speculative bug seems to have bit Riley. Sometime in the mid-1880s, he purchased two lots in Los Angeles, the hot spot of real estate speculation on the West Coast. Riley described the property as "very choice!" He toyed with the idea of constructing a winter residence in Los Angeles but abandoned this in favor of constructing an estate in Washington's suburbs. In 1887, he offered the two California lots for sale at $4,000 per lot.[48] By 1889, the Riley investment in real estate development totaled around $40,000, or approximately ten times the annual salary of a cabinet-level official.

Riley enjoyed professional pride, if not substantial financial gain, for his teaching, writing, and what amounted to royalties for use of his Missouri-era insect illustrations. As state entomologist, Riley received modest honoraria for articles in farmers' almanacs, encyclopedias, and other publications. As chief of the Entomological Commission and the Division of Entomology, Riley continued to write articles for encyclopedias, and he served as entomological editor for the *Rural New Yorker*, the *American Agriculturist*, and the *American Naturalist*.[49]

As previously noted, Riley taught courses at colleges in Kansas and Missouri. In 1873, Riley proposed to Charles Bessey that he teach a course on economic entomology at Iowa Agricultural College, suggesting six lectures at fifty dollars each, nine lectures at forty dollars each, or twelve lectures at thirty-five dollars each.[50] In 1880, when the Entomological Commission was financially strapped and Riley's reinstatement as federal entomologist was in doubt, Riley proposed the establishment of a combined state entomologist and chair of entomology at Iowa Agricultural College with himself as head.[51] Following his reinstatement as federal entomologist, Riley found little time for teaching; however, in 1891–92 he delivered the Lowell Lectures in Cambridge, Massachusetts, and in 1895 he delivered a series of eight lectures at the University of Missouri at Columbia.[52]

Like payment for articles and teaching, Riley's income for the use of his insect illustrations was primarily a source of professional pride and satisfaction. Early in his tenure as Missouri state entomologist, Riley began selling electrotypes of his woodcut illustrations to colleagues for their use in publications. In October 1873, for example, he sent Bessey a bill for nine electrotype figures "at the usual rate," for a total of $36.75, or about $4.00 per figure.[53] Some of Riley's colleagues, however, duplicated and resold his illustrations at reduced rates, though they had purchased the illustrations on the condition that they use them for specific publications.[54] During his tenure as chief of the Division of Entomology, Riley continued selling electrotypes of his early illustrations. Because the appropriation for Missouri state entomologist contained no separate items for illustrations, equipment, assistants, postage, or other expenses, he considered his illustrations as his personal property.[55] His practice of selling Missouri-era illustrations was never questioned and, as federal entomologist, he consistently maintained that anything produced by federal employees belonged to the government. Leland O. Howard continued this policy.[56]

Riley's correspondence in 1887 provides a glimpse into his day-to-day administrative and financial activities. While on vacation in Gloucester, Massachusetts, Riley instructed Howard to pay the Brown Stone Company and to inform Mrs. Riley that he had done so, to instruct Theodore Pergande on how to proceed with museum assignments, to keep a daily journal of activity at the division, to get examples of insecticide implements and remedies, and to send silkworm eggs to Miss Carrie Klingel in St. Louis.[57] A few days later, Riley wrote to J. C. Pearsall at the City National Bank of Washington, DC, enclosing a note for $500 for payment on the Saverly note (apparently a second mortgage on some property) in case his account should be overdrawn before his return to Washington. Incredibly, Riley traveled from Gloucester to Washington for *one day* to take care of personal financial and administrative

FIGURE 10.4 Sketch of House for Professor C. V. Riley, Washington Heights, by Hornblower & Marshall, Architects, March 30, 1889 (courtesy of the National Agricultural Library Special Collections, Beltsville, Maryland).

matters before taking the train to New York, arriving a few days prior to his planned departure for Europe.[58] From his New York hotel room, Riley instructed Howard to take his checkbook from his desk drawer in the office and deliver it to Eugene Schwarz with instructions to place it in his desk at home, to send him a statement of his bank balance with a list of checks and deposits, and to deliver the check for Otto Lugger to his wife, who would turn it over to him.[59] Meanwhile, Riley corresponded with bankers in Washington and New York regarding mortgages and other matters, and he authorized David W. Coquillett in Los Angeles to offer two lots for sale.[60] Amid this tangle of personal and administrative detail, Riley negotiated with Comstock and J. D. Putnam to schedule him for a talk at the Entomological Club when the AAAS met in New York.[61] Riley delivered two papers at the AAAS and even joined an excursion by ship along the New Jersey coast on the very day he was scheduled to sail for Europe.[62]

Despite Riley's chaotic administrative and financial style, the family prospered, making possible construction of a new house in Washington Heights, an exclusive real estate development on high ground within the northwest quadrant but just outside the city's historical northwest boundary.[63] Financing the house required some consolidation of assets, including the sale of the Riley residence at Thirteenth and R.[64] Construction proceeded during Riley's absence, while directing the American Agricultural Exhibit at the Paris Exposition in 1889–90.[65]

The house plans by Hornblower and Marshall depict a frame structure faced with brick and stone, with an elaborate assemblage of towers, turrets, dormers, and porches on a jigsaw ground plan of bays, angles, and curves, with nary a straight line. Employing a Romanesque Revival architectural style, Hornblower and Marshall followed the lead of the prestigious architects McKim, Mead, and White, who designed houses for wealthy patrons in resort communities along the East Coast. Harking back to English rural forms, the rambling, irregularly angled, multi-gabled compositions were intended to convey age and permanence.[66] Riley considered naming the residence "Vedalia," in honor of the lady beetle that had recently controlled the Cottony Cushion Scale but chose instead the name "Sunbury" after the English estate near the village of Walton-on-Thames of his childhood. In November 1890, the Rileys moved into their new home; two months later, Emilie gave birth to Cathryn Vedalia.

In Washington, Riley's health and vigor deteriorated dramatically. Stamina and concentration that once had come effortlessly now failed at critical times. Riley's health problems began in Missouri, but in Washington he suffered a mounting cycle of nervous agitation, upset digestion, headaches, insomnia, and exhaustion.[67] These problems were no doubt aggravated by strife with LeDuc and financial uncertainty during his separation from the Agricultural Department. Once reinstated, Riley buried himself in his work, not taking a vacation for the first three years. In 1884, Loring granted him permission to travel to Europe. In Montpellier, he spoke on phylloxera; in Lyons, he gathered information on silk spinning technology, and in Edinburgh he participated in the International Forestry Convention.[68] The break from office pressure and the change of surroundings provided only temporary relief. Back at work, Riley confessed to Lintner, "I am a badly used up man . . . I may have to give up entirely before long."[69] On his doctor's advice, in 1886 he secured Colman's permission to take a three-month leave without pay and embarked on a trip to Europe; however, in Europe he again busied himself with phylloxera and silkworms, and he corresponded regularly with Howard regarding departmental matters.[70]

Travel to Europe or to stateside resorts provided Riley's sole relief from mounting ills. In 1887, he spent four months in Europe. From Paris he wrote

FIGURE 10.5 Sunbury, the Riley family residence, ca. 1891 (Riley Family Records, National Agricultural Library Special Collections, Beltsville, Maryland).

to Howard, "My health for the most part is perfect and I sleep on an average some six hours nightly . . . but I have relapses of the old trouble whenever my spirits tempt me to do too much, and at times the over active, throbbing, wakeful singing brain, discourages me and makes me almost wish I had none."[71] Riley's sudden disappearances created stress at the office and at home. In January 1894, he left suddenly for the Hygeia Hotel in Fortress, Virginia. Eleven-year-old Mary, distraught that she had not been able to say goodbye, wrote that she hoped he would get plenty of rest.[72] Riley's absences provided only temporary relief. The headaches and associated pains increased relentlessly. In January 1888, he confided to Baron Carl Robert Osten Sacken that he might be forced to give up his position.[73]

Riley was uncomfortable with orthodox religion.[74] Although he and George were baptized as infants in the Anglican Church, there is no indication that they attended church as youngsters. In Kankakee and Chicago, Riley attended church services sporadically. In St. Louis, he came under the influence of Emilie and her father, who were both staunch Unitarians. Gottlieb Conzelman had rejected orthodox Lutheranism prior to immigrating to the United States.[75] Riley presumably accompanied the Conzelmans to the Church of the Unity in St. Louis. Reverend J. C. Learned announced proudly that his church welcomed the views of such heterogeneous thinkers as Humboldt, Agassiz, Gray, Washington, Jefferson, Emerson, and Longfellow.[76] In Washington, DC, the Rileys were associated with a Unitarian-style congregation called the Peoples' Church, whose pastor later conducted Riley's funeral.[77]

Like most of his white contemporaries, Riley assumed that human beings were divided into a hierarchy of races in which the Anglo-Saxon race presumably had the greatest potential for improvement, while African Americans, American Indians, and others had exhausted their evolutionary potential.[78] On occasion, he attempted to imitate African American dialect when ridiculing uninformed notions about insects.[79] Riley never went out of his way, however, to offend African Americans or other racial or ethnic groups. Although he worked closely with congressman Van H. Manning, who was notorious for his racist views, he did not share Manning's racism. On the contrary, he related to African Americans and other minorities on a personal, human level. For example, a black porter who accompanied Riley and others on an excursion to the Rocky Mountains in 1873 said of Riley, "Smart fellow . . . he knows more about bugs than any chap in these yer [*sic*] diggings! If he can't tell a caterpillar from a chip-monk, then you can scalp *me*!"[80] The Division of Entomology report for 1883 depicts a young, neatly dressed African American operating an insecticide sprayer, a possible indication that Riley considered African Americans valuable members of society but not destined for the managerial elite.[81] Riley's association with Isidor Bush, a prominent vintner and Jewish scholar, indicates his acceptance of Jews.[82]

With respect to the use of liquor, tobacco, and other "sins of the flesh" that agitated his generation, Riley practiced moderation. In his Chicago youth, Riley abstained from drinking alcoholic beverages in accordance with his Good Templar pledge to abstain from the production, sale, or consumption of alcoholic beverages. Riley later recalled, however, that he took his first drink around 1865 at age twenty-two, that is, during his Templar years.[83] Nevertheless, Riley apparently remained committed, at least in principle, to Templar-style abstinence even after moving to St. Louis. In a lecture given during his first year as state entomologist in which he encouraged young men to study entomology, Riley closed his talk by saying that those who studied insects would have no need to frequent saloons.[84] In St. Louis, Riley did not join a Good Templar lodge, although the Templars were active there (the sixteenth international session was held in St. Louis in 1870).[85] As noted above, in St. Louis, Riley came to approve of moderate social drinking. At the AAAS meeting in Dubuque in 1873, when Riley's colleague George Swallow, who supervised the university's vineyard and wine-making facilities, delivered a paper entitled "Good Wine a Social and National Blessing," a delegate named Dr. Palmer, who described himself as a "claw hammer" prohibitionist who represented the temperance society, attacked him vigorously. Riley was torn between his loyalties to Swallow and to Charles W. Murtfeldt, secretary of the Missouri State Board of Agriculture, who was a strict prohibitionist. For once, he took no position.[86] In later life, he occasionally drank wine, sherry, whisky, or brandy in moderate quantities. He disliked American beer but occasionally, when in England, he drank beer and ale.[87] Emilie agreed with Charles's tolerance of alcohol in moderation. According to family tradition, she allowed sherry but nothing stronger in the Riley house.[88]

Riley accepted the use of tobacco, and he smoked Cuban cigars of the brand Reina Victoria, named in honor of Queen Victoria.[89] Riley's proposal for double-decker buses in St. Louis included a separate men's smoking compartment. Both the Cosmos Club and Riley's residence provided separate smoking rooms.[90]

Riley found little time for recreational pastimes, but he did have one consuming passion besides entomology: flying machines. Riley recalled how, as a lad, he constructed mechanical toys propelled with rapid screw propellers that ascended to the ceiling of his room and remained there until their source of propulsion (probably rubber bands) was exhausted. As a youth he studied the flight of birds and insects, and as federal entomologist he compiled an extensive file on heavier-than-air flight.[91]

By the 1890s, heavier-than-air flight proponents had coalesced into two schools, one associated with Samuel Pierpont Langley, secretary of the Smithsonian Institution, and the other with English-based American-born Hiram S. Maxim, inventor of the mousetrap and the Maxim gun. Whether Riley consulted with Langley on flying is not known; however, he was an intimate of Maxim. In August 1894, when Riley was attending the British Association for the Advancement of Science in Oxford, England, Maxim invited him to visit him at Baldwyn's Park, Baxley, Kent. Soon thereafter, Riley reported Maxim's experiments and offered his own views on flying in an article in *Scientific American*. Riley described the recent ascent of Maxim's machine, which, although it flew only a few hundred feet, was one of the first documented ascents of heavier-than-air vehicles. Maxim's immense structure weighing eight thousand pounds, manned by three men, and powered by gigantic steam boilers, ascended briefly but came crashing down when one the boilers exploded, seriously injuring the crew members. Maxim, sobered but undaunted, explained to Riley that the machine was being repaired for the next flight. Riley closed his account with his own suggestion for an alternate design. Riley contended that both Langley and Maxim erred by basing their concepts on bird's flight, which relied on inclined planes and rapid motion. He proposed to imitate fish that swam freely upward, downward, and forward with a minimum of effort. Riley's fish-shaped vessel (of which no drawings are known) would be buoyed upward by gas-filled compartments and thrust in various directions by propellers. In 1894, Riley was reported to be negotiating with a stock company to finance his design. When he died a year later, nothing came of his proposal.[92]

In September 1888, having just celebrated his forty-fifth birthday, Riley prepared his will and testament.[93] He may have been prompted by his deteriorating health to dictate his will. A year earlier, he indicated to Colman that he might soon die, and he didn't care whether he went "up or down." A year later, in 1889, in his presidential address before the American Association of Economic Entomologists, he told his listeners, "[I have] the feeling that my own labors . . . are . . . about to end."[94]

In his will, Riley expressed his wish to be buried in the family plot in Glenwood Cemetery "with as little expense and . . . ostentation as possible."[95] Riley bequeathed most of his personal and real property to his "dear wife," and he entrusted her with the care of their "beloved children." To his half-sister, Nina, he willed certain unspecified "personal property" plus $10,000 and two lots with improvements (houses) at 1303 and 1305 R Street.

To George B. Goode, director of the National Museum, Riley willed all entomological material, specimens, and collections not already included in the insect collection he donated in 1885 (see chapter 13). To Howard he bequeathed his books and pamphlets on Hymenoptera as a "trifling token of friendship and long official association," and to Schwarz he gave his books and pamphlets on Coleoptera.

The most surprising beneficiaries named in Riley's will were Harry Wheatley, Riley's boyhood friend, now living on Church Street, Walton-on-Thames, who was to receive $1,000, and Riley's "ward," Charles Fewtrell Wylde, currently a pupil in a private school in Southborough, Kent, who was to receive $5,000 upon his coming of age in 1900.[96] It is noteworthy that Riley honored the memory of his biological father, whom he probably never knew personally, by assuming responsibility for his "ward," who had his father's name. Riley requested that Harry Wheatley and Nina support Wylde's education up to his majority, with any extra expenses involved in his education to come from the bequest made to Nina. The prominent place of Nina, Wheatley, and Wylde in Riley's will reflects Riley's lifelong ties to his English roots. His Aunt Mira and Uncle Tim were presumably well-to-do and therefore not included in the will.

Finally, Riley named Theophilus Conzelman, Emilie's brother and director of the Conzelman family enterprises in St. Louis, as guardian of his children, should Emilie die while the children were in their minority. Emilie was named "sole executrix," in charge of executing the will. Riley may have had premonitions of eminent death, yet he still had much ahead professionally, including at least one stunning success.

II

"A Great Big Silk Farm"

There can be no good reason . . . why silk-culture may not become one of the industries of this country.

—C. V. Riley, *Fourth Missouri Report* (1872), 138.

Two prominent programs during Riley's tenure as federal entomologist were his campaign to promote the silk industry in the United States and his victory over scale insects that threatened the California citrus industry. Addressing beneficial silkworms and pernicious scale insects both called for international scientific cooperation and foreign language skills, two of Riley's strengths; however, the two campaigns ended with dramatically different results. Silk culture, although heavily subsidized, limped along for more than a decade before being abandoned as a total failure, while the brief and minimally financed scale insect campaign brought about the dramatic suppression of the Cottony Cushion Scale through the introduction of its natural enemy, the Vedalia Beetle.

As Missouri state entomologist and federal entomologist, Riley was America's most prominent advocate of silk culture.[1] Raising silkworms had been a favorite boyhood pastime, recorded in his sketchbook from Dieppe.[2] In Chicago, Riley learned that the Ailanthus silk moth, a large and attractive moth native to China, had been introduced in France and England with the intention of expanding the silk industry.[3] Introduced into the United States in 1861, the Ailanthus silk moth soon established populations around Baltimore, Philadelphia, New York, and Chicago. The lepidopterist John G. Morris in Baltimore, along with commissioner of agriculture Isaac Newton and his

FIGURE 11.1 Polyphemus Moth (Riley's American Silkworm), a native saturniid (giant silk moth) of interest for alternative silk sources. Produced by Riley when the moth appeared in the Chicago vicinity; scanned by Donald C. Weber, October 2016, from *American Entomologist* 1 (March 1869).

entomologist Townend Glover, recommended the Ailanthus Silkworm as an alternative to the mulberry silkworm. In 1866, Riley performed an experiment that demonstrated the remarkable ability of Ailanthus silk moth males to find the female by means of scent. Having raised Ailanthus moths from eggs, the species not yet having established itself around Chicago, he placed a female on a tree and released a male, marked with notched wings for identification, a mile away and downwind from the female. The next morning the notch-winged male had joined the female on the tree.[4]

In the following years, Riley obtained eggs of various native and foreign silkworms—including the true silkworm, *Bombyx mori* (L.)—from the Agricultural Department, from colleagues in Europe, and his own scouting of the countryside around Chicago and St. Louis. From these, he raised silkworms for experimentation to determine their commercial potential as silk producers. By 1871, Riley was convinced that Americans, like Europeans, could rear silkworms commercially.[5] While in Europe that summer, he investigated silk manufacture at Lyon, the center of French silk production.[6]

In his fourth Missouri report (1872), Riley argued that despite numerous failed attempts since colonial times, conditions now favored silk culture.[7] In support of his argument he noted, (1) that women, now freed from spinning and weaving wool, could supplement family incomes with silk culture; (2) that

Featured Insect

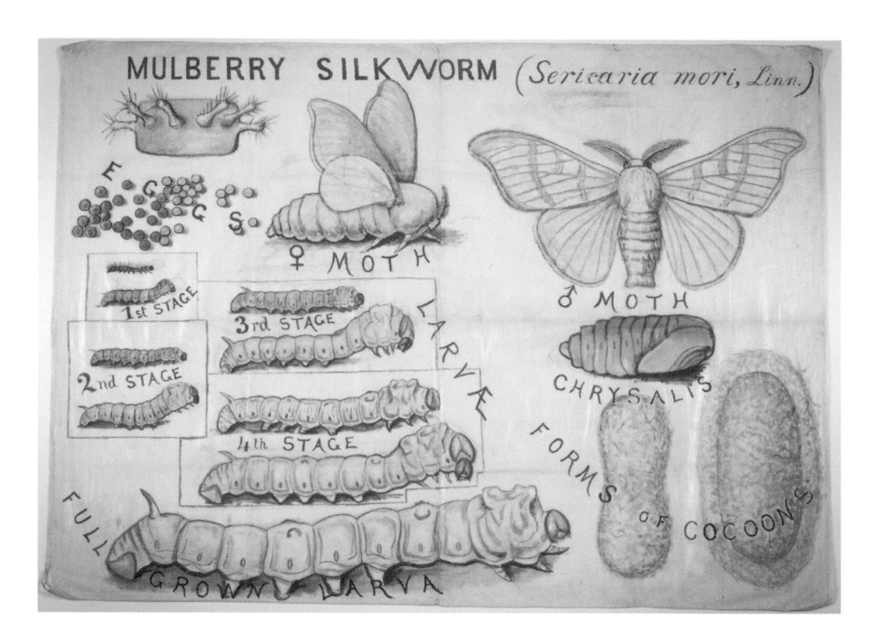

Silkworm [*Bombyx mori* (L.) (Lepidoptera: Bombycidae)]. Riley Insect Wall Chart (Courtesy of Morse Department of Special Collections, Kansas State University Libraries). The failure of silk production in the United States in Riley's time has continued to the present day. Synthetic fibers, starting with the semisynthetic Rayon in 1904, supplanted much of the potential market for true silk. However, silk is still popular, and silk production has doubled in the past twenty years to over 600,000 metric tonnes worldwide. Production is almost exclusively (over 99 percent) in Asia, with China (66 percent) and India (26 percent) the leading producers, and Brazil, Romania, and Egypt the only significant other producers, all less than 0.5 percent of world production, outside of Asia.* Three species of giant silk moths (family Saturniidae), including Riley's "American Silk Worm" [old name *Telea polyphemus*; current name "polyphemus moth," *Antheraea polyphemus* (Cramer)], are sources for silk in negligible quantities, compared to *Bombyx mori*. Many other insects spin silk, most of them as part of the pupation process, and of course spiders spin silk for various purposes, including prey capture. Silk of all sorts remains a source of fascination for its unique qualities, and as an aspiration for synthetic chemists and genetic engineers.

*FAOSTAT 2016, latest statistics on unreeled cocoons from 2013, http://faostat.fao.org/ (accessed October 31, 2016).

American ingenuity could overcome the European and Far Eastern advantage of low wages paid to silkworm workers; (3) that the current importation of spun silk indicated a strong domestic market for American-grown silk; (4) that the construction of intercontinental railroads facilitated the importation of silkworm eggs from the Far East; (5) that America's mild climate favored silkworm production; and (6) that numerous indigenous and acclimated silkworm species already flourished in the United States. Also, Congress had signaled its willingness to establish protective tariffs for domestic industries like silk culture by its protection of sorghum sugar producers.

Riley cautioned that America's high wage scale prevented direct competition with silk producers in Europe and Asia; however, a 10 percent tariff on "raw silk" (silk thread that had been "reeled" or spun, from cocoons) would allow American producers to supply silk for the domestic market. The central problem, Riley declared, was the lack of both trained operators to unwind the fine thread from cocoons and recombine these into strands of silk thread ("raw silk") and the reeling machines for them to operate.[8] In Europe, female specialists operating steam-powered machines carefully monitored the silk filament (measuring up to one thousand feet for a single cocoon) as it was unwound, and then combined these filaments into strands of ten or more that constituted finished ("raw") silk thread. Riley argued that "superior [American] intelligence and the advantage of [a superior] climate" would offset high American wages compared to Europe and Asia.[9]

Riley identified eight moth species extant in North America whose cocoons contained usable silk filaments, of which only one, *Bombyx mori*, the traditional silkworm, was currently utilized for silk production. Riley contended that investigation of the other indigenous and introduced species in America would indicate those species that could be improved through the application of "Darwinian principles" in their selection and breeding.

Riley's reference to "Darwinian principles," based probably on his recent review of Darwin's *The Variation of Animals and Plants under Domestication*, confused the distinction Darwin made between selective breeding (which had been practiced for centuries) and the production of new species through natural selection. Darwin, in fact, derived his initial concept of selection from farmers' selective breeding of "domestic productions."[10] Darwin's theory that similar selection in nature (natural selection) had produced new species required long stretches of geologic time, a concept he derived from Charles Lyle's *Principles of Geology* (1830–33). Riley pointed out, correctly, that selective breeding of *Bombyx mori* over four thousand years had produced the world's only domesticated insect: the true silkworm was now unknown in the wild.[11] However, this was far too short a time to produce new species of silkworm moths. His assertion that "Darwinian principles" could guide American silkworm breeders in the selection and breeding of native and foreign silkworms to produce more and better quality silk was not really "Darwinian" but a restatement of the traditional practice of selective breeding.

It is unclear whether Riley attempted to hybridize different species of silk-producing moths himself, but he consulted at least one person who did.[12] Étienne Léopold Trouvelot, an astronomer employed at Harvard and living in Medford, Massachusetts, was one of America's leading silkworm experts in the 1860s and 1870s. A colleague of Louis Agassiz, Samuel H. Scudder, and Alpheus Spring Packard, he became famous in the 1870s for his illustrations of the heavens as viewed through the telescope at the Harvard Observatory. In 1869, while Trouvelot was working with Gypsy Moths, a small number of the insects escaped near his Bedford, Massachusetts, home. Trouvelot knew enough about Gypsy Moths at that point to be concerned about their escape and reported the breakout to local authorities, but no one seemed concerned about a few tiny caterpillars. By 1881, the year Trouvelot returned to France, the moths had quietly multiplied until they reached outbreak numbers.[13]

Trouvelot shared Riley's conviction that Americans could utilize native and introduced silkworms. During the Civil War, when the supply of cotton for New England mills was cut off, Trouvelot experimented with *Telea polyphemus*, a large brightly colored native American moth that he called the "American Silk Worm," now called the Polyphemus Moth. By 1865, Trouvelot claimed to have solved the problem of breeding Polyphemus Moth in captivity and to have developed an efficient method for reeling silk from its cocoons. On a five-acre woodlot behind his house, enclosed by an eight-foot wooden fence to prevent caterpillars from escaping, Trouvelot bred (according to his own account) over 1 million of these caterpillars. The silk from his Polyphemus Moths, however, was still inferior to that of the true silkworm.[14] Somehow, he had to improve his stock.

At a meeting of the Boston Society of Natural History in 1867, Trouvelot announced that he had developed a method for interbreeding different species of moths. He confined two males and one female of the same species together with one female of an allied species at the appropriate time for mating. Following copulation by a male and female of the same species, he said, the remaining male and female of different species were induced to mate.[15] Unlike

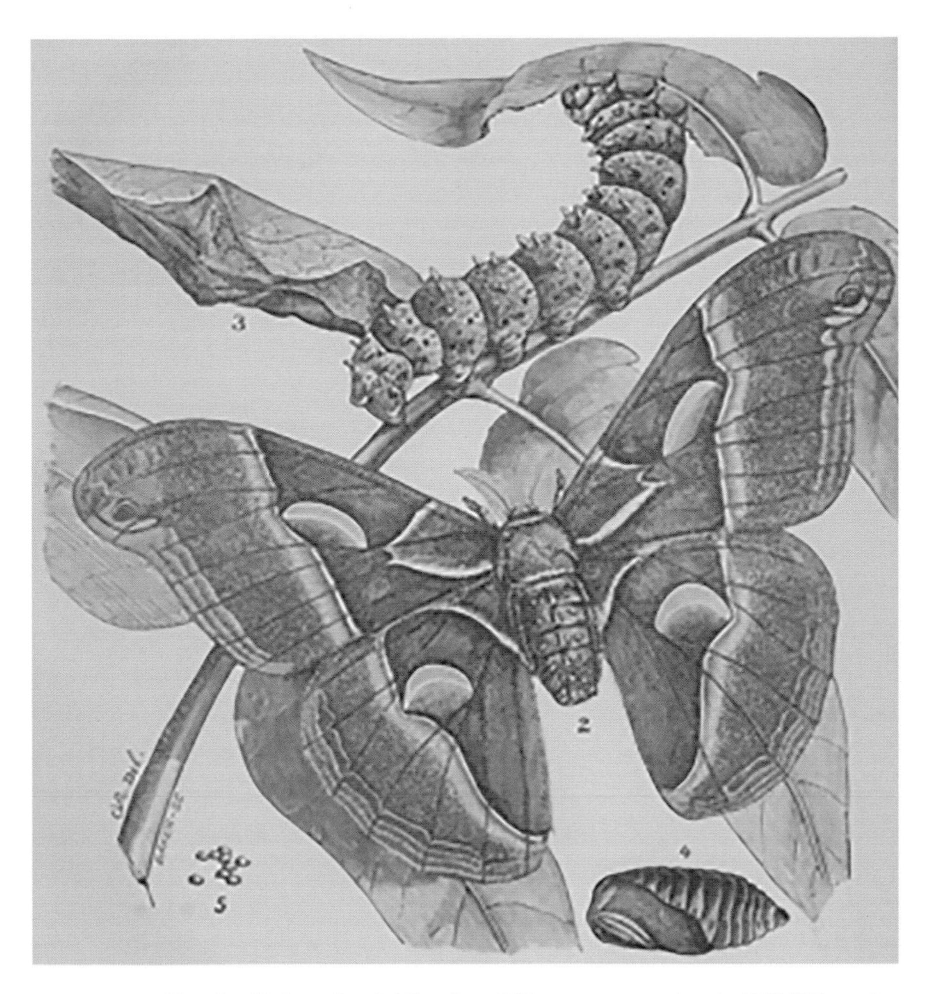

FIGURE 11.2 Sketch of life cycle of Ailanthus Silkworm, appearing in C.V. Riley, *Annual Report of the Missouri State Entomologist* (1872), v. 4, between pp. 112 and 113.

Riley, he made no reference to Darwinism. In 1870, most of Trouvelot's caterpillars died of disease and he turned back to astronomy, but he shared his knowledge of silkworm breeding with Riley.

Trouvelot's Polyphemus Moth cocoons contained hard sticky matter that made reeling difficult, but Riley expressed his hope that selective breeding would correct this deficiency.[16] Riley preferred *Attacus* [now *Hyalophora*] *cecropia*, another large indigenous moth whose worm was a voracious feeder.[17] Originally confined to the American interior, the Cecropia Moth was now being successfully raised on the Atlantic coast. Citing heretofore unreported larval changes, Riley regarded this modified form of Cecropia Moth as an example of rapid evolution. Based on its potential as a silk producer (even though no silk had actually been produced from its cocoons), Riley classified the species as "beneficial," though in the wild it sometimes multiplied to the extent that it defoliated fruit trees.[18]

Among the introduced silkworms, Riley first choice was the *yamamai* silkworm, the caterpillar of the Japanese oak silk moth that Japanese breeders had long utilized for silk production.[19] Riley had high expectations for the Japanese oak silkworm because of the vast variety of North American oak species. Unlike the Ailanthus Silkworm, however, it did not thrive in North America, and Riley's efforts to breed it in captivity were only partly successful. Nevertheless, Riley hoped that selective breeding would improve the species, and he urged Glover and Commissioner Horace Capron to acquire eggs from Japan for further experimentation.[20]

Riley's second choice of introduced species was the *pernyi* silkworm (now called the Chinese tussar moth, *Antheraea pernyi*), another oak-feeding Asian species that had been introduced in France and England in the 1850s and later in the United States.[21] Riley reported that Alexander Wallace, a correspondent in England, had sent him some *A. pernyi* eggs in 1869, but they all died. Riley rated it as the third most promising of the eight "American" species, after the true silkworm and *A. yamamai*.

Riley admitted that the three exotic established species (*A. yamamai*, *A. pernyi*, and the Ailanthus silk moth) had only limited potential until they were improved through selective breeding. On the other hand, he saw great potential in Osage Orange, a vigorous tree in the mulberry family that was native to the American Midwest, as a source of food for silkworms. This hardy, fast growing tree was, by midcentury, familiar in the Midwest, where it was planted in hedgerows. The tree produced large, rough-skinned grapefruit-sized fruit, hence the common name Osage Orange. Introduced in France in 1820, the leaves were fed to silkworms, but with limited success. In 1870–71, Riley experimented with Osage Orange as larval food for the true silkworm and reported obtaining a large number of cocoons. In 1872, Samuel Cornaby of Spanish Fork City, Utah, informed Riley that he had raised up to fifty thousand silkworms on an exclusive diet of Osage Orange.[22] Riley reported, however, that the worms Cornaby sent him belonged to a Syrian race he had seen years before in Covent Garden Market, London, a race that had no commercial value because of its inferior silk. Nevertheless, Riley took some of these cocoons to France in 1871 to demonstrate the potential of Osage Orange as a food source. Osage Orange was more frost-hardy than mulberry, its leaves resisted wilting, and it produced an abundance of foliage. Riley admitted that worms fed on Osage Orange produced inferior cocoons and that the premium races of domesticated silkworms did not thrive on an exclusive diet of Osage Orange. He hoped, nevertheless, that through selection the Syrian race could be improved, and he anticipated the day when an improved race would be bred outdoors on Osage Orange hedgerows.[23]

Interspersed with his writings about the biology of silkworms, Riley related various American success stories. In 1866, the Oneida Community in western New York State employed one hundred fifty silk operators and reported a profit of $25,000 in its first year.[24] By the early 1870s, the number of looms in and around New York City had increased from fifty to five hundred, surpassing Boston, the traditional American center of silk production. In Paterson, New Jersey, the Dale Manufacturing Company had recently built the largest American silk plant, where three hundred fifty employees operated the latest equipment from Lyon to produce silk braids, cord, and dress trimmings.[25]

California presented a special case. In the years 1860–68, California silkworm breeders reaped enormous profits through the sale of silkworm eggs to Europe, where the silkworm disease *pébrine* had decimated the silkworm population.[26] In San Bernardino County, Louis Prevost, a French immigrant, produced silkworm eggs in large quantities on a ten-thousand-acre mulberry orchard. In 1865–70, Louis Pasteur identified the microorganism that caused silkworm disease and prescribed sanitary measures that allowed French and Italian producers to supply Europe with disease-free eggs. Simultaneously, silkworm disease appeared in California. At the time Riley wrote in 1872, the once-thriving California silkworm egg industry was in ruins.[27]

In the Midwest, Riley found more promising activity. In 1869, Ernest Valeton de Boissière, a wealthy French philanthropist, purchased 3,500 acres in southeastern Kansas, where he planned a cooperative colony based primarily on silk culture. When Riley visited "Silkville" in November 1871, most of Boissière's holdings were still devoted to cattle grazing, but he was experimenting with broom making, canned meat, sorghum syrup, castor oil, potato starch, Morocco leather, and safety matches. Silk culture, his primary project, was beginning to take shape. Some eight thousand mulberry trees furnished food for the growing silkworms. Boissiere anticipated that silk production would commence in two years. In the meantime, members of the cooperative had woven silk ribbons and trimmings from Japanese raw silk on looms imported from France.[28]

As state entomologist, Riley lacked a legislative mandate or executive agency to implement his proposals.[29] This changed in 1878 when he replaced Glover as entomologist in the Agricultural Department. Now speaking as federal entomologist, he outlined his concept at the AAAS meeting in St. Louis in August 1878.[30] That year he issued a manual on silk culture, experimented with Osage Orange as silkworm forage, and called on Congress to appropriate funds for reeling stations.[31] Perhaps at Riley's suggestion, William LeDuc urged Congress to pay a bounty of fifty cents per pound for American-produced cocoons as a means of improving rural life.[32] LeDuc ordered silkworm eggs from Japan that Riley and his assistants hatched and raised, then distributed the worms free of charge.[33]

While waiting for funding from Congress, Riley outlined three options for those interested in silk culture: produce cocoons for sale in Europe, reel silk from the cocoons raised at home, or produce eggs for sale. Regarding the second alternative, Riley warned that only organized communities with skilled personnel like those at Silkville, Kansas, the French-Italian colony at Fayetteville, North Carolina, or the Italian settlement at Vineland, New Jersey, could operate reeling stations capable of competing with Europe.[34] His third alternative, to produce eggs for sale in Europe, is puzzling. He estimated that a married couple raising silkworms for the production of eggs could earn $900 per year. Considering that the average midwestern farm income was $700 annually, this was certainly appealing.[35] Yet Riley noted earlier that *pébrine* disease had rendered silkworm eggs from California unwelcome in Europe. Riley claimed that, in 1877, buyers in France had paid 1,691,400 francs for eggs from the United States, but he cited no examples of Americans who had made profits from the sale of eggs.[36]

Following Riley's removal as entomologist in May 1879, Comstock inherited the silk project, but he displayed little enthusiasm for it. In his report for 1879, he noted that silkworm eggs ordered from France had hatched and died during transit, thus frustrating any attempt to experiment or to supply applicants.[37] Meanwhile, Riley, as chief of the Entomological Commission, continued to speak and write in support of silk culture. His silkworm manual that he distributed during Comstock's tenure served to keep silk culture in the public eye.[38]

By the time Riley was reinstated in 1881, popular enthusiasm for silk culture was sweeping the country. Fueling the boom was national economic expansion, the accelerated mechanization of textile manufacturing that freed women from spinning and weaving wool, women's fashions that featured silk garments, and a federal budget surplus that could be tapped for the promotion of silk. In addition, Russian Mennonite immigrants with practical experience in raising silkworms established centers of silk production in the Midwest and West.[39]

Silk cooperatives, associations, and for-profit enterprises sprang up across the country. In 1882, the Women's Silk Culture Association of Philadelphia organized an exhibition featuring a dress designed for Lucretia Garfield, widow of the slain president, that was spun from silk raised in twelve states.[40] At the fair, Riley exhibited twenty yards of silk ribbon manufactured from silk from Osage Orange–fed worms.[41] That year, the American Silk Exchange in New York City exhibited silk products from twelve states and the District of Columbia and announced that it would buy cocoons for shipment to French manufacturers in Lyon.[42] At the New York fair, Riley exhibited a "very neat jar of pale cocoons" produced from Osage Orange–fed worms.[43]

That year, the Mississippi Silk Company, represented by Mrs. M. B. Hillyard of Mobile, Alabama, organized a network of silk centers in Alabama, Georgia, and Tennessee. The Mississippi Silk Company also bought cocoons but, unlike the New York Exchange, reeled its own silk rather than sending the cocoons to France.[44] In New Orleans, a group of women chartered the Southern Silk Association, which offered silkworm eggs and mulberry trees for sale.[45] The Mississippi Valley Silk Culture Enterprise Company, in Holden, Missouri, issued a silk directory and offered to furnish half-grown silkworms on consignment. The Boy's Silk Culture Association of Philadelphia, operating as an agent of French silk manufacturers, offered to buy cocoons.[46]

In California, women took the initiative in 1882 by organizing the California Silk Culture Association. The next year, the California legislature established the California State Board of Silk Culture, which offered premiums for the best silk cocoons and purchased cocoons for reeling at the state reeling school. The year following (1884) private investors incorporated the California Silk Culture Development Company.[47]

Silk boomers claimed twenty-five thousand active participants in 1882 and predicted the number would double within a year.[48] William B. Smith of the American Silk Exchange pointed out that Americans consumed $85 million worth of silk per year, of which $35 million was imported, and $50 million was manufactured in the United States from imported raw (reeled) silk. Only a negligible amount of silk thread was actually produced in the United States, but Smith foresaw the day when all silk consumed in America would be produced domestically.[49]

Riley contributed to the silk boom by releasing an updated version of his silk manual, experimenting with varieties of silkworms, and distributing silkworm eggs.[50] At the World Cotton Exposition in New Orleans in 1884, Riley and his staff exhibited eggs, larvae, moths, and cocoons of native and foreign silkworms, as well as samples of silk reeled in Kansas from a strain of silkworms that Riley had raised on a diet of Osage Orange.[51] William Saunders, the department's horticulturist, distributed mulberry slips and seeds. In departmental publications, Riley highlighted success stories like that of J. Herbelin of New Orleans who produced 2,500 pounds of cocoons and who hired European reelers to operate fifteen reels of his own design.[52] Although Riley was more circumspect about lobbying Congress following his conflict with LeDuc, he nevertheless publicly advocated duties on imported silk.

In June 1884, at Commissioner George Loring's urging, Congress appropriated $15,000 for silk culture.[53] Loring named Riley as director of the new Silk Division, and Riley in turn chose Philip Walker, an "enthusiastic believer in sericulture" with contacts to the French silk industry, as deputy director. At the time the appropriation was passed, Riley was on rest and recovery in Europe.[54] He conferred with Benjamin Franklin Peixotto, the US consul in Lyon, who shared Riley's conviction that Americans could compete with low paid European labor if protected by a tariff on imported reeled silk.[55]

Beginning with the fiscal year 1884–85, Congress appropriated $15,000 each year until 1890, when the appropriation was raised to $20,000. In 1886, Congress appropriated an additional $10,000 to erect a reeling station at the Agricultural Department. For at least two years, Congress passed special appropriations of $5,000 to subsidize the reeling station of the Women's Silk Culture Association at Philadelphia, $2,500 to subsidize a reeling station of the Ladies Silk Culture Association at San Francisco, and an indeterminate amount to subsidize a reeling station in New Orleans. The appropriations for silk culture for the years 1885–90 approximated the annual appropriation for investigation of harmful insects, which was about $20,000.[56] In terms of correspondents and requests for publications, silkworm eggs, and other matters, the Silk Division, during its seven years of existence, had more public contact than any other division.[57]

A singular oddity of the silk program was its placement under the entomologist. Riley recognized that the primary problems with establishing silk culture in the United States were the wage differential between Europe/Asia and America and the lack of trained reelers and modern reeling machines. These were not biological but, rather, economic and technological issues. One wonders why Riley and the commissioners didn't seek the advice of Jacob R. Dodge, the department statistician, when planning and implementing the silk program. Dodge was a veteran of the department since Patent Office days and a colleague of Riley's since at least 1874, when they both addressed the National Agricultural Congress in Atlanta. In 1876, Riley considered Dodge a suitable candidate for commissioner of agriculture.[58] Riley's failure to consult Dodge may have had to do with the fact that economics, as a discipline, was just finding its place in the American academic and administrative system. The American Economics Association was first organized in 1885, and agricultural economics as a subdiscipline came along later.[59] Dodge, one of the first practitioners of what eventually became agricultural economics, might have contributed useful economic analysis. In 1883, he presented a paper at the recently organized section of Economics and Statistical Science at the AAAS in which he argued that agricultural production increased in value in direct proportion to the availability of railroad transportation. Dakota wheat, he said, would be practically worthless if there were no railroads to transport it to market.[60] In 1891, as the nation slid into a wrenching depression, Dodge became a whipping boy for the national press because of a paper he delivered at the Economics and Statistics Section.[61] Journalists criticized Dodge for his pessimism regarding farmers' economic prospects, a view that didn't coincide with the popular image of self-sufficient, industrious American yeomen. The good rapport evident between Riley and Dodge until 1877 apparently soured once they worked as colleagues on the same floor. In 1887, Riley wrote to Dodge that he hadn't answered a letter Dodge had addressed to him because "I felt the tone unworthy of you . . . it illustrated a petty rankling feeling which I have long felt influenced your attitude toward me."[62] For whatever reason, Riley did not solicit Dodge's help, or anyone else's, regarding the economics of domestic silk production. This is all the more surprising because, in his locust and Cotton Worm campaigns, Riley solicited input from experts in other disciplines, such as ornithologists, botanists, and astronomers. For personal reasons, he seems to have abandoned the problem-centered, interdisciplinary approach that he otherwise utilized so effectively.

Riley's limitations in economics became apparent in his assessment of American labor and technology. In 1872, he asserted that Americans could rear silkworms as well as Europeans, but by the mid-1880s, after reviewing thousands of applicants for silkworm eggs, Riley was less sanguine regarding Americans' abilities as silkworm breeders.[63] Riley likely groaned as he read the letter from Mrs. C. Thompson of Kamm, Illinois, in February 1883: "The silkworm eggs . . . was a good many hatched out which most all died. Most of the balance of eggs hatched but as I did not receive manual for about two weeks after receiving eggs I did not know how to take care of them so there was only about twenty worms came to this spinning fr[om?] which spin cocoons which was very nice. There was six of them [finally?] which laid eggs which I have saved and intend to hatch out this season. The spring was very cold and wet which I think was bad for them. Would be pleased to have some more eggs and your manual for 1883 as I wish to experiment further."[64] Riley seems to have screened many or all of the applications for silkworm eggs.

In 1885, Riley offended American silkworm breeders when he announced that, because of the uneven quality and diseased condition of American silkworm eggs, all eggs for distribution that year would be purchased in Europe. When American suppliers protested, Riley replied that he didn't sympathize with them. Twenty years earlier, before Pasteur had solved the mystery of silkworm disease, Americans had reaped substantial profits selling eggs in Europe. Now, the situation was reversed. Silkworm breeders in France and Italy, employing Pasteur's methods, produced disease-free eggs, whereas testing by department specialists in 1884 revealed that out of thirty-nine batches of American-produced eggs, thirteen were diseased.[65] Stung by critics who questioned his patriotism, Riley relented and purchased some eggs from American producers in 1886, a decision he soon regretted. Most of the American eggs were contaminated or diseased. An exasperated Riley remarked that the majority of Americans hadn't the slightest idea of the care necessary in silkworm egg production. In the following years, the division purchased all eggs for distribution in France and Italy.[66]

Silk proponents frequently argued that silk culture was an ideal occupation for underemployed or idle women. Peixotto proposed silk culture as a means for women to avoid unnamed evils associated with idle time and as a way to supplement family income within the safety of the home.[67] Walker lamented the "vast amount of idle hen time" wasted on American farms.[68] A writer in the *Farmer's Review* observed that thousands of unlucky young women were doomed "to paint bad pictures, write poor sketches, drag their lives out teaching, or marry uncongenial husbands for the sake of a living."[69] Norman J. Colman embellished the argument by praising the superiority of American women. He bragged that he employed only American women to operate the reeling machines at the department (without specifying whether immigrants or daughters of immigrants were "American"), and he asserted that his operators could do twice as much work as women in France.[70] Female silk boosters frequently agreed that silk culture was particularly suited to women, though they avoided references to idleness and unsupervised activities outside the home.

Riley, sobered by screening thousands of requests (mostly from women) for manuals, eggs, and mulberry trees, revised his initial optimism regarding women's special aptitude for silk culture. He now warned against entrusting silk culture only to women, the aged, invalids, and children. Only capable, experienced adults, both men and women, he insisted, were suitable for practicing silk culture.[71]

Similarly, Riley revised his sanguine expectation that American technology would compensate for low labor costs in Europe and Asia. His change of mind grew out of his disappointment with the Serrell reeling machine. Edward W. Serrell was born in England and immigrated with his family to New York, where he became a prominent civil engineer. Prior to the Civil War, he designed railroad, bridge, and tunnel projects. In the Union army, he developed long wire, armor plate, impromptu gun carriages, and iron viaducts. About 1883, Peixotto and the American secretary of state Frederick Frelinghuysen notified Riley that Serrell was perfecting a new silk reeling machine in southern France.[72] Serrell recognized that inventions should be flashy as well as serviceable. He pointedly announced that, by using *electricity*, his machine produced a superior quality of thread while reducing labor costs. In an era when Thomas Edison's incandescent lights were illuminating cities across the country, Serrell capitalized on the notion that any machine run by electricity must be superior. Peixotto gushed, "What the cotton gin has done for cotton . . . the Serrell invention may . . . do for silk."[73] Here was a sure-fire recipe for success: Americans across the continent would produce cocoons to be reeled by efficient women operating Serrell reeling machines, so that America would outstrip traditional silk centers in Europe and Asia.

While the Serrell reeler was receiving final modifications, Loring, probably at Riley's request, subsidized the Women's Silk Association of Philadelphia's reeling station and Jules Herbelin's operation in New Orleans. The record of both operations for the season of 1884–85 was dismal. In Philadelphia, the cost of raw materials outstripped proceeds from the sale of finished reeled silk; in New Orleans, Herbelin's operation barely broke even, with materials costing 94 percent of the proceeds for finished silk.[74]

In May 1885, Serrell received a patent for his invention, and in 1886 the first machines were installed at the Agricultural Department.[75] Riley had hoped to install twelve Serrell machines, but Congress appropriated only enough for six. Due to various mechanical defects, these machines were soon replaced by new and improved Serrell machines. Colman expressed hope that the output of raw silk would double.[76] In 1886–87, the station, located in an annex, produced 143 pounds of first rate silk and 120 pounds of second rate silk.[77] The reeling station attracted considerable public attention, but Riley remarked that the season's output was only "as satisfactory as could have been expected under the circumstances."[78] He explained that the inferior silk was due to inexperienced producers who overcooked the cocoons, giving them a brown tinge. In order to correct this problem, Riley established a network of purchasing stations that purchased fresh, living cocoons to be "chocked" [cooked] at the central depot in Washington.[79]

In 1887, at the invitation of Cincinnati officials, Riley and his staff moved the departmental silk station to the Cincinnati Industrial Exposition for several months. The move prompted favorable publicity, but it disrupted production.[80] Walker, who appears to have been an inventor-mechanic in the same mold as Barnard, announced that he was modifying the machines to correct mechanical deficiencies. He also developed an apparatus for washing silkworm eggs and a device for boiling cocoons (boiling was necessary to loosen the threads for reeling).[81] By 1887, Riley had doubts regarding the Serrell reelers. In his report for that year, he wrote, "I am free to confess that I have little hope of [a] final favorable result so far as the main object of the experiment is concerned."[82]

Despite his doubts, Riley invited Walker to accompany him to the Paris Exposition in 1889. After investigating French reeling technology at Lyon, Walker designed a machine he said would correct the problems with the Serrell machines. Walker predicted, "The cocoons [would be] put in at one end [and] come out raw silk at the other."[83] Walker, impatient with the French, who "won't learn any better," still dreamed of a Whitney-style technological revolution that would turn the United States into a "giant silk farm."[84]

By the spring of 1889, the silk program was being sustained primarily by Walker's optimism and Riley's tenacity. At that point, national politics intervened in the person of Jeremiah McLain Rusk, President Benjamin Harrison's new secretary of agriculture.[85] A brawny six-foot-two farmer, horse breeder, stockman, and popular governor of Wisconsin, Rusk frequently delighted crowds at fairs by rolling up his sleeves, taking on all comers, and winning every thrashing contest. Back-slapping, wisecracking "Uncle Jerry," the "jolliest secretary" in Harrison's cabinet, soon transformed his office into a clubhouse where boisterous companions of all political persuasions gathered to swap yarns and savor fine tobacco, utilizing Jerry's extensive collection of corn-cob pipes. An advocate of tariff protection for domestic sugar from sorghum and sugar beets, Rusk believed that silk production in the United States could succeed only if protected by a tariff or supported by a bounty.[86] One of Rush's first decisions was to remove the silk program from Riley's Division of Entomology. What lay behind this decision was not clear, but there was likely a personality clash between the fastidious, cerebral Riley and the boisterous extrovert Rusk. Riley was clearly not pleased by Rusk's abrupt removal of silk from his portfolio. Returning from the Paris Exposition in September 1889, Riley informed his new chief that he had investigated the latest automatic reeling machines in Lyon, then curtly remarked, "[S]ince you have decided to remove the silk program from my Division, I refrain from further comment."[87] This was Riley's last official statement on silk culture.

With Riley's exit, the silk program soon expired. In 1891, Rusk declined to recommend continuation of the silk appropriation, and Congress dropped the item from Agriculture's budget.[88] The silk program that Riley had conceived and directed over the past dozen years and that Congress had subsidized for the past six years produced only minimal quantities of mediocre silk at a net financial loss. In place of the anticipated economic gains, it resulted in delusion and embarrassment for the commissioners, the Division of Entomology, and Riley.

12

Vedalia the "Wonder Beetle" and Biological Control

The history of the introduction of [the Cottony Cushion Scale] . . . and its final reduction to unimportant numbers by means of an apparently insignificant beetle . . . will always remain one of the most interesting stories in . . . practical entomology.

—C. V. Riley, *Annual Report of the Secretary of Agriculture* (1889), 334

Following the Civil War, horticulturists in Florida introduced improved orange stocks to replace feral orange trees planted by Spanish colonizers. William Saunders, the horticulturist at the Agricultural Department, supplied them with Italian Navel Orange stock. Horticulturists in California followed suit so that, by the 1880s, growers in the two states were supplying most of the domestic market. In the 1870s, Florida production was three times that of California, but by the early twentieth century California growers produced three times as much as Florida.[1]

Production of citrus fruits came with a complex of native and foreign insect pests. During Comstock's first year as federal entomologist, he received more inquiries about orange tree pests than about any other insects except the Cotton Worm. His investigations in Florida in January and February 1880 revealed half a dozen insect pests of oranges. The Cotton Stainer, notorious for sucking the juices of the developing cotton seeds and permanently staining the cotton lint, had recently begun sucking the juices of oranges and staining the fruit, thus acquiring the name "Cotton Stainer on Orange." "Red Rust" was a mysterious malady that discolored and spoiled oranges. The Orange-Leaf Nothris caterpillar consumed leaves and buds and spun webs that suppressed the growth of young orange trees. The Orange-Leaf Notcher, a beetle, was named for its habit of eating notches in orange tree leaves. The most notorious and least understood citrus pests were the scale insects. Named for the scale-like covering that protected them from weather and natural enemies, these tiny pests were impervious to all known insecticides.[2] In the summer of 1880, Comstock and his wife traveled to California, where they encountered additional species of scale insects and other pests of orange trees.[3] California fruit growers had initiated a State Board of Agriculture to halt the importation of plants infested with Grape Phylloxera and other pests, but the makeshift quarantine failed to protect their vineyards and orchards.

Entomologists had paid little attention to scale insects because they were small and unattractive compared to butterflies, moths, and beetles. Victor Antoine Signoret, Riley's partner in the phylloxera campaign, apparently enjoyed the challenge of creating order in obscure and unattractive insect groups. In addition to his studies of phylloxera, he published the first monograph on scale insects in 1869. Signoret assigned the scale insects to the family Coccidae, in the order Homoptera, a puzzling group in which the outward appearance and metamorphosis diverged from the normal course of insect development. In the Homoptera, especially scales and mealy bugs, males and females of the same species sometimes differed so radically that the casual observer might assign them to different genera.[4] Homoptera produced offspring at times by normal sexual union of male and female, at other times through parthenogenesis from females alone. Immature Homoptera often regressed from complex and mobile immature stages to simplified later stages, often, in the case of females, loosing their legs and becoming ultimately little more than egg packets. Signoret's monograph treated 277 species; however, he discussed mostly European species and did not include those attacking citrus groves in America. Asa Fitch had described about half of the approximately thirty known American species, but he focused on those indigenous to the Eastern woodlands rather than those threatening citrus groves.[5]

The life cycle of the scale insects, John Comstock explained, was unusual and complex and remarkably adapted to the utilization of its host species and the perpetuation of its own.[6] Newly hatched scale insects, tiny creatures equipped with legs, antennae, and piercing-sucking mouthparts, attached themselves to plants, sucked the juices, and excreted a protective wax covering

Featured Insects

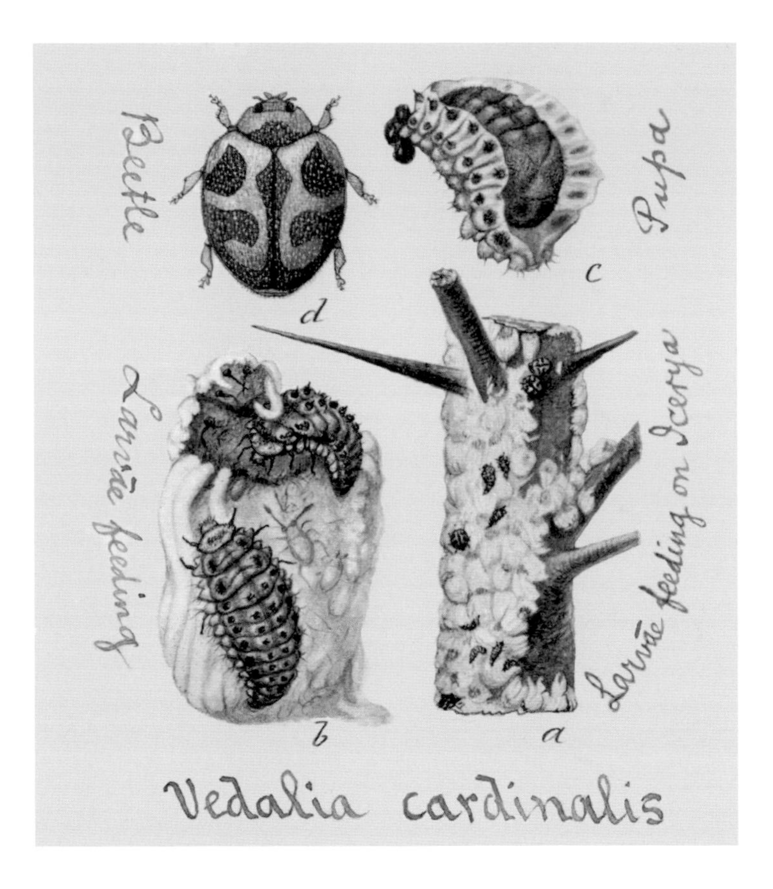

Featured Insects: **Vedalia Beetle** [*Rodolia cardinalis* (Mulsant) (Coleoptera: Coccinellidae)] and its prey, **Cottony Cushion Scale** [*Icerya purchasi* Maskell (Hemiptera: Margarodidae)]. Riley pencil sketch of life cycle of *Vedalia cardinalis* (USDA ARS National Agricultural Library Special Collections, with special thanks to Diane Wunsch and Mike Blackburn) and color photograph of Vedalia Beetles preying on Cottony Cushion Scale (Jeffrey W. Lotz, Florida Department of Agriculture and Consumer Services, Bugwood.org). The introduction of the Vedalia Beetle from Australia to California was the premier of classical biological control and launched this method as a spectacular solution to the Cottony Cushion Scale outbreak that was threatening the California citrus industry. Ironically, the parasitoid fly *Cryptochetum iceryae* (Williston) (Diptera: Cryptochetidae) was introduced before the Vedalia Beetle, and today provides suppression of the scale during the California winter and in cooler climates, where Vedalia is not as effective. As an internal gregarious parasitoid, its immature stages are hidden, leading casual observers to overlook or underestimate its efficacy.* In California, continuing changes in citrus pest management, especially new insecticides, threaten to provoke scale outbreaks by jeopardizing established natural enemies, including Vedalia and *Cryptochetum*.** These two species, especially Vedalia, have been distributed, mainly from California, to many additional invasion sites of Cottony Cushion Scale.*** The most recent introduction is the first and so far the only purposeful classical biological control program on the Galapagos Islands, to quell the Cottony Cushion Scale's devastation of native vegetation there, and thus to protect "Darwin's laboratory of evolution." This program, using Vedalia beetles collected in Queensland, Australia, has been a spectacular success.****

*Robert F. Luck, "Notes on the Evolution of Citrus Pest Management in California," in *Proceedings of the Fifth California Conference on Biological Control*, Riverside, July 25–27, 2006, ed. Mark S. Hoddle and Marshall W. Johnson, 1–7 (Riverside: University of California, Riverside, 2006); Caltagirone and Doutt, "Vedalia Beetle Importation."

** Elizabeth E. Grafton-Cardwell and Ping Gu, "Conserving vedalia beetle, *Rodolia cardinalis* (Mulsant) (Coleoptera: Coccinellidae), in citrus: a continuing challenge as new insecticides gain registration," *Journal of Economic Entomology* 96 (2003): 1388–98.

*** See world map (Figure 2) in Caltagirone and Doutt, "Vedalia Beetle Importation."

****Mark S. Hoddle et al., "Post release evaluation of *Rodolia cardinalis* (Coleoptera: Coccinellidae) for control of *Icerya purchasi* (Hemiptera: Monophlebidae) in the Galapagos Islands," *Biological Control* 67 (2013): 262–74.

that dried to a permanent "scale." The larvae (in modern usage, nymphs) then shed their legs and antennae, and in an apparent reversal of normal insect development, transformed themselves into legless sacs, still firmly attached to the plant host. From this amorphous entity emerged tiny adult males with wings, or egg masses from the sessile female. Males typically had no mouthparts for taking in nourishment, their only function being reproduction.[7]

Although Comstock clarified important aspects in the classification and life histories of American scale insects, the only remedies he could offer consisted of isolating orchards, inspecting trees prior to planting, and discarding contaminated fruit boxes. Arsenic-based insecticides like Paris Green and London Purple were useless, because the scale insects sucked internal plant juices rather than chewing the foliage. In theory, contact poisons like kerosene emulsions promised some relief, but Comstock concluded that further testing was necessary before he could recommend the Riley-Barnard milk-kerosene emulsion that Riley's team had developed for the Cotton Worm.[8] Other contact substances like tobacco, sulfur, lye, alcohol, ammonia, and carbolic and sulfuric acids proved ineffective, too expensive, or harmful to plants. Pyrethrum was effective only when applied in massive dosages, it cost ten to fifteen times more than kerosene mixtures, and it destroyed many beneficial insects.[9]

Riley had noted the increasing threat of pests to Florida citrus groves in his 1878 report, but he had no opportunity to investigate.[10] While Riley was out of office (1879–81), Comstock gained considerable support among growers in Florida and California through his investigation of citrus pests. When Riley was reappointed in 1881, he added scale insects to his other top priorities, the Cotton Worm and the silk program. The Cotton Worm and scale insect investigations overlapped in significant ways. For example, kerosene-emulsion was developed as a control for the Cotton Worm but proved more effective against scale insects. The same was true of the cyclone nozzle. Also, among the personnel Riley assembled in the course of his Cotton Worm campaign, two agents (Henry Hubbard and Albert Koebele) proved to be outstanding investigators of scale insects.

Riley appointed Hubbard as a special agent to the Entomological Commission in 1880 to investigate the Cotton Worm and also pests of orange trees. As noted in chapter 9 regarding the Cotton Worm, Hubbard soon developed the kerosene-soap emulsion that was an improvement on the Barnard-Riley kerosene-milk emulsion.[11] That same year (1880), Riley met Koebele, a German immigrant, at the Brooklyn Entomological Society. Impressed with Koebele's skill at mounting insect specimens, Riley hired him as a special agent of the Entomological Commission in 1881 and sent him south to conduct Cotton

Worm investigations. In 1882, he sent Koebele to Brazil to study cotton insects and other insect pests (including scale insects). On his return, Koebele, recovering from an unhappy love affair, requested to be transferred and Riley sent him to San Francisco to collect insects for the National Museum and to investigate locusts and other insect pests in California.[12]

Daniel W. Coquillett, Riley's third investigator of scale insects, was not involved in the Cotton Worm program. A farm lad from Pleasant Valley, Illinois, he was recruited by Cyrus Thomas (as Illinois state entomologist) to report on injurious moth species. In 1882, Coquillett developed tuberculosis, prompting his family to move to Anaheim, near Los Angeles, where the climate proved beneficial to his health. In 1883, Riley appointed Coquillett as a field agent for the Division of Entomology, and that year Coquillett produced his first report on scale insects.[13]

In April–May 1882, Riley joined Hubbard in Florida. They confirmed the discovery by William H. Ashmead, a journalist and entomologist in Florida, that "Red Rust" was caused by a mite and they demonstrated spraying with kerosene emulsion could control it.[14] In May 1882, Riley reported this and other discoveries regarding citrus pests before the National Academy of Sciences.[15] Hubbard's report, *Insects Affecting the Orange* (1885), incorporated their discoveries and control measures and became the Bible of citrus pest control in Florida.[16]

While Riley and Hubbard were making progress in Florida, the situation in California worsened dramatically. Whereas in the early 1880s, the Red and Black Scale constituted the main threats to orange groves, in 1883 the Cottony Cushion Scale began multiplying rapidly. By 1886, it threatened to destroy orange culture in California.[17] In April 1886, Riley traveled to California for a firsthand look, and he also transferred Koebele to southern California.[18] He had time only for a quick visit because he was scheduled to travel to Europe in May. While in Europe from May to September, he directed Hubbard to send him specimens of the scale, and he corresponded with naturalists in Australia, New Zealand, and the Cape Colony concerning its origin, life history, and possible controls.

Riley had first learned of the Cottony Cushion Scale in 1872, when Richard Harper Stretch, an entomologist in San Francisco, sent him a specimen, known then as *Dortheasia characias*.[19] Riley speculated that George Gordon, the San Francisco sugar baron, had introduced it to California on acacia trees imported in 1868–69.[20] William Miles Maskell, a naturalist in New Zealand, described the species as *Icerya purchasi* in 1878.[21] Riley explained that its alarming spread in California was due to its tremendous fecundity; its dispersal by wind, animals, and on nursery stock; its diversity of food plants; its imperviousness to insecticides; and the absence of natural enemies.[22] By rearing broods from eggs and nymphs sent by Koebele and Coquillett, and by examining specimens in the National Museum collection (presumably including specimens added by Comstock), Riley identified an intermediate form, between the first and second instars, establishing that there were three molts, not two as reported by Coquillett, Koebele, and Comstock. He discovered a nymphal stage that he tentatively described as the second stage of the male, and he supplied the first detailed description of the adult male. Riley made the first calculations of the rate of growth through various stages from egg to adult.[23] He later calculated that a single Cottony Cushion Scale, with a life cycle of thirty days from adult to adult, was theoretically capable of giving rise to twenty-two trillion offspring within six months.[24] Riley bred the scale's first known North American native parasite; however, it had failed to control the current outbreak and thus offered no solution. The same was true of a newly discovered species of earwig predator.[25]

In April 1887, in response to an invitation from California Fruit Growers Association president Ellwood Cooper, Riley addressed the fruit growers at their semi-annual convention in Riverside, California. Introducing Riley, Cooper recalled that, a few years before, he himself had traveled to Europe in search of controls for grasshoppers, olive tree scale insects, and other pests of California crops. His European contacts referred him to his own countryman, Riley, as the best authority. Riley, drawing on his 1886 report (which at that time was in press), reviewed what was known of the Fluted Scale (the name he preferred for Cottony Cushion Scale) and recommended spraying kerosene-emulsion with the cyclone nozzle sprayer.[26] He then called upon the fruit growers to sponsor a mission to Australia to identify and import natural enemies of the scale. As Congress had forbidden him and his staff to travel outside the United States on official business, Riley suggested that growers raise $1,500 to $2,000 to send an independent agent to Australia.[27]

In the 1850s and 1860s, many entomologists such as Asa Fitch, Benjamin Walsh, and William LeBaron had urged the transplantation or propagation of predators and parasites as controls, but it was Riley who had the most practical experience in these early experiments in biological control.[28] As Missouri state entomologist, he had advised farmers to collect parasite-infested pupae of the Rascal Leaf-Crumpler (a pest of fruit trees), the Common Bagworm, and the Oyster Shell Bark Louse of Apple and then confine them in screen-covered boxes or in exposed locations where the emerging host species would perish, but the parasites would spread.[29] In 1869–70, Riley distributed parasites of the Plum Curculio to correspondents in Missouri. Riley had limited experience in what is now known as classical biological control: transporting natural enemies from the continent of origin of a pest to control the target pest in its newly invaded area. In 1873, he and Jules Planchon had introduced an American parasite of Grape Phylloxera to France, but it had proved ineffective as a control. In 1884, Riley and Otto Lugger, after previous futile attempts, succeeded in importing parasites of the European Cabbage Butterfly from Germany. They sent batches of the parasite to Iowa, Nebraska, and Missouri, where the species multiplied so rapidly that naturalists in the Midwest surmised that it had been introduced much earlier.[30]

Riley and other naturalists assumed that the Cottony Cushion Scale originated in Australia and had been spread from there to California, New Zealand, and South Africa. In the spring and summer of 1887, however, the *Pacific Rural Review* published correspondence between New Zealand naturalist William Maskell and Waldemar G. Klee, a professor of botany at the University of California and state inspector of fruit pests, that prompted Riley to have second thoughts.[31] In this exchange, Maskell indicated that he had not compared specimens of *Icerya purchasi* with *Icerya sacchari*, a scale insect pest of sugar cane that was indigenous to the island of Mauritius, in the Indian Ocean. This opened the possibility that the two scale insects might be identical, in which case the home of *Icerya purchasi* might be Mauritius, not Australia. Riley could not rest until he had compared the specimens himself. What if it turned out that the government entomologist was searching for enemies of the scale on the wrong continent?

Riley's entomological sleuthing during the summer, fall, and winter of 1887 reveals his global command of entomological information. In June, prior to his departure for Europe, he requested Koebele to obtain specimens of the "sugar louse" from a sugar refinery in San Francisco.[32] While in Europe, ostensibly for rest, he appeared at a meeting of the British Association for the Advancement of Science in Manchester, where he discussed the question of *Icerya purchasi*'s origin with his British colleagues. He then traveled to Paris to compare the type specimens of *I. purchasi* and *I. sacchari* in Signoret's collection. He had to enlist his French entomological colleagues to open Signoret's cabinet, as Signoret was confined to bed rest. His examination of Signoret's specimens indicated that the two species were distinct and supported an Australian origin for Cottony Cushion Scale.[33]

Nicholas Pike, a former neighbor of Dr. E. Icery, the physician-naturalist on Mauritius for whom the genus *Icerya* had been named, had supplied

Signoret with the specimens of "pou blanc," which Signoret named *Icerya sacchari*. Riley learned that in 1874, Pike had also sent a specimen of "pou blanc" to the MCZ. At Riley's request, Hermann Hagen sent the specimen to him in Europe. Riley also learned that Pike had sent a specimen to Maskell in New Zealand. Riley persuaded Maskell to send the specimen to Howard in Washington, DC, who forwarded it to Riley in Europe. Riley examined it and concluded that it also was distinct from *Icerya purchasi*. Upon his return to Washington, he placed the specimen in the departmental collection, commenting wryly that it was "probably the most traveled insect in any collection."[34] Riley's examination of *Icerya* specimens housed in Paris, Cambridge, and New Zealand convinced him beyond doubt that Australia was the home of the Cottony Cushion Scale.[35]

While in Europe, Riley requested Frazier S. Crawford in Adelaide to send Australian parasites of the scale to his agents in California. In November 1887, Crawford mailed living specimens of a large parasitic fly to Coquillett and Koebele; however, most of the parasites were dead upon arrival. Koebele forwarded some insects to Klee in Alameda, but Klee reported that he was unable to retrieve any living parasites.[36] Riley, meanwhile, sent specimens of the parasite to Samuel Wendell Williston, the leading American authority on Diptera, who erected a new genus for it and named it *Lestophonus iceryae*.[37]

Meanwhile, at Riley's urging, Henry Harrison Markham, congressman from Los Angeles, sponsored a congressional resolution to send an entomologist to Australia. Riley also urged Markham and other Californians to work for removal of the travel prohibition.[38] When Congress rejected Markham's appeal, Riley conceived the idea of utilizing the Melbourne Exposition, scheduled to open in 1889, to circumvent the restriction on department travel. With the approval of Colman and Edwin Willits, Colman's assistant for scientific affairs, Riley enlisted George L. Rives, assistant secretary of state, to include his project in the Melbourne Exposition. Riley toyed with the idea of going himself but concluded that his preparations for the Paris Exhibition, opening in May 1889, prevented him.[39] He assigned Koebele to accompany the State Department's delegation, with responsibility to collect natural enemies of the scale and to send these to California. Koebele's salary was paid by the Agricultural Department, but his travel expenses were paid from the exposition budget.[40] Riley also appointed Francis M. Webster, a division agent in Indiana, to accompany and assist Frank McCoppin, former postmaster of San Francisco, who was responsible for the US exhibition in Melbourne.[41]

When Koebele sailed for Australia in September 1888, Riley instructed him to first search for *Icerya* parasites.[42] When Koebele arrived in Adelaide, Crawford supplied him with living specimens of *L. iceryae*, the parasite he had sent earlier and that had (presumably) perished, and Koebele shipped a package containing the insects to San Francisco. The ship's crew handled Koebele's American-bound shipment carelessly. Even worse, customs officials in San Francisco held it for weeks, during which time the ice that was intended to prevent the rapid maturing of the parasites melted and most of the hosts and parasites perished.[43] Incensed at the delays in customs, Riley complained to the secretary of the Treasury, who instructed his California agents to forward future shipments without delay and without charge.[44] The San Francisco customs agents then delivered Koebele's third shipment of *Lestophonus*-infested *Icerya* promptly to Klee, who forwarded them to Coquillett in Anaheim. Koebele predicted that *Lestophonus* would be worth millions of dollars to California orange growers. He warned, however, that Crawford had recently discovered that *Lestophonus* was known to have its own parasite. Crawford urged Klee and Coquillett to take steps to prevent the establishment of this hyperparasite in California.[45] Riley, in his 1888 report (completed January 1889), expressed confidence that the parasite would control the *Icerya* as it had in Australia.[46]

In 1887, prior to Koebele's sailing to Australia, Crawford sent some living specimens of *L. iceryae*, which he had renamed "*Cryptochaetum iceryae*," to Klee, who released some flies in San Mateo County.[47] Over time, *Cryptochetum* proved to be an effective control of *Icerya* in cooler, moister northern California. However, it offered no relief to citrus growers in dryer, warmer southern California, where groves were succumbing to huge *Icerya* infestations.[48]

While Riley focused his attention on *Lestophonus* and other parasites, he took little notice of small batches of lady beetles that Koebele included with his shipments. On October 15, 1888, practically as an afterthought to his discussion of *Cryptochetum iceryae*, Koebele also reported finding a lady beetle, *Vedalia cardinalis*, feeding on *Icerya*.[49] Koebele sent twenty-eight beetles with his first shipment, which arrived at Coquillett's testing grounds on November 30, 1888. His second shipment included forty-four beetles that arrived December 29, 1888, and his third shipment of fifty-seven beetles arrived on January 24, 1889.[50] When Koebele informed Riley that he had included the lady beetles in his first shipment, Riley responded off-handedly, "the sending of the . . . Coccinellids [Vedalia] is of course desirable[,] but I think we have much more to hope from the *Lestophonus* [*Cryptochetum*]."[51]

Coquillett released the beetles on a scale-infested tree on the Wolfskill orchard (located in modern downtown Los Angeles) and enclosed the tree in a tent.[52] In April 1889, when he opened the tent, he was astonished to find the tree practically free of scale. Equally astonishing was the multitude of progeny some 129 beetles had produced. When Coquillett allowed the beetles to spread to adjoining scale-infested trees, they stripped scale from one tree after another in a widening circle, multiplying as they spread. Vedalia Beetles bred so rapidly that Coquillett was soon able to supply nearby growers with colonies of these lady beetles.[53]

News of the miracle beetle spread quickly. Commercial growers and individual gardeners arrived daily to fill pillboxes with the wonder beetles. By the time Riley arrived at the Paris Exposition in May 1889, reports of victory over the scale blazed from newspaper headlines. Colonel J. R. Dobbins, a grower in San Gabriel Valley, described the euphoria: "The Vedolia [*sic*] has multiplied . . . and spread so rapidly that every one of my thirty-two hundred orchard trees is literally swarming with them. . . . People are coming here daily, and by placing infested branches upon the ground . . . can secure colonies of thousands of the Vedolia [*sic*] . . . Over fifty thousand have been taken away to other orchards during the present week . . . I feel positive . . . that the entire valley will be practically free from Icerya before the advent of the new year."[54]

The colonel's prediction was fulfilled even earlier. From April to August 1889, Coquillett distributed 10,555 beetles to 208 orchard owners. By August, practically all orange trees in the Los Angeles area were free of *Icerya*; by October, the remaining orchards in southern California were reported free of scale.[55] William F. Channing (son of a noted Unitarian minister) declared "[T]he deliverance was more like a miracle than anything I have ever seen."[56] In his report prepared in December 1889, Riley assured growers that the Cottony Cushion Scale was no longer a threat to orange culture.[57]

Riley recognized that Vedalia represented a special case, but only later did entomologists recognize its tremendous genetic flexibility. The gene pool of the original colonized stock represented about 514 individual beetles. Its adaptability was demonstrated in the course of its subsequent colonization by California stock in Egypt, Cyprus, the Soviet Union, Portugal, Puerto Rico, Venezuela, Peru, Chile, Hawaii, the Philippines, Guam, Uruguay, Argentina, Taiwan, and Peru. Vedalia is one of a very few species to successfully colonize such widely diverse habitats through the progeny of a few individuals.[58]

Though pleased by the Vedalia success, Riley regretted that other natural controls of scale had not become established in California. He preferred a range of controls, including predators and parasites in conjunction with insecticides, rather than reliance on one predator. He warned that Vedalia was known to feed only on the Cottony Cushion Scale; it was not a cure-all for scale infestations or other pests. California orange growers, on the other hand, had only praise for Vedalia. By 1890, the beetle had so nearly eradicated the scale that Coquillett had difficulty maintaining a colony of Vedalia for release, should the scale reappear.[59]

When Koebele sailed for Australia, Riley instructed him to concentrate on parasites (now called parasitoids) rather than predators. This instruction and his later caution against relying exclusively on predators like Vedalia suggest that he had reconsidered his earlier conclusion, with Forbes in the 1870s and 1880s, that parasitoids were less effective than predators as control agents.[60] By 1888, in his search for control of the Cottony Cushion Scale, Riley placed his hopes on parasites (parasitoids) rather than on predators. Whatever influenced this change in Riley's evaluation of parasitoids, subsequent experience tends to corroborate his emphasis on parasites (parasitoids) as control agents. A review of classical biological control worldwide published in 1976 rated parasitoids as overall more effective than predators. At that time, approximately 84 percent of insect control agents employed worldwide were parasitoids. Vedalia, the predator with which classical biological control effectively began, is clearly an exception.[61] More important in modern classical biological control are considerations of nontarget effects of introduced natural enemies, with the emphasis on specialized candidates for introduction that will not attack hosts or prey other than the targeted invasive pest. There are far more likely to be specialized parasitoids available for introduction than predators, but Vedalia as a very specialized predator is again unusual in this regard.[62]

While the Vedalia drama unfolded, Coquillett developed a fumigation process for the control of scale insects that had long-range implications for pest control. The use of gas to control greenhouse pests had long been practiced, but this procedure had not been successfully applied in orchards. Production of fumigant gas was time-consuming and costly, enclosing full-grown trees was difficult, and heavy doses of poisonous fumigants destroyed foliage.

From January to August 1886, Coquillett tested insecticides against the Cottony Cushion Scale on the orange grove of the Wolfskill Ranch in Orange County, one of the largest groves in southern California. Alexander Craw, the ranch foreman, had had some success fumigating with bisulphate of carbon (carbon disulfide in modern usage). Trees exposed to fumigant for three hours were freed from scale insects, but this process took too long to be economical.[63]

From outward appearances, the California operations of the Division of

Entomology appeared to be functioning harmoniously, when, at the end of August 1886, Riley suddenly released Coquillett from government service. Some participants attributed Riley's action to personal conflict between Koebele and Coquillett. According to this version, Koebele (whom Riley regarded as one of the most talented entomologists on his staff) convinced Riley to release Coquillett.[64] Riley, in public statements, insisted that he was forced to drop Coquillett because $10,000 of the division's appropriation for that year was diverted to the new Division of Ornithology and Mammalogy.[65]

Now out of a job but committed to the control of the scale, Coquillett joined forces with Craw at the Wolfskill Ranch. Within a month, in September 1886, Coquillett developed a fumigant based on hydrocyanic acid gas (now hydrogen cyanide, or HCN) that proved to be quick, effective, and inexpensive. Coquillett's improved method involved drying the HCN so that it killed the scale insects but left the foliage undamaged.[66]

As a private contractor, Coquillett, possibly in partnership with Wolfskill and Craw, applied for a patent for the HCN process. He attempted to keep his formula secret, but A. S. and A. B. Chapman and L. H. Titus of San Gabriel learned of Coquillett's discovery and appealed to Eugene W. Hilgard at the University of California in Berkeley, who sent his chemist, F. W. Morse, at the California Experiment Station, to assist the San Gabriel growers. By June 1887, Morse developed an HCN fumigation process that was similar, perhaps even identical, to Coquillett's.[67]

Riley had given low priority to fumigation, but now he apparently had second thoughts because he rehired Coquillett in July 1887, instructing him to continue development of HCN gas fumigation. There is no record of what transpired between Riley and Coquillett, but Riley presumably gave Coquillett a dressing down for applying for a patent (though this was clearly his right as a private citizen), and Coquillett, eager to rejoin the Division, accepted Riley's terms, dropped his patent application, and returned to his post.[68]

Coquillett and Morse both claimed credit for developing HCN fumigation and making it known to growers. Morse published the results of his experiments and his formula in the University of California Experiment Station Bulletin in June 1887. He and his supporters (including his colleague Charles W. Woodworth at the experiment station) insisted that he, not Coquillett, should be credited for making the process known. In June 1887, at the request of A. B. Chapman, the Los Angeles Board of Agriculture pronounced Morse's "killing" the best method and invited him to return to southern California, all expenses paid.[69] Four months later, in October 1887, at a gathering of Los

Angeles fruit growers, Craw praised Coquillett for developing the HCN process. Coquillett, in the *Report of the Commissioner of Agriculture* (submitted January 20, 1888), claimed that he developed the process about six months before Morse. He noted that, soon after his reappointment in July 1887, the Los Angeles County Horticultural Commission withdrew its offer of $1,000 for a method "for perfect extermination of *Icerya*" because the "government agent" [Coquillett] had solved the problem.[70]

Growers in southern California quickly adopted the HCN fumigation process. In 1887, Los Angeles County, as well as the cities of Riverside and Anaheim, provided growers with county or municipally owned fumigators at minimal cost.[71] In Los Angeles County, nurseries were required to fumigate trees. Growers, meanwhile, developed tree-sized tents, some patented, others not, that facilitated the fumigation of orchards.[72]

In late 1887, A. D. Bishop, a grower in Orange County, asked Coquillett (now reinstated with the Division of Entomology) to conduct fumigation experiments against the Red Scale (*Aonidia aurantii*) on his orange trees. They discovered, among other things, that fumigating at night appeared to be more effective than by day (although they were not sure why).[73] Bishop and several other growers applied for a patent based on the innovations of night fumigation, circulation of the fumigant by a fan, and other modifications of HCN fumigation.[74] Riley, incensed at their attempt to patent a process that a division employee had developed, complained to Willits, who protested to the Patent Office. When officials in that office denied the patent, Bishop et al. hired an attorney and appealed directly to the commissioner of patents, who, in January 1891, reversed his employees' decision and granted the patent.[75] Bishop et al. insisted they had learned of the HCN process from Morse, not Coquillett.[76] Riley, who was engaged in a running battle with the Patent Office over the Barnard patents, advised growers to ignore all fumigation-related patent claims and to proceed with fumigating as they saw fit.[77] In 1893, Bishop and his partners convinced a law enforcement officer to arrest a fruit grower who had used HCN fumigation, and the case went to court. In 1894, the presiding judge overruled the commissioner of patents, declaring that "An old process does not become a new and patentable process by being used at night. . . . Of course, night excludes the light. Everybody knows that. Nor is the night patentable."[78] His ruling most certainly cheered Riley.

Overall, Riley's opposition to patenting helped pave the way for patent-free fumigation as practiced in California and elsewhere. The HCN fumigation process was soon overshadowed by the Vedalia success, but a few years later it was used to control the Red Scale and the San Jose Scale, both of which were not preyed upon by Vedalia.[79] From the late nineteenth century on, HCN fumigation remained a primary control of citrus pests, and in the early twenty-first century, hydrogen cyanide is still considered an important fumigant.[80]

In late 1889, Riley was enjoying public recognition for his work at the Paris Exposition and for the Vedalia triumph. His enjoyment was suddenly interrupted when McCoppin and his supporters published articles in the *Pacific Rural Press*, claiming that McCoppin, rather than Riley, deserved credit for importing Vedalia.[81] As the controversy over credit heated up, it became apparent that the arrangement between the Department of Agriculture and the Department of State contained weaknesses that could be exploited by those so inclined. For example, McCoppin and his supporters alleged that Riley had submitted his report first to the Department of Agriculture rather than to the Department of State, which they insisted was the lead agency.[82] Some McCoppin supporters wanted to grant McCoppin patent rights to Vedalia, others proposed that a statue of him be erected in his hometown, and some even proposed that the Vedalia Beetle be renamed "*Vedalia McCoppiensis*."[83]

The dispute over credit for Vedalia's success widened the rift between Riley and California horticulturists, who insisted that the importation of natural enemies was the primary if not the sole effective control measure for all injurious insects. The editor of the *Pacific Rural Press* remarked, "[T]he unprecedented success of the Vedalia has caused fruit growers . . . to expect immediate and similar results from all new insects."[84] Byron M. Lelong, secretary of the California State Board of Horticulture, and Ellwood Cooper (the Santa Barbara horticulturist who had praised Riley at the fruit grower's convention in 1887) led the insect-importation campaign. In early 1891, upon Lelong's urging, the California legislature appropriated $5,000 for the Board of Horticulture to send an entomologist to Australia, New Zealand, and other destinations in the South Pacific to collect beneficial insects. The board called upon Rusk, the new secretary of agriculture, to place Koebele in charge of the mission.[85]

Riley considered a second expedition to Australia unnecessary. He preferred that entomologists like Crawford in Australia send promising insect predators and parasites by steamer, and he cautioned against exclusive reliance on control by insect enemies of pests. Rusk ordered Riley to cooperate with the Californians, and Riley reluctantly authorized Koebele to sail, advised him where to look, and advised him to take along native American parasites and predators that might benefit farmers and horticulturists in the South Pacific. Koebele sailed in August 1891.[86]

Coquillett and Koebele depended on the cooperation and good will of local state officials and horticulturists. They suffered under Riley's sometimes tactless manner and his lack of sensitivity to their dependence on local support. Riley's abrupt furloughing of Coquillett, for example, led Coquillett to ally himself with growers. Now Koebele was caught between his duty to carry out Riley's instructions and his dependence on the Californians who paid for his expedition. Riley wrote Koebele prior to sailing, "There will be no questions as to whose instructions you will go under, or as to whom you will make your reports."[87]

Lelong countered by assuring Koebele that Californians considered him "the greatest benefactor to our state."[88] Reminding Koebele that Rusk supported him (Lelong) and the Californians rather than Riley, he proposed that Koebele ignore Riley's instructions and press on to China and Japan to search for beneficial insects.[89] Lelong insinuated that Riley and Coquillett hoped that Koebele's mission would fail, and he sweetened his appeal to cooperate with the Californians by providing Koebele with generous financial support. Apparently the California Board of Horticulture was not bound by strict accounting procedures like those of the Agricultural Department.[90]

In the course of Koebele's second mission, his relations with Coquillett, never very good, deteriorated further. Koebele complained to Riley that Coquillett purposely mishandled insects sent to him in order to discredit Koebele's mission.[91] When Coquillett defended Riley against the Californians, Riley insisted that Koebele follow suit. He scolded Koebele for sending insects directly to Lelong and wondered aloud whether Koebele had "lost his head."[92]

As Riley predicted, Koebele's yearlong expedition failed to repeat the Vedalia success.[93] Cooper praised Koebele's discovery of a predator of the Black Scale, a pest on olive and citrus trees, but it proved only moderately effective as a control. Of the forty species Koebele brought back, only four species became established, and these proved to have little significance as controls. Riley summed up Koebele's mission as "moderately promising."[94] By contrast, Lelong and Cooper celebrated Koebele as a conquering hero. California fruit growers awarded Koebele a gold watch and gave his wife a pair of diamond earrings.[95]

Riley's feud with the Californians involved some comical episodes. In May 1890, Lelong was prowling around the Department of Agriculture grounds in Washington, DC, when, he claimed, he discovered lady beetles feeding on aphids directly under the window of the federal entomologist. On his return to California, he held a press conference in which he predicted that his captures would lead to the control of all remaining scale and aphid pests on the West Coast. Back in Washington, a Department of Agriculture employee by the name of Wilkinson filed the clipping announcing Lelong's "discovery," commenting on the margin, "Oh My."[96] Riley referred to Lelong and supporters as "political entomologists."[97] In 1895, the year Riley died, he and Lelong were still quarrelling. Riley was reportedly leading an effort to abolish the California State Board of Horticulture (Lelong was president) and to incorporate its functions into the University of California.[98] Lelong, following a head injury when he fell from a streetcar in San Francisco, committed suicide in May 1901 on the steps of the California statehouse.[99]

By 1893, Riley's relations with the California horticulturists had become so embittered that he closed the California office. He transferred Coquillett to Washington, where Coquillett soon resigned to assume charge of the Diptera in the National Museum collection, becoming a world authority in that order and naming over one thousand new species.[100] Koebele resigned to accept a position as entomologist (with a much higher salary) for the Hawaiian provisional government.[101]

Despite Riley's caution that the Vedalia success was a special case, he continued experimenting with imported parasites and predators. In 1891, Riley reported the successful importation of a parasite of the Hessian Fly, *Entedon epiogonus*, that he distributed to Francis Webster (the agent who worked at the Melbourne Exhibition) in Indiana, Stephen Forbes in Illinois, A. J. Cook in Michigan, and James Fletcher in Ontario. Riley and Forbes agreed that the parasite showed promise, but that it would take years before its effectiveness could be assessed.[102]

Riley and other entomologists involved in importations in the 1880s and 1890s frequently expressed concern that insect pests might be inadvertently introduced into new environments. As noted above, Riley and Koebele warned that a parasite of a scale parasite might become established in California. Riley rejected Koebele's plan (on his second mission) to ship parasites of Coccids on their hosts because injurious species might be introduced to California.[103] He also advised Koebele to ship potentially beneficial insects only in the "quiescent state" [cold storage] in order to maintain strict control over their release.[104] In 1890, Riley warned that a recently discovered species of Icerya from Montserrat and Sonora, Mexico, might become established in Florida, Texas, and California, in May 1901 and he recommended quarantining plants from those areas.[105]

The control of the Cottony Cushion Scale by Vedalia is often cited as *the* classic case and effective beginning of classical biological control.[106] Willits, scientific director of the Department of Agriculture, termed the Vedalia story an "entomological romance."[107] Riley was the central figure in the Vedalia story. He undertook the careful investigation that confirmed beyond doubt that the Cottony Cushion Scale originated in Australia, and he engineered the arrangement whereby the Agricultural and State Departments utilized the Melbourne Exposition to send Koebele to Australia. Vital to this undertaking was Colman, Riley's long-term friend, now his supervisor. A less flexible or trusting commissioner might well have nipped Riley's scheme in the bud. Although Riley conceived of, led, and organized the program, his agents Koebele and Coquillett, in cooperation with Riley's Australian contacts, carried through the identification, transfer, propagation, and distribution of insect predators and parasites as natural controls.

Riley's success with Vedalia contrasts sharply with his silk program. Throughout his tenure as federal entomologist, Riley devoted a major portion of his time to silk culture. The Cottony Cushion Scale/Vedalia project, by contrast, occupied Riley's partial attention for about five years, from 1884 to 1889. During those years, appropriations for the silk program (which were listed separately from division funds for the investigation of injurious insects) increased from $15,000 to $23,208. In 1889, the year of the Vedalia triumph, the budget for silk culture outstripped the total appropriation for the division's investigation of injurious insects. From 1878 to 1891, the silk program gobbled up funds totaling $128,253 in special appropriations, plus a considerable portion of the Division of Entomology's regular expenditures.[108] The Vedalia project, by contrast, had no special appropriation, and its total cost, estimated at $1,500 in the currency of that era, involved a moderate item in the division's budget.[109] The results of the two projects were the inverse of the time and resources devoted to them. The silk program produced nothing, except perhaps the negative conclusion that silk culture in the United Sates was uneconomical. The Vedalia project, on the other hand, rescued California citrus growers from injury estimated in the millions of dollars and initiated the practice of classical biological control.

Riley's promotion of silk culture went astray because it involved, essentially, an economic problem for which he was not particularly well qualified. He seems not to have fully understood that, when economics was the primary issue, he needed the assistance of others. The Vedalia project, by contrast, called upon Riley's biological acumen, his extensive experience with predators and parasites, his worldwide contact with naturalists, his skillful choice of a capable exploration and introduction team, and his mobilization of resources in Washington. Riley's strengths lay in biology and the skillful management of agencies to carry out biological investigations. Riley's genius shone most brightly in his capacity as a naturalist, his first and foremost calling.

13

Creating a National Insect Collection

I make this donation [my insect collection] in the belief . . . that the National Museum . . . must inevitably grow until it shall . . . surpass other institutions . . . as a repository of natural history collections.

—C. V. Riley to Spencer F. Baird, October 23, 1885

EARLY AMERICAN ENTOMOLOGISTS regarded their collections as a means of bringing order to the New World insect fauna but also as a declaration of America's scientific independence from Europe. Their collections survived only when they were entrusted to institutions. Thomas Say's collection wasted away following his death in 1834.[1] Frederick Ernst Melsheimer's collection of American Coleoptera was spared a similar fate when it was purchased by the MCZ. Similarly, Thaddeus W. Harris's collection passed into the care of the Boston Society of Natural History.

When Riley arrived in Washington in 1878, the transition from individual to institutional care of collections was well under way; however, the focus was in Philadelphia and Cambridge, Massachusetts, rather than in Washington. In the reconnaissance of the West, entomology had fared rather badly. Plant specialists like John Torrey, Asa Gray, and George Engelmann had utilized western expeditions to map the continent's botany, and vertebrate zoologists like Spencer F. Baird, Elliott Coues, and Robert Ridgeway had assembled impressive collections of birds, mammals, and other quadrupeds, but entomologists had not been as successful. Say had made a promising start in the 1820s when he accompanied the Stephen H. Long expedition to the Great Plains, but thereafter entomology in federal surveys had languished. John LeConte complained to Baird in 1859, "The Vertebrata of the interior seem to monopolize attention, to the entire exclusion of the Articulata."[2] In the 1870s, Ferdinand Hayden engaged Cyrus Thomas, Alpheus Spring Packard, and Samuel H. Scudder, who reported on the Rocky Mountain Locust and other insect species, but Hayden's emphasis on entomology ceased in 1879 when the western surveys were consolidated into the US Geological Survey.[3] John Wesley Powell, who orchestrated the consolidation, focused exclusively on geology, explicitly excluding botany and zoology.[4]

More serious was the lack of a central depository for insects collected on federal expeditions. The Smithsonian Institution had the greatest potential, but Joseph Henry, the Smithsonian's first secretary, believed that, at least initially, the Smithsonian should avoid overhead costs associated with a library or museum. He preferred to sponsor original investigations by independent naturalists who published their findings under the Smithsonian name, while allowing those associates to retain the natural history specimens used in their research. When Congress nevertheless mandated a library, Henry responded by firing the librarian.

Baird, who in 1850 became Henry's assistant, proceeded to modify Henry's policy of not establishing collections at the Smithsonian.[5] Like Riley, Baird valued direct observation over book learning. As professor of natural history at Dickinson College, in Carlisle, Pennsylvania, he organized field trips where students observed animals in their natural habitat.[6] Baird considered collections absolutely necessary to his specialty in vertebrate zoology and other natural history branches, but he knew nothing was to be gained by confrontation with Henry. Working behind the scenes, in 1857 he facilitated the transfer of the Patent Office's natural history collection to the Smithsonian. Henry, intent that curation of specimens not deplete Smithsonian funds, insisted that Congress appropriate an additional $4,000 per year for the upkeep of collections.[7] The collections from the Patent Office formed the basis of the US National Museum collection, which was funded separately from the Smithsonian Institution.

In 1858, Baird declared, "The greatest deficiency in American natural history is . . . in . . . entomology."[8] Faced with the problem of what to do in a field outside his specialty, Baird sponsored a series of taxonomic publications covering the insect orders. The Smithsonian monographs that appeared during the following decade constitute classics in American entomology: Baron Carl Robert Osten Sacken (Diptera, 1858), LeConte (Coleoptera, 1861–62), John

Gottlieb Morris (Lepidoptera, 1860 and 1862), Hermann August Hagen and Philip Uhler (Neuroptera, 1861), Hermann Loew and Osten Sacken (Diptera, 1862), and Scudder (Orthoptera, 1868).

Baird continued Henry's policy of consigning specimens collected by federal agents to specialists in various orders, e.g., LeConte for Coleoptera and Scudder for Orthoptera. Although technically subject to recall by the Smithsonian, in practice the specimens remained on permanent loan.[9] This was an improvement over previous practice, when the Patent Office served as a catch-all for specimens collected by federal agents.[10]

One reason Baird continued loaning specimens to outside collaborators was that Townend Glover, who served as entomologist at the Patent Office (1854–59) and at the Agricultural Department (1863–78), had decidedly heretical views regarding insect specimens. Leland O. Howard recalled Glover's impact on collections under his control: "When I first came to Washington [in 1878] . . . there was no National Collection of Insects . . . Glover, the entomologist . . . up to June 1878, was not in favor of preserving collections. . . . His idea was to make a colored drawing of a species, and then throw the specimen away. The drawing, he argued, is practically indestructible, whereas the specimen is subject to . . . destruction by museum pests, it fades when exposed to light, and is easily damaged by a jar to the [collection cabinet]."[11]

From the 1840s to the 1870s, with entomological collections stalled in Washington, the American Entomological Society in Philadelphia and the MCZ in Cambridge took the lead. The American Entomological Society's collection fared well for a time. In 1864, the society sponsored a collecting expedition to the Rocky Mountains and secured donations from leading entomologists like Uhler and Osten Sacken. That year, Henry agreed to send duplicate Smithsonian specimens for the society's cabinet. The society's hopes to establish a national insect collection collapsed in 1865 when Thomas Wilson, its main benefactor, died. The society, now strapped for funds, reorganized as a subsection of the Academy of Natural Science and transferred its collection to the academy building.[12]

Following the Philadelphians' failure to establish a national insect collection, the initiative passed to the MCZ. From the inception of Agassiz's museum in 1859, he appointed expert entomological curators for the museum's insect collection: Scudder (1859–63), Packard (1863), Uhler (1864–66), and Hagen (1867–92). Hagen, formerly at the University in Königsberg, Prussia, acquired key private collections, including many from Philadelphia naturalists.[13] Thus, in the 1870s, while the American Entomological Society was retreating to the academy and Hagen was quietly assembling a national insect collection, Baird was still searching for an entomological partner for the National Museum.[14]

With Riley's arrival in Washington in 1877–78, Baird suddenly had an entomologist who not only had expertise in all insect orders but possessed what was probably the most extensive general collection of North American insects. In any case, it was the most complete collection of economically significant species. Riley was a product of the Midwest, with no particular debts to Agassiz and the Harvard-Cambridge elite. He agreed with Baird that the national insect collection should be at the National Museum.

It is not clear when the two met and agreed to work together, but by June 1878, Riley, who had recently replaced Townend Glover as entomologist in the Agricultural Department, was assisting Baird by identifying insects and answering letters relating to entomology. On July 2, 1878, Riley apologized to Baird that he had not met the seven-day turnaround for answering correspondents because he had been out of town for a few days getting married.[15] In May the following year, having been dismissed by William LeDuc, Riley assured Baird that he would continue to handle entomological correspondence. He expected no pay at the present, but he suggested a paid curatorship for himself upon completion of the new museum building in 1881.[16] Baird replied three days later, accepting Riley's offer of assistance, but making no promises regarding pay.[17] Having advanced to secretary of the Smithsonian following Henry's death in early May 1878, Baird now had more freedom to carry out his plans for the National Museum, but in a continuation of Henry's frugal policy with respect to the Smithsonian budget, he preferred to engage experts from federal agencies to curate the collections on an honorary basis.

This exchange of letters marks the beginning of the National Insect Collection.[18] The contrast between the two partners is striking. Baird, reared in a prominent Pennsylvania family, followed an academic career, first as a student, then later as a professor at Dickinson College, before his appointment as assistant secretary in 1850. Fifty-five years of age in 1878, Baird was Riley's senior by twenty years. With his low-key and nonconfrontational manner, Baird preferred to work behind the scenes and seldom gave public addresses. He rarely attended AAAS or other scientific gatherings, but he did serve as president of the Cosmos Club in 1881.[19]

Riley, on the other hand, came from an unstable family and had no formal training in natural history. A joiner by instinct, he relished every opportunity to speak before groups and to serve in organizations. Their respective familial circumstances contrasted as well. Mary Churchill Baird was an invalid whose dependency Baird bore with tenderness and patience. They had one child, a daughter.[20] Emilie Conzelman Riley, by contrast, was an efficient manager of the Riley household during Riley's extended absences, and she bore him seven children.

Despite these contrasts, the two formed an effective pair. Riley's expertise in invertebrates complemented Baird's expertise in vertebrates. Despite their differences in age, Baird and Riley functioned as equals. Unlike many young naturalists, Riley had won his spurs on his own, independent of Baird. Prior to Riley's move to Washington, the two had indirect contact through the Chicago Academy of Sciences. Baird arranged for Smithsonian support for Robert Kennicott's exploration of Alaska and the Yukon (1860–63) and for Kennicott's scientific corps that accompanied the Western Union Telegraph expedition to Alaska (1865–66). Kennicott was a role model for the youthful Riley. He treasured his small collection of insects from Kennicott, and he displayed Kennicott's portrait in his St. Louis office.[21]

During the years 1879–81, when Riley directed the Entomological Commission from his house, Baird continued to consult him on entomological matters. Apparently Baird never called upon John Comstock for assistance. Following Garfield's election in November 1880, Riley, expecting to regain his position in the Agricultural Department, urged Baird to formalize their partnership. Baird responded, in January 1881, by appointing Riley honorary curator of insects for the United States National Museum (at that time, Comstock was federal entomologist). Riley naturally accepted.[22] Three months later, in April 1881, Riley was reinstated as federal entomologist. With this appointment, Riley held the three central entomological positions in the federal government: chief of the Entomological Commission, entomologist at the Agricultural Department, and honorary curator of insects at the National Museum.[23] Riley's appointment as honorary curator formalized the partnership between the Agricultural Department and the National Museum. Later that year, Baird wrote of this partnership: "The department of entomology . . . has . . . been little cultivated in the National Museum . . . [A]lthough . . . the Smithsonian paid much attention to gathering material . . . The necessity of a department of systematic and economic entomology has . . . always been recognized . . . [In] the present reorganization . . . Professor C. V. Riley, the entomologist of the Department of Agriculture, has been appointed honorary curator, and has deposited his own private collection of insects . . . as a nucleus for [a national insect collection]."[24]

The agreement between Riley and Baird was the Magna Carta of the National Insect Collection. Both partners recognized the necessity of reliable, long-term curation of the collection, and Baird entrusted Riley with this responsibility. Riley and Baird also agreed that the collection belonged at the National Museum and not the Agricultural Department, despite the fact that entomologists in the Agricultural Department, and later at the state experiment stations, were the primary donors and users of the collection. They recognized that the museum had a clear focus on accession, preservation, and classification of specimens, whereas the Agricultural Department was subject to competing administrative and political pressures. Fortunately for Riley, the new museum building was ready for occupation in 1881, the same year he was reinstated as federal entomologist. That year, Baird announced, "[T]he Museum is now prepared to properly care for such collections, under the direction of Professor Riley . . . and it is hoped that . . . a truly national exposition of the insect fauna of the country will be brought together."[25] This was good news to entomologists who remembered the destruction of Benjamin D. Walsh's collection in the Chicago fire of 1871 and the wasting away of Say's collection decades earlier. In 1886, Baird explained the relationship between the Department of Insects at the National Museum and the Division of Entomology at the Department of Agriculture: "As the Museum . . . will . . . receive a great deal of its best material through the Department of Agriculture, one of the chief aims of this national collection should be to reciprocate, not only by preserving all systematic material and thus aiding said Department of Agriculture in necessary determinations, but by giving particular attention to the biological side of the collection."[26] The success of this unusual arrangement depended ultimately on the individuals who initiated and continued the agreement.

The two partners agreed that the collection should emphasize both systematic and applied entomology. Riley, who established the Entomological Commission in response to locusts and other insect pests, found a soul mate in Baird, who organized the United States Fish Commission (1872) in response to the decline of commercial fish stocks in the Atlantic.[27] They further agreed that systematic and applied entomology should include public instruction. Riley, who had prepared displays of insects at the office of the *Prairie Farmer*, in the state entomologist's office, at agricultural and horticultural societies, and at state fairs, now prepared an exhibit at the museum "for the instruction and edification of the public."[28]

Riley and Baird envisioned a federal-state network of investigators devoted to agricultural science, a concept that came into being in 1887, when Congress passed the Hatch Act establishing federally supported state experiment stations. Aware that most entomologists at that time were collectors and systematists with little inclination toward agricultural applications, they stressed the opportunity to augment the national collection through the donation of private collections.[29]

Riley's placing of his collection in the National Museum and the pivotal role his collection played in the growth of the national insect collection raises the question, how did his collection come into being? The sources on this point are sketchy. When Riley crossed the Atlantic in 1860, he apparently did not bring an insect collection with him, but during his first year at Kankakee he traded a heifer for a microscope to examine insect specimens. A few weeks later, he obtained a more powerful model out of Chicago. When he moved to Chicago in 1863, he took along a small collection. Through his own collecting, augmented by specimens farmers sent to the *Prairie Farmer*, his collection soon grew to substantial proportions.[30] In 1866, for example, a New York State correspondent sent Riley two worms for identification, explaining that he had sent them to the *Rural New Yorker* and the *Country Gentleman*, but neither had an entomologist able to identify them. Perhaps on the basis of immature insect specimens preserved in alcohol in his collection, Riley identified both species, which were injurious to apple trees, and explained their natural history.[31]

At the Illinois State Fair in 1866, Walsh presented a single case of insect specimens representing the most injurious insects. Riley exhibited five cases of insects at the same fair, complemented by oil paintings of caterpillars.[32] The following year (1867), Walsh (by then Illinois state entomologist) displayed forty cases of insects embracing ten thousand species and sixty thousand specimens.[33] It is quite likely that Riley's collection was at that time comparable to Walsh's.

There are only fragmentary references to Riley's collection during his Missouri years. In 1873, a reporter described a walnut cabinet in Riley's office filled with insect specimens.[34] That same year, another reporter described two walnut cabinets (one of which had sixty drawers) containing a total of fifty thousand "bugs" in the state "Bug Shop."[35] As state entomologist, Riley continued displaying insect specimens at the state fairs.[36] In 1867, Riley and Otto Lugger prepared a cabinet of sixty drawers for the University of Missouri at Columbia. Riley's own collection was by then certainly larger than the collection he prepared for the college.[37]

Riley regarded the collection at the state entomologist's office as his personal property in the same way as Walsh considered his collection his personal property.[38] Apparently no one questioned his transfer of the collection to Washington. Following his appointment as federal entomologist in May 1878, Riley apparently moved the collection from his home (or perhaps the office of the Entomological Commission) to the Agricultural Department. When a year later he was forced out of the Agricultural Department, he transferred all or parts of the collection back to his home. Sometime following his appointment as honorary curator of insects at the National Museum in January 1881 (possibly after his reappointment as entomologist in the Agricultural Department in April 1881), he transferred his collection to the new museum building. In 1885, he offered to deed his collection to the museum under the condition that either Congress grant him a salary as curator (in addition to his salary as federal entomologist) or that the federal government purchase the collection.[39] No price was mentioned, but in 1867 the federal government had purchased Glover's collection and drawings for $10,000, and Riley likely thought of that as a minimum.[40] In October 1885, he wrote to Baird again, now offering his collection as an outright gift, his only condition being that, should he move out of Washington, he be permitted to withdraw the "purely biological" material. Riley noted that the collection represented over twenty-four years of labor (which would place its inception in 1861) and the expenditure of many thousands of dollars. He estimated its market value to foreign museums seeking a general collection of North American insects at $20,000.[41] Baird accepted, and Riley's collection officially became the nucleus of the National Insect Collection.[42]

Although Riley held his collection as the trump card, he felt under pressure to play it. In 1885, Riley, his wife, and all the children were ill. Their five-month-old son died that year. At age forty-three, he was concerned about the disposition of his possessions, his collection being perhaps the most valued item. (In 1888, he prepared his last will and testament.) Riley was possibly concerned about Baird's health as well. Although only sixty-two, Baird was showing signs of professional and personal pressures. In fact, Baird died less than two years after Riley made his donation in 1885.

As part of his donation, Riley described the contents of his collection and other material then in the "National Museum Collection of Insects."[43] He described four components: the Riley Collection, the Department of Agriculture Collection, the National Museum Collection, and the Exhibit Collection of Economic Entomology. The fourth component included little that was of enduring taxonomic value and will be omitted from the present discussion.

The Riley Collection consisted of 766 boxes of pinned specimens representing seven orders of insects plus Arachnida, insect galls, insect architecture, and miscellaneous items "not yet arranged." This totaled 115,058 specimens representing 15,328 species. In addition, there were nineteen boxes of alcoholic material containing 2,850 vials "chiefly of adolescent states of insects." Very small insects, numbering approximately 3,000, were mounted in balsam on slides. Riley also described the boxes he had designed for the collection. He reported the total number of species at 20,000, roughly twice the number when he had moved from Chicago to St. Louis.

The Department of Agriculture Collection consisted of 623 boxes, and a large assortment of slides and alcoholic specimens accumulated "during the past seven years," i.e., since 1878, the year Riley and Baird initiated their partnership. The 623 boxes represented specimens from three sources: 300 boxes from Comstock's tenure (1879–81); 323 boxes "containing all of the more recent material collected for or reared at the Department during the past four years" (i.e., during Riley's tenure since 1881); and the Brazilian collection of J. C. Brenner, the collection of Koebele, and several purchased collections. No mention was made of material consigned by the Smithsonian on "permanent loan" to outside collaborators.

Riley described the material from Comstock's tenure as consisting of Coleoptera and Lepidoptera "in rather poor condition . . . looked upon as duplicates . . . and rarely used in the work of the Division."[44] This assessment reflects Riley's animosity toward Comstock and LeDuc rather than an objective appraisal. Although Comstock lacked Baird's support during his tenure as federal entomologist, he enlarged the Agricultural Department collection by three hundred boxes, an impressive showing for two years in office. In 1879, upon being asked to explain his plans for the Division of Entomology to the Entomological Club, Comstock replied that he had assigned a full-time assistant to the rearing and mounting of insects, that he planned to establish a biological collection, and that he had invited specialists to arrange specimens in the different orders.[45] The full-time assistant was most likely Theodore Pergande, Riley's assistant dating back to his Missouri days.[46] Comstock's plans to establish a biological collection may have been in response to Riley's removing his collection from the department. Comstock's significant innovation was engaging outside specialists to work up the various sections of the collection. During his short tenure, he engaged six specialists to work on a volunteer basis.[47]

The National Museum Collection, Riley reported in 1884, had been accumulated "during the past three years or since the Department of Insects has been in existence" (1881–84). It consisted of fifty boxes of pinned specimens and one hundred bottles of material preserved in alcohol. The latter included an assortment of older material turned over from the Agricultural Department. Specialists had removed the best specimens, the "residuum" being of no great value. Riley estimated that it totaled about twenty thousand specimens, representing two thousand species. Riley's statement indicates the relative size of the three components that formed the initial National Insect Collection:

Collection	No. Boxes	Percent of Total	No. Specimens	Percent of Total	No. Species	Percent of Total
C. V. Riley	766	53%	115,058	62%	15,328	69%
Dept. Ag.	623	43%	50,000	27%	5,000	22%
National Museum	50	4%	20,000	11%	2,000	9%
Total	1,439	100%	185,058	100%	22,328	100%[48]

In 1881, Baird reserved Riley several rooms for entomology in the new museum building. From that time on, Riley spent a few hours at the museum every afternoon, identifying, mounting, and arranging specimens.[49]

Riley still hoped for a paid position as curator. Such appointments at the Smithsonian were rare but not unknown. For example, in 1882, when Thomas resigned his professorship at Southern Normal University and his position as Illinois state entomologist to become curator of Native American artifacts for the Bureau of Ethnology at the Smithsonian, his position was funded through Powell's Bureau of Ethnology. Baird agreed to this arrangement because Powell's bureau paid Thomas's salary and because he feared that artifacts from the mound builders in the Mississippi Valley might go abroad rather than to the museum.[50] Unfortunately for Riley, there was no comparable danger of Europeans absconding with American insect specimens. Baird saw no reason to pay Riley as entomological curator when he was already on the payroll in the Agricultural Department, and he quietly ignored Riley's broad hints at a paid curatorship. When Baird died in 1887, Samuel Pierpont Langley, an astronomer and physicist, became secretary of the Smithsonian, and George B. Goode, an

Agassiz-trained ichthyologist who had functioned as director of the museum without benefit of title, was named museum director.[51] Goode and Riley continued the partnership Baird and Riley had initiated.

Riley considered it his prerogative to select an assistant to carry on the day-to-day work of the Department of Insects. He narrowed his selection to three candidates: Herbert Osborn, professor at Iowa State University; Otto Lugger, his former assistant in Missouri, now at the Maryland Academy of Science and Literature in Baltimore; and John B. Smith, recently recruited by him for the Division of Entomology and highly regarded for his systematic work at the Brooklyn Entomological Society. Riley considered Lugger best qualified for routine work, but he thought Smith more likely to make original contributions.[52] Goode and Baird concurred in Riley's choice of Smith, who was appointed as assistant curator on August 1, 1885. Smith proved to be an excellent choice. He was a recognized authority on Coleoptera and Lepidoptera, but more important, he had an ebullient personality and a contagious sense of humor. Colleagues and students delighted at his uninhibited enthusiasm on field trips as he pursued flying insects in the bushes and over fields, his short legs scarcely raising him above the meadow grass.[53] He had the light touch Riley lacked. He was popular among his peers and was elected secretary of the AAAS Entomological Club in 1884 and in 1888.[54]

While Smith's colleagues were pleased at his appointment, Riley set standards that challenged even Smith's equanimity. Riley's written communications reveal his often oppressive style as well as Smith's forbearance. On one occasion, after a review of Smith's performance, Riley demanded that Smith make no additions to his personal collection, that he sell his personal collection to the museum or other parties at a reasonable price (Smith sold it to the museum in 1887),[55] that Smith recognize that his continued employment was at Riley's pleasure (although he was officially appointed by the museum director), that he recognize Riley's position as head of the Department of Insects, that he was to resign his position as editor of *Entomologica Americana*, that all correspondence regarding museum matters (whether addressed to Riley or not) was to be considered official, that Smith was to avoid invading specialized areas in which Riley was working, and that the monograph of Noctuidae on which they were working jointly was to take first claim of Smith's time. Having reviewed this dismal list, Smith, with utter magnanimity, expressed his pride in the progress of the collection and his pleasure in being associated with Riley.[56] Despite Riley's overbearing, even unfair treatment, Smith remained unperturbed; he and Riley maintained a productive working relationship.

Smith pressed Riley on occasion to recommend him for pay raises and advancement in rank, but during four years as assistant curator, Smith's salary increased by only $300.[57]

On the other hand, Riley wrote chatty letters to Smith from Europe and, as though confiding to a friend, reported on his own health: "In proportion as my head gets better and I get more rest, in that proportion, I get homesick and want to be with my family and at my work again."[58] On one occasion, he wrote to Smith with detailed instructions concerning work to be done and then concluded, "I congratulate you on your matrimonial venture and would beg you to give my very best regards to Mrs. Smith."[59] Here was the paradox of Riley's personality: always the grand plan in mind, relentless in control of his subordinates, complaining of his precarious health, yet on the other hand socially cordial, with no hint of animosity. In 1889, Smith resigned to accept a professorship at Rutgers College.[60] As a scientist in the expanding experiment station system, Smith became renowned for his work on the biology and control of mosquitoes.[61]

Specimens for the National Insect Collection came primarily from federal government agents, who were required to deposit their material with the museum. Most of the specimens came from Department of Agriculture employees over whom Riley had supervisory control, as well as extension agents at state experiment stations.[62] By 1894, the collection had more than doubled in size, from approximately 20,000 to 40,000 species and from approximately 350,000 to 610,000 specimens.[63] Although not happy with the relationship of the experiment stations to the Division of Entomology, Riley nevertheless promoted the National Insect Collection as a central component of the experiment station system. Smith, as assistant curator, coordinated the work with the experiment stations.

Other federal agents, like those with the Fish Commission, the army, and the navy, contributed specimens more sporadically. One of the most prolific contributors was Thomas L. Casey, scion of a distinguished military family and a colonel in the Army Corps of Engineers. He rose to fame entomologically by describing over nine hundred species of Coleoptera, in the process creating consternation among coleopterists by his extreme "splitting" of species. Strangely enough, this seems to have been overlooked by Riley, who commended Casey and accepted his collection of 20,000 species and 9,200 types, his splitting of species notwithstanding.[64]

Another important source of specimens was the donation or purchase of private collections. Riley's placement of his collection in the museum prompted others to follow suit. Donated private collections included the Texas insects of T. W. Belfries, Lucien M. Turner's insects from Labrador, the Alaska insects of Edward Burgess and C. L. McKay, as well as collections from C. H. Dally (India), Jean F. Perez (Mexico), William Beutenmüller, and Samuel Wendell Williston (Kansas).[65] Riley often diverted Agricultural Department funds to purchase biological material, but even he had limits. In 1884, Titian R. Peale, Philadelphia artist and naturalist, offered to sell one hundred colored plates of Lepidoptera "with full descriptions from original observations and notes of transformation" for five dollars per plate. Baird urged Riley to purchase them with Agricultural Department funds, but Riley demurred because the work had no significance for economic entomology.[66]

Amateur collectors, prompted by patriotism and/or scientific curiosity, sent specimens to the Smithsonian where, they were assured, their captures would be curated in secure cabinets alongside other natural history treasures. Unfortunately, amateur collectors' aspirations sometimes exceeded their entomological competence. Baird, who was even more solicitous than Riley of amateur cooperation, insisted that specimens, regardless of source, be processed through the system, entered in the accession file, and acknowledged with personal thanks from the Smithsonian secretary. Riley chafed under what he considered Baird's excessive patronizing of donors, pointing out that a letter of acknowledgment over the signature of the secretary implied that the specimen was a valuable addition when in fact it might be worthless. Somewhat irritated, Riley summarized the take for 1886: "There were altogether 103 accessions during the year, which came to the department in the ordinary course, a large proportion useless for any purpose."[67] Baird prevailed, however, and the flow of specimens from various outside sources continued. From a vessel at Galveston, Texas, came a single centipede for which the notation on the accession record read "irreconignizable [*sic*]."[68] By contrast, a single specimen, *Carabus truncaticolis*, from S. Applegate in Alaska, was noted as "a rarity, the most important accession of the month."[69]

In typical ad hoc mixing of responsibility, Riley charged Division of Entomology staff with processing and cataloging specimens for the National Insect Collection. Division procedures were based on guidelines developed by Riley dating back to his Chicago days. Pergande was a key figure in taxonomic work. His twenty-two notebooks of observations on insects reared over three decades comprise a masterpiece of the genre.[70] Lugger, Howard, Eugene A. Schwarz, George Marx, and others contributed according to their expertise. Riley operated like an entomological gadfly, buzzing about in virtually all orders (arachnology included), naming or renaming specimens, challenging experts on their own turf, and resolving questions through access to specimens in the collection.[71] Remarkably, in most cases this self-trained entomologist was proven right. One notable exception, early in his career, was his judgment that ticks were *not* the carriers of Texas fever in cattle.[72] A Riley portrait from the 1870s, commissioned by him, depicts him seated, examining an insect with a hand lens. Riley, the field naturalist par excellence, was also Riley the taxonomist, identifying and cataloging specimens.[73]

Inevitably, Riley's expansion of the National Insect Collection conflicted with Hagen's project of establishing a central entomological collection at the MCZ. In December 1884, (a year before Smith's appointment as assistant) the editor of *Science* charged that an honorary curator at the National Museum without a full-time assistant was "worse than useless," and he warned that collections sent to the National Museum were not properly cared for.[74] The next month Charles H. Fernald, professor of natural history at Maine State College, insisted that the MCZ was the only repository in the United States that provided dependable curation. He concluded that, because the curator at the National Museum was fully occupied elsewhere, it would be better for Americans to sell their collections to European museums than to consign them to destruction in a museum without proper staff.[75]

These were fighting words. Riley responded that the National Museum, not the MCZ, held first claim as the central repository of the nation's insect fauna.[76] Citing the statutory requirement that federal employees deposit their specimens in the National Museum, he asserted that in the past three years the National Museum had curated more material than any other institution, including the MCZ. He concluded, "Washington is fast becoming the chief natural history centre of the country; and the national museum . . . offers . . . as secure a repository for collections as any other institution."[77] The exchange reflected the longstanding rivalry between Riley and the Harvard academic elite. At the 1880 AAAS Entomological Club meeting in Boston, Scudder pointedly described Comstock, who was in charge of the Agricultural Department collection, as a curator "who was in earnest," implying that Riley was inexperienced as a curator.[78] Riley bristled at such innuendo from Cambridge patricians and readily returned the criticism. That same year, he characterized Hagen's insecticide prepared from beer mash as "entirely useless."[79] In the mid-1880s, Riley lost two critical rounds to the MCZ when LeConte's Coleoptera and Osten Sacken's Diptera went to Cambridge.[80] By the late 1880s, however, the ascendancy of the National Insect Collection seemed assured, and Riley and his Boston-Cambridge competitors discontinued their public posturing.

The collection steadily outgrew its allotted space; however, during the depression of the 1890s, Riley's pleas for more space and assistance fell on deaf ears. Following Smith's departure, Goode hired an aide at a lower salary to handle routine maintenance while relying on entomologists in the Division of Entomology in taxonomic matters. Riley urged Goode to hire a second aide but to no effect.[81] Distressed by conflict with J. Sterling Morton, the new secretary of agriculture, and budgetary cuts at the museum, Riley predicted "a period of comparative stagnation" with regard to the collection.[82] When Goode solicited his recommendations for the Department of Insects in 1893, Riley replied that, considering the poor reception of his previous recommendations, he lacked the spirit to make any additional requests.[83]

Riley's spirit reflected the state of his health and his escalating conflict with Morton (see chapter 15). In this atmosphere, he resigned as federal entomologist on April 23, 1894. He expressed relief at terminating his "slavedom [*sic*] as Government Entomologist" and voiced his desire to become the full-time curator of insects at the National Museum.[84] Once again, Riley departed for Europe. Following his return in the fall of 1894, he proposed to Goode that he be paid a salary, that he be relieved of routine taxonomic duties, and that he be allowed to carry out original investigations in three areas: the inheritance of acquired characteristics, the role of insects in the food chain of fishes, and the nature and development of flight in insects.[85] He hinted that if a salaried position were not forthcoming, he would be forced to look elsewhere. Although Goode was sympathetic to Riley's proposal, his hands were tied by budgetary constraints. Whether Riley was actually in financial need is not clear. In 1885, he had told Eugene W. Hilgard that he was financially independent and therefore free to serve or resign from the Agricultural Department at his pleasure.[86] Stressed and ill, Riley spent little time at the museum in the winter of 1894, and in the summer of 1895 he sailed to Europe again.

After Riley's death in September 1895, Goode appointed Howard as honorary curator, continuing the tradition of dual appointment. Howard served as curator of the National Insect Collection until his death in 1950; thus, Riley and his appointee, Howard, guided the National Insect Collection through its first seventy years. While the two differed in style, their vision for the collection and for the partnership between the two agencies was essentially the same. Howard's appointment coincided with a new era of civil service and college-educated entomologists. In recognition of the changing times, Howard appointed four custodians: David W. Coquillett (Diptera), Eugene A. Schwarz (Coleoptera), William H. Ashmead (Hymenoptera), and O. F. Cooke of the Bureau of Plant Industry (Myriapods).[87] These appointees, Cooke excepted, were handpicked members of Riley's cadre.

In 1976, entomologists from around the world gathered in Washington for the Fifteenth International Congress of Entomology. Riley's influence was still evident. Presiding over the conference was Curtis W. Sabrosky, a Dipterist at the Department of Agriculture who was assigned to the National Insect Collection. In the early twenty-first century, the institutional partnership between the Department of Agriculture and the National Museum is still functioning. The expansion of the collection and the increase in personnel and facilities fulfilled Riley's vision of a world-class collection. Until 1940, the museum had only two positions in entomology. In 2016, the staff consisted of forty Department of Agriculture positions and about eighty museum positions. The collection is now the second largest worldwide.[88] Riley provided the expertise, energy, leadership, and his own collection; Baird provided the institutional framework. The National Insect Collection stands as an enduring tribute to their partnership.

14

Unfinished Business

[I have] the feeling that my own labors . . . are . . . about to end.

—C. V. Riley, Presidential Address, American Association of Economic Entomologists, 1890, *Insect Life* 3 (January 1891), 209

During the last decade and a half of his life, Riley was not only widely recognized as the nation's leading applied entomologist, he was also the most prominent interpreter and popularizer of entomology in America. At the Entomological Society of Washington, at the "Saturday Lectures" of the Biological Society of Washington, and at the National Museum lectures, Riley spoke on Yucca Moth fertilization and other popular topics.[1] In his Lowell Lectures in Boston in 1891–92 he utilized "Stereopticon Views" of hand-drawn illustrations, speaking before ever-increasing audiences on the topics of social insects, gall insects, the interactions of insects and plants, and (appropriate to his New England listeners) the Gypsy Moth.[2]

In 1888, near the end of Norman J. Colman's commissionership, Riley initiated *Insect Life*, an Agriculture Department publication devoted to the investigation of injurious insects. *Insect Life* incorporated the ideals of an entomological publication Riley had sought since writing for the *Prairie Farmer* a quarter of a century earlier. An official publication of the Agricultural Department, *Insect Life* was distributed free of charge to universities, experiment stations, and other entomologically oriented institutions. As editor, Riley controlled content, style, and layout. *Insect Life* was an immediate success. Agricultural editors praised its timely and informative articles, and they copied liberally from its pages.[3]

In addition to his prestige as lecturer and author, Riley was widely praised for his organization of the American agricultural exhibit at the Paris Exposition in 1889.[4] The exhibit featured the Hessian Fly, Army Worm, Rocky Mountain Locust, and other insect pests that were displayed on plants they attacked, along with recommended control measures. The exposition jury awarded Riley grand prizes for his phylloxera and viticulture displays.[5] Four years later, in 1893, he prepared a special exhibit on entomology for the Columbian World's Fair in Chicago.[6]

Although Riley enjoyed his reputation as economic entomologist and as popular interpreter of insect form and behavior, he considered his work in some respects unfinished. He had failed to gain membership in the National Academy of Sciences.[7] Riley was likewise thwarted in his efforts to achieve a professorship. Although he insisted on being called "Professor Riley" on the basis of his honorary doctorate from the University of Missouri, he never gained a regular professorship. His 1877 proposal to assume a combined position as both a professor at Iowa Agricultural College and the Iowa state entomologist was sidetracked when he was appointed chief of the Entomological Commission. In December 1886, in a letter to John O. Westwood, Riley indicated his wish to succeed Westwood as professor of zoology at Oxford University. Westwood admired Riley and reportedly would have been pleased to have Riley follow him, but following Westwood's death in 1893, his Oxford colleagues chose Edward B. Poulton, one of their own, to succeed him.[8] Combined with Riley's wish for a professorship was his wish to produce a synthesis on entomology as an update to Packard's *Introduction to Entomology*, for use as a textbook in schools and colleges. He often expressed this wish but never found time or opportunity to carry out the project.[9]

Another item of "unfinished business" was the matter of patent rights to insecticides and insecticide machinery. In the 1870s and 1880s, commercial firms frequently claimed that Riley, Charles E. Bessey, and other government officials endorsed their products.[10] Riley not only refused to give official endorsements but also energetically opposed government officials endorsing commercial products or accepting compensation for endorsements. In 1879, responding to suggestions in the press that he stood to gain personally from the sale of insecticides, he denounced these charges. Although during his tenure as Missouri state entomologist he had recommended Paris Green, he had rejected offers from commercial firms to promote sales of their products. Likewise, in the early 1870s, when Missouri nurseries were shipping millions of grapevine stocks to France, he turned down offers to recommend specific commercial products or to invest in the sale of vines.[11]

Riley often voiced his fundamental objection to the American patent system. Having invented a spraying device for Paris Green, a locust catcher, and having coinvented the Riley-Barnard cyclone nozzle, he considered himself a qualified judge of patents.[12] He supported the National Grange in its efforts to revise the patent laws so that farmers would not be prosecuted for buying a product some firm claimed to have patented. Riley likewise advised farmers and orchardists to ignore patent claims to standard products like Paris Green, London Purple, and orchard fumigators.[13]

From his office across the Mall from the Patent Office, Riley frequently feuded with the patent commissioner. When William Barnard's heirs pressed Barnard's patent claims to the cyclone nozzle, Riley requested that Secretary of the Interior L. Q. C. Lamar (whose department included the Patent Office) allow D. A. McKnight from the attorney general's office to represent the Agricultural Department.[14] Riley wanted McKnight to argue the case because he had been involved in the Barnard claims from the beginning. When the case dragged on and a new judge was appointed, Riley wrote to Assistant Attorney General John B. Cotton in October 1890, urging him to engage McKnight (now a private attorney rather than an officer of the Justice Department) to represent the Department of Agriculture. Riley felt the case so important that he was prepared to pay McKnight's expenses if necessary.[15] With Riley's guarantee of payment, McKnight took on the case.[16]

At about the same time Riley brought McKnight in on the Barnard case, Californians renewed their patent claims to the hydrogen cyanide fumigation process. Riley advised California growers to ignore these patent claims because, he argued, the fumigation process had been developed for the public, was now common knowledge, and was therefore not subject to patent.[17] In June 1891, when the HCN fumigation case was in process, Riley stated publicly his views on the American patent system and his prescription for reform. The basic problem, Riley asserted, was that the Patent Office regularly granted patent rights for minor modifications to existing machines or processes. In his view, the free market could replace the entire patent system: "The economic entomologist points out the want and the means of meeting it; the cultivator creates the demand and the manufacturer supplies it."[18]

In July 1891, as if to snub Riley, the commissioner of patents granted the Californians a patent to HCN fumigation. Coinciding with this decision, and with the Barnard case apparently headed the same way, the *New York Sun* published an article charging Riley with patenting the Riley nozzle and using his official position to advertise it. Riley responded indignantly, "I have never taken a favor for information given [;] have never attempted to control any . . . discoveries for my own benefit [;] have never applied for a patent on any of them and have strenuously opposed all attempts to do so . . . whether by those employed under me or by those not so employed."[19] Norman J. Colman came to Riley's defense, pointing out that, far from profiting from his position, Riley had often subsidized staff members from his own pocket. The *Sun* admitted its mistake and issued an apology.[20]

Smarting at the suggestion that he had used his office for personal gain, Riley appealed to his scientific colleagues gathered in August 1891 for the AAAS meeting in Washington. Riley's public arraignment of the Patent Office, especially its infringement on other departments as in the Barnard case, created a considerable stir among the scientists. His protest, however, failed to bring about substantial changes in the patent system.[21] Riley's battle with the Patent Office persisted after his resignation and, in a sense, even after his death; in 1897, the Patent Office issued patents for the kerosene emulsion and the cyclone nozzle to Barnard's widow.[22]

Riley's free market concept seems to have prevailed with regard to the cyclone nozzle despite the posthumous granting of patent rights to Barnard's heirs. By the mid-1880s, various versions of the cyclone nozzle were being produced by American manufacturers.[23] The Vermorel nozzle, patterned after the cyclone nozzle but patented and manufactured in France, represented a special case. By 1889, various firms offered Vermorel nozzles in the United States, apparently without an American patent. In the early 1890s, with the advent of all-purpose sprays (fungicide and insecticide), Vermorel nozzles became widespread.[24] In 1894, Vermorel donated fifteen sprayers to the Department of Agriculture for display at the Columbian Exposition.[25] When Secretary of Agriculture J. Sterling Morton asked Riley whether the cyclone nozzle had been patented outside the United States (i.e., in France), Riley replied, evasively, that he didn't know.[26] In fact, he had to have known, having visited the Vermorel establishment in 1889.[27] The question of patents thus remained a major item of unfinished business.

A third item of unfinished business was organizing the emerging profession of economic entomology. Organization of the profession became critical in the 1890s as college-educated scientists filled positions in the experiment stations.[28] In his position as Missouri state entomologist and through his participation in the United States Agricultural Society and its successor the National Agricultural Society, Riley had advocated federal funding for experiment stations. College administrators, however, feared federal domination of the station network. As a result, they insisted on a combined state and federal system.[29]

In December 1879, Riley addressed a gathering of agricultural scientists who were organizing as the American Agricultural Association. Speaking as chief of the Entomological Commission, having been released as federal entomologist a few months previously, Riley discussed agricultural education in England, France, and Germany as possible models for the United States.[30] Riley held that England, with its aristocrat-dominated agricultural schools, was too small and homogeneous to serve as a model for the United States. In France, the Department of Agriculture (located within the Department of Commerce and Public Works) operated an extensive system of agricultural schools and experiment farms. This provided a better model. Riley, however, preferred the German model where the minister of agriculture held a cabinet-level position, and each province maintained a model farm supervised by a prominent farmer with training in agricultural science. Riley admired the national agricultural school in Berlin and the Ministry of Agriculture that distributed agricultural bulletins (Amtstblätter) gratis to farmers. He applauded the German government for giving free passes on the national railroad to those attending agricultural fairs, for requiring the planting of fruit and ornamental trees along public roads, and for mandating that farmers destroy caterpillars and other pests. All this, he asserted, was administered most efficiently and economically.[31] Riley's extolling the virtues of German agricultural education despite his strong emotional ties to France is an indication of his international and cosmopolitan orientation in matters of science and learning.

Turning to the United States, Riley began by pointing out that the agricultural commissioner was appointed on the basis of party politics with no regard to scientific qualifications or the wishes of farmers (Fredrick Watts and William LeDuc being two prime examples). He called for the establishment of a national agricultural advisory board comprised of division heads (Riley anticipated returning as chief entomologist) and other scientifically qualified individuals. The board would then choose a qualified person to serve as agricultural commissioner. Riley was unclear as to whether this person would be chosen from the board or from a wider selection of candidates. He also proposed the establishment of a central experiment station in Washington, with branches in each state.[32]

Following his address at the American Agricultural Association in 1879, Riley contributed little to the debate leading to the passage of the Hatch Act in March 1887. Why he maintained a low profile on this issue is unclear and not a little surprising. He may have sensed that the establishment of state experiment stations inevitably shifted influence away from the federal government and his division. His fellow Missourians, Congressman William Hatch, who was chairman of the House Committee on Agriculture, and Commissioner Colman, played key rolls in the final formulation of the Hatch Act, but Riley remained outside the immediate circle that orchestrated its passage.[33]

Although Riley played no substantial role in the passage of the Hatch Act, he was well aware that this legislation marked a basic change in the organization of agricultural science. Under the Hatch Act, each state was authorized to establish a department devoted to agricultural research (i.e., an experiment station) at the agricultural college or university of its choosing. Stations were funded at $15,000 per year from combined federal and state funds. Initially the federal government funded 85 percent and the states 15 percent.[34]

Hatch Act funding fed the growth of the land-grant universities. After struggling in the postwar years with meager funding and enrollments, the universities began assembling agricultural faculties, establishing departments of agriculture, and turning out students trained in agricultural science. By 1887, when Riley's staff reached a plateau, the agricultural colleges were producing increasing numbers of graduates who anticipated careers in agricultural science. The change was evident at Cornell, where the experiment station, established in 1880, had been crippled by lack of funding. Suddenly Hatch Act funds facilitated a doubling of the station's staff and John H. Comstock's construction of an insectary (something Riley longed for in Washington).[35] After 1887, each state's annual allotment of $15,000 was roughly the equivalent of the budget for the Division of Entomology. Although this amount was for the entire program (not only entomology), funding at this level meant that each state hosted a program comparable in scope to Riley's division. Following the Hatch Act, an expanding cadre of experiment station employees with college degrees rapidly replaced the older generation of self-trained and (often) foreign-educated scientists like Riley.[36]

The immediate issue was how to organize the expanding ranks of economic entomologists into an effective professional organization. Existing entomological organizations held limited appeal for experiment station professionals. The American Entomological Society, after experimenting with the *Practical Entomologist* in the 1860s, had now returned to systematics. William Henry Edwards, a lepidopterist, complained that Ezra T. Cresson, the society's secretary, devoted so much of the society's proceedings to "long papers on Hymen[optera] [Cresson's specialty] and Coleop[tera]" that no one else had a chance of getting published.[37] Comstock, who visited the Entomological

Society of America in the 1880s, noted that the society's collection included no biological material.[38]

The Entomological Club of the AAAS was more appealing. Here, entomologists discussed outbreaks of injurious insects, the relative merits of insecticides, and rules governing insect taxonomy. Riley, however, favored the establishment of a new organization that would have close ties with the agricultural colleges as well as with the AAAS. He also believed membership should be limited to professional economic entomologists, that is, those whose primary task was controlling insect pests.[39] Riley proposed that the society hold its annual meetings in Washington and semi-annual meetings with the AAAS and/or winter gatherings in various regions so that experiment station workers would not have so far to travel.[40]

In the first volume of *Insect Life* (1889), Riley called for a national organization of economic entomologists. He cited the increasing number of economic entomologists, their need for communication and exchange, and the need to speak with a unified voice on insecticides and related matters.[41] From his post at the Paris exhibition during the summer of 1889, Riley shared his ideas with James Fletcher, dominion entomologist of Canada, and Leland O. Howard. During that summer, Fletcher joined Howard in Washington, and together, with Riley's input from Paris, they sketched out a blueprint for the society. In one of his letters, Riley outlined to Howard his vision of the new organization, as well as some tactical planning to get it launched:

> I am glad that Fletcher is so favorable to the idea of a National Association of Economic Entomologists. All things considered, I believe it will be much better for him to issue the call as president of the Entomology Club, and go ahead under that call and organize at the forthcoming meeting of the Association. I shall . . . not be there, and that will make it all the easier probably if the members . . . choose to put me in as the first presiding officer. This, however, I don't much care, though of course I should appreciate the honor. My own belief is that it is better to confine it to the working and practical economic entomologists, as there is hardly a necessity for a National Association of any other kind. I hope that if Fletcher comes . . . you will draft very carefully a constitution and by-laws, and set forth the object of the Association fully and plainly. If you care to send me a rough copy of such a draft I will try to do what I can in offering suggestions just in time for you to receive it before the meeting.[42]

Riley wanted Fletcher to direct the organizational meeting in order to mollify critics who charged that he was dominating the undertaking. Fletcher, operating on home territory in Toronto where the Entomological Club held its meeting, could direct the proceedings while Riley was in Paris.

In June 1889, Fletcher met with Howard in Washington, where the two, with the aid of Riley's "Dear Howard" letters, drafted the constitution.[43] Professional entomologists appeared in gratifying numbers in Toronto in August. The plan moved like clockwork. The constitution and by-laws were adopted, and officers were elected: Riley, president; Stephen Forbes, vice president; and John B. Smith, recently resigned as Riley's assistant at the National Museum, secretary.

The objectives of the new association were broad and inclusive, but the membership was confined to workers in economic entomology, which was reinforced in the name: "Association of Official Economic Entomologists."[44] Among the twenty-three charter members (including Riley, not in attendance), seventeen were extension scientists, two were USDA employees, and four were in the Canadian federal service.

The first regular meeting of the new association was held November 12, 1889, at the National Museum. As honorary curator and head of the Department of Insects at the museum, Riley presided over the meeting on his own turf. By-laws were approved and "Official" was dropped from the name, making it the Association of Economic Entomologists. The second meeting was held in connection with the Association of American Agricultural Colleges and Experiment Stations. The new entomological society thus sought the sponsorship of the agricultural schools in preference to the AAAS, the latter incidentally having no funds to allocate.

The society's proceedings were published in *Insect Life*. Riley thus held positions as federal entomologist, president of the Association of Economic Entomologists, editor of *Insect Life*, and curator of the National Insect Collection. This concentration of offices did not please everyone. Forbes, for example, warned Riley against abusing his power.[45] Despite disclaimers that he sought the presidency, Riley was the obvious choice. He was the most articulate advocate of economic entomology, and he held the key positions in the discipline.

At the early meetings, members discussed whether experiment stations should publish new species names and other "non-economic" matter in station bulletins. The members unanimously agreed that such material should not be included. A second issue was whether station scientists should recommend patented insecticides and machinery. The members concluded, no doubt to Riley's disappointment, that experiment station officers could recommend patented products, provided that analysis and tests demonstrated their worth.[46]

The decision to allow extension scientists to test and certify insecticides could be interpreted as a safeguard to growers, in line with the Riley-Walsh practice of exposing unscrupulous salesmen and dubious products; however, by assuming responsibility for testing insecticides, extension scientists took on a time-consuming task that tended to forge a partnership between the experiment stations and producers of insecticides. The potential for conflicts of interest was thus built into the system.

From Riley's perspective, the most pressing organizational problem was how the Division of Entomology and the station entomologists would share responsibility for investigating injurious insects and advising growers. Prior to the Hatch Act, Riley sought to place his agents to best advantage both politically and geographically. Recognizing the political clout of the state of Illinois with its expanding grain industry, Riley directed his agent, Francis M. Webster, to investigate grain insects. Forbes, the strong-willed proponent of insect ecology, took exception to Riley invading his turf.[47] The new regime established under the Hatch Act strengthened Forbes's hand considerably. Riley hastened to reassure state officials that he wanted to work in harmony with them. In 1891, he observed that "the men who . . . shaped the . . . Hatch bill were careful that the Department's function should be to indicate, not dictate, to advise and assist, not govern or regulate."[48] Despite such reassurances, the problem of USDA personnel melding harmoniously with station personnel persisted.

By coordinating their meetings primarily with the agricultural colleges rather than the AAAS, professional entomologists faced the danger of losing touch with the mainstream of science. In 1892, the secretary of the Entomological Club announced that the club would meet concurrently with the Association of Economic Entomologists at the meeting of the AAAS in Rochester, New York. He encouraged club members to present papers of an applied nature at the meeting of the economic entomologists, reserving for the Entomological Club meeting matters of "purely scientific interest."[49]

Competition for the loyalties of entomological professionals mounted when the Association of American Agricultural Colleges and Experiment Stations assumed official oversight of station entomologists. Forbes, serving as chairman of the Entomological Committee of the association, set the agenda and meeting dates for experiment station entomologists. Riley bristled at the suggestion that the organization he had so carefully crafted might operate separately from the AAAS, where he had decisive influence. He also objected strenuously when the Entomological Committee restricted membership in the Association of Economic Entomologists to experiment station agents from the United States, excluding Fletcher and the Canadians. The Entomological Club, weakened by the exit of station entomologists, met at the AAAS meeting at Rochester in 1892. New officers were elected for the following year, but the "next" meeting was never held. Thus ended the Entomological Club that from its founding had served as the primary forum for American entomologists.[50]

The problem of a fragmented entomological discipline, divided along arbitrary lines of "economic" and "non-economic," soon became evident. In his presidential address before the Association of Economic Entomologists in 1892, Riley avoided the issue, stressing instead the progress of economic entomology over the past two decades (the years of Riley's career in Missouri and Washington). Riley noted with gratification the new esteem in which the public now held his discipline. What change had taken place since the days of William Kirby, William Spence, Thaddeus W. Harris, and Benjamin D. Walsh, when entomologists were characterized as eccentric bug hunters! Riley now took the offensive: "Applied entomology . . . when pursued with unselfish enthusiasm born of love of investigation and the delight in benefiting our fellow men is inspiring [and] deservedly so, considering the vast losses to our farmers from insect injury. . . . Our work is elevating in its sympathies for the struggles and sufferings of others."[51]

Riley's euphoric remarks regarding the status and future of economic entomology sidestepped matters of fundamental discord, in particular the fragmentation of entomologists into aesthetic (or pure) and applied camps. In this sense, his efforts in organizing the new profession remained unfinished business.

Another matter of unfinished business involved Riley's ambitions for advancement. The elevation of the commissioner of agriculture to secretary of agriculture in the last month of Grover Cleveland's first administration (March 1889) provided for the appointment of an assistant secretary to handle increased administrative duties. Riley was unhappy at the prospect of an additional level of administration between him and the secretary, but he was pleased with Jeremiah McLain Rusk's appointment of Edwin Willits, president of Michigan Agricultural College in Lansing, as assistant secretary in charge of scientific work.[52] Neither Riley nor Willits harbored grudges from a decade earlier, when Willits, who was Anna Comstock's cousin and then a junior congressman, escorted the Comstocks around Washington and introduced them to influential Washington personalities.[53]

The rapport between Riley and Willits continued during the Harrison-Rusk administration (1889–93). In 1892, Grover Cleveland was elected a second time and, as was customary, he prepared to appoint a new secretary and assistant secretary of agriculture. Riley made it known that he wanted to replace Willits as assistant secretary (he did not oppose an additional level of bureaucracy when he filled the position). The time seemed propitious. Riley considered himself on good terms with Cleveland (though it is not clear whether they ever met personally), and Cleveland had announced that he would not reappoint officials from his first cabinet.[54] Riley, now settled with his family at Sunbury and enjoying the prestige accorded a successful Washington bureau chief, considered the assistant secretary position a natural next step in his career. When Cleveland named Julius S. Morton, former governor of Nebraska, as secretary of agriculture, Riley's expectations rose accordingly.[55] Known widely as the founder of Arbor Day, Morton would likely share Riley's midwestern frame of reference concerning agricultural science. The *Chicago Times*, referring to Riley as "an old friend and admirer of Morton," quoted Riley on Morton's qualifications: "Mr. Morton is admirably fitted [as] Secretary of Agriculture, being a college man, a statesman of the highest character and thoroughly identified with the agricultural . . . interests of the country . . . His appointment is hailed with satisfaction by those interested in the department regardless of party."[56] When Morton arrived in the capital in March 1893, Riley invited prominent Washington scientists to Sunbury for a lavish reception in Morton's honor.[57]

Cleveland and Morton, however, had other plans. Morton asked Willits to remain in office until the end of the Columbian Exposition in January 1894.[58] More importantly, Cleveland directed Morton to name a southerner to the position. Cleveland noted that for decades northerners had headed the department, Rusk from Wisconsin, Colman from Missouri, Loring from Massachusetts, Willits from Michigan, and now Morton from Nebraska. Southern interests needed to be recognized. Morton reviewed twenty-three candidates (including Riley), but in consideration of Cleveland's mandate to choose a southerner, he rejected them all.[59] Morton eventually chose Charles W. Dabney Jr., a native of Virginia and president of the University of Tennessee. Dabney's credentials were impeccable. He was a southerner, the son of a Presbyterian clergyman who had served as General "Stonewall" Jackson's chief of staff and later his biographer. Before being named president of the University of Tennessee, Dabney had served as director of the North Carolina state agricultural experiment station.[60] If the president insisted on a southerner, Dabney was the man. When it became clear that Riley was no longer a contender, he insisted he didn't really want the appointment; however, he demonstrated his displeasure in tactless ways.[61]

Coinciding with Cleveland's assumption of the presidency in March 1893, an economic depression set in that dominated government and life during his administration. In the months following his inauguration, businesses and banks failed by the hundreds, throwing an estimated one-fifth to one-third of factory workers out of work. Prices for farm commodities plummeted, and banks foreclosed on mortgages. The Cleveland administration responded with draconian cuts in government programs.[62] Morton, a strict fiscal conservative, believed that federal and state government, including the Department of Agriculture, was bloated and inefficient. He directed division chiefs to submit a list of individuals to be dropped from the payroll.[63] Riley replied that no one could be dropped, and in fact, the division had vacancies that should be filled.[64]

Morton's cuts struck directly at Riley's accustomed free hand. Reversing Colman's policy of sending Riley to the AAAS and other professional gatherings at government expense, Morton informed Riley that he would have to pay his own travel and expenses to the Association of Economic Entomologists, even though he was a featured speaker there.[65] Morton also forbade Riley to take leave, even for health reasons.[66] Even more demeaning was Morton's interference in departmental publications, in particular the establishment of an editorial committee to edit forthcoming reports. Riley, author of the famed Missouri Reports and thousands of articles in agricultural and scientific journals, rankled under the requirement that he submit his reports to a committee to correct his prose with regard to capitalization, punctuation, parentheses, and other matters of form.[67]

Although direct documentation is lacking, Riley likely resented Dabney's zealous application of civil service regulations. When Cleveland took office in 1893, only 698 out of 2,497 departmental employees were classified according to the civil service merit system. Three years later (after Riley's death in September 1895), only the secretary, assistant secretary, chief of the Weather Bureau, and common laborers were still outside civil service. Although Riley never openly complained about civil service regulations, he most certainly would have bristled at the suggestion that Dabney, or anyone, might rule on his scientific credentials or propose that he submit himself to civil service exams.[68]

Thus, the beginning of the Cleveland-Morton administration boded ill for Riley and his program. Considering that up until then he had operated as a power broker of agricultural science, this represented another area of unfinished business.

A final item of unfinished business carried a name: John Henry Comstock. During his two years out of office (1879–81), his relations with Comstock had so soured that any reconciliation appeared unlikely. At one point Riley referred to Comstock as "a little sneaking liar."[69] In 1881, Comstock returned to Cornell, where he became the nation's preeminent entomological teacher and scholar. Rising each morning prior to classroom lectures, Comstock constructed his *Introduction to Entomology* (1888) that became the Bible for generations of entomological students, the kind of reference work Riley had envisioned writing himself.[70] Comstock's entomology department at Cornell flourished and, at the invitation of David Starr Jordan, his former student, now president of Stanford University, he helped organize the entomological department at Stanford.[71]

In 1893, Comstock presented seminal works on the taxonomy of Lepidoptera that were hailed as the basis of insect taxonomy according to evolutionary principles.[72] Riley was keenly aware of his rival's accomplishments. He may also have known, or suspected, that Comstock's pay as a professor and academic administrator was, by the 1890s, on a par with his, or even higher.[73]

On at least two occasions subsequent to Riley's reinstatement as federal entomologist in 1881, Comstock and Riley met personally. In August 1887, Comstock and Riley attended the AAAS in New York City. There they discovered that they both opposed Cope's bid to become president of the association. Late one night, following apparently successful maneuvering by the anti-Cope contingent, Comstock came to Riley's room to thank him for his support. Riley wrote to Howard regarding his visit with Comstock, "It is always pleasant to forget and forgive past wrongs so generously, if so tardily acknowledged."[74] Anna Comstock recalled years later that on that occasion Riley was in a conciliatory mood, that Comstock met him half way, and they buried the hatchet.[75]

Three years later, in 1890, Riley stopped in Ithaca to inspect Comstock's insectary. Anna Comstock, rising to the occasion, invited him to tea. She recalled later that she and Riley kept up a conversation but that John Henry remained noticeably quiet. Riley indicated privately to Anna Comstock that her husband was probably better off having returned to Cornell rather than remaining in Washington.[76] On this uncertain note, the relationship between two founding fathers of the American entomological profession remained stalled. Riley's problematic relationship to Comstock was another, painful, matter of unfinished business.

15

A Valuable Career Cut Short

In January 1894, nine months after Grover Cleveland's cabinet (including Julius S. Morton and Charles W. Dabney) assumed office, Riley accepted an invitation from the Montserrat Company to investigate scale insects infesting citrus plantations on Montserrat Island, a British colony in the Lesser Antilles. In typical unconventional fashion, Riley took leave of absence without pay for two weeks and arranged to have Henry Hubbard accompany him (Hubbard was not on the government payroll). Riley arranged with Morton (apparently without consulting Dabney) to hire Hubbard on commission for two months, his primary duty being a revision of his citrus insect report. Riley planned to have Hubbard accompany him to Montserrat, gather information relevant to his revision, and return home to finish his report. The Montserrat Company paid Riley's salary and travel expenses for Riley and Hubbard. Riley anticipated benefits for citrus growers in Montserrat and in Florida. He even entertained the possibility that they might duplicate the Vedalia victory by introducing a lady beetle from Florida to the island of Montserrat.[1]

Due to transportation problems in the West Indies, Riley and Hubbard missed their return passage. Upon Riley's return in late April, Dabney, apparently miffed at Riley's unorthodox arrangements and his failure to inform him of his plans, presented Riley with a letter charging him with disregard of travel regulations and unauthorized leave of absence. He charged Hubbard with performing private, commercial work on government time, and he informed Riley that Hubbard would not be paid.[2] Riley, harried from lack of sleep, attempted without success to see Morton, and failing this, sent Morton a letter explaining their hurried preparations for the trip and the problems related to unreliable transportation and communication in the West Indies. He charged Dabney with impugning his and Hubbard's honor and intentions.[3] Morton replied that Dabney's conduct was appropriate: Riley had instructed Hubbard in a manner contrary to Hubbard's commission (the commission made no reference to Hubbard traveling to Montserrat), and Riley had failed to inform Dabney of these arrangements. In light of Morton's response, Riley saw no possibility of reconciliation. Following a meeting with Morton, Riley submitted his letter of resignation on April 26, 1894.[4]

In a public statement, couched in the form of a press release, making no reference to the Montserrat controversy, Riley stated that he was resigning in the interests of his health and his family. When, in the following months, speculations circulated that strife with Morton had led to his resignation, Riley urged Morton to make a public statement that he had not been forced to resign, and Morton complied.[5] Meanwhile, Riley circulated a letter among his colleagues, reiterating that he had long contemplated his resignation, that his official duties had been burdensome, and that his health was the most important consideration in his decision to retire. Alpheus Spring Packard and others urged Riley to reconsider and to return to his office, but Riley replied that his decision was final.[6] Leland O. Howard organized a dinner and reception for Riley on June 1, 1894, the day his retirement became effective.[7]

Riley's initial reaction was one of relief. He spent June and part of July resting and working at the National Museum. In July, he departed for England, complaining that he felt ill and weak.[8] By fall, Riley was back, but his outlook was little improved. His English sojourn had failed to relieve old health problems. Financial worries compounded his gloom. He considered applying for reinstatement with the Department of Agriculture, but Morton informed Riley that Dabney had responsibility for hiring and that Dabney declined to consider the possibility. In August 1894, prior to Riley's departure for England, John D. Richie, the New Zealand secretary of agriculture, asked him to suggest someone to serve as government entomologist. Riley replied that he knew of no qualified candidates who were available, then added (much to Richie's surprise) that, if the salary were adequate, he would consider applying himself. Negotiations continued through the next year, and the issue was pending at the time of Riley's death.[9]

Affairs at division headquarters deepened his despondency. Morton, in his annual report for 1894, boasted that he returned $600,000 (amounting to 23 percent of his budget) back to the treasury.[10] Howard's report as entomologist was condensed to 3 pages, down from 120 pages the year before, just enough space to announce the removal of the entire field staff and the discontinuation of most of the division's publications, including *Insect Life*. Thus ended Riley's dream publication and his hopes to place a field representative in every state. The single positive sign at the department was Howard's appointment as Riley's successor. Riley had urged Howard's appointment, but Morton had acquiesced only after Dabney polled leading American entomologists, who overwhelmingly endorsed Howard.[11]

In January 1895, Riley attended the meeting of the American Pomological Society in Sacramento, California. There, Byron Lelong and Ellwood Cooper revived old charges regarding the introduction of the Vedalia Beetle, and they circulated statements that Riley had been "fired" by the secretary of agriculture.[12] From California, Riley traveled to Columbia, Missouri, to deliver a series of lectures at the University of Missouri. He evidently hoped that his lectures would lead to a professorship.[13]

In April 1895, Riley escorted Alice and Mary to Paris to continue their education. Riley arranged for stops in England to visit Sister Nina, Aunt Mira, and Uncle Tim; in Bonn, where he had studied art; and in Denmark, where Emilie's relatives awaited them. In June, Riley attended the meeting of the Entomological Society of London, which had elected him an honorary fellow in 1889.[14] Riley returned in August to participate in the Association of Economic Entomologists and the AAAS annual meeting. He then joined his family for a vacation in Nova Scotia before returning to Sunbury.[15]

On Saturday, September 14, 1895, the sun rose clear and warm with high humidity and fair easterly winds, a beautiful September day in Washington. Shortly after 9 a.m. Riley and his fourteen-year-old son, William Garfield, pushed off on their bicycles from Sunbury. Upon reaching Columbia Road, they started down the steep incline with Riley leading. They were coasting rapidly, in contrast to the prudent speed usually assumed by Riley along this route. The incline gave way to level road at the intersection of Connecticut and "S" Street. Suddenly the front wheel of Riley's bike struck a piece of granite that had fallen from a wagon hauling broken stone for cobblestone pavement. Upon striking the stone, the bicycle wheel turned at a right angle, catapulting Riley headfirst over the handlebars and onto the pavement. Dr. A. B. Cline, manager of Taylor's Drugstore, located at the intersection, rushed to aid the stricken man. The alarm went out and soon Drs. Playter and Clark arrived to attend to Riley, who was unconscious. A deep laceration rent the skull over the left eye and across the nose. Blood oozed from Riley's left ear and collected in a pool on the pavement, indicating that the base of the skull had been fractured. The knuckles of both hands were badly abraded, indicating that he had gripped the handlebars tightly despite the fall. A horse-drawn police ambulance transported the unconscious man a quarter of a mile back to his home. Surrounded by his wife, children, and friends, Riley lingered on, with no hope of recovery, until he died shortly before midnight.[16]

Reverend Alexander Kent of the People's Church conducted the funeral at Sunbury on Tuesday, September 17, 1895. Alice and Mary, summoned by telegraph, departed Paris immediately but arrived after the funeral. Although many Washingtonians were still on summer vacation, mourners filled the mansion. Floral tributes covered the casket and overflowed into the living room. At Glenwood Cemetery, honorary pallbearers drawn from Washington's scientific elite carried the casket to the Riley family plot. Army commander A. W. Greeley and meteorologist Cleveland Abbe had worked with Riley in the locust investigations. Otis T. Mason, curator of ethnology at the National Museum, was a longtime associate and neighbor of Riley. William T. Harris, now US commissioner of education, had known Riley since their association in the St. Louis circle of metaphysicians. Theodore Gill, ichthyologist at the National Museum, had helped organize the Biological Society of Washington. Harvey W. Wiley was Riley's colleague as chief of the Division of Chemistry.[17] Riley's gravestone bears the inscription, "And in short measure—life may perfect be."[18] With death coming four days short of his fifty-second birthday, life was indeed short, but, measured against his grand ambitions and considering his precarious health, it was far from perfect.

In the following months, Emilie gathered the children about her and maintained the home as she had during Riley's absences. She was economically well situated due to the resources left by Riley, her father, and her younger brother, Theophilus Conzelman. As directed in Riley's will, the latter managed her finances. The family resided at Sunbury until 1898, when they moved to a more modest house at 1754 S Street NW. In 1914, a new house was built for the family at 2141 Leroy Place.[19] Alice and Thora lived there with their mother until her death in 1947. Mary lived with them until her death in 1939. William and Cathryn moved out sometime in the intervening years. William died in 1963, Alice in 1968, Thora in 1972, Helen in 1973, and Cathryn in 1978.[20]

Alice, the statuesque eldest daughter with her father's dark hair and coloring, assumed management of the household. Her autocratic manner created tension among the siblings, who nevertheless remained together for many years. Remarkably, only one of the seven Riley children married. Helen, the third youngest, fell in love with an Englishman. They married in England and had one child, Emilie, who also married an Englishman.[21]

FIGURE 15.1 Riley family plot, Glenwood Cemetery, Washington, DC (Riley Family Records, National Agricultural Library Special Collections, Beltsville, Maryland).

An assessment of Riley's life and contributions must begin with Howard's interpretation of Riley. Howard was Riley's assistant for sixteen years (1878–94), and he succeeded Riley as USDA entomologist, continuing in this position for thirty-three years. While Riley's early death deprived Riley of the opportunity of writing his own memoirs, Howard lived to age ninety-three and wrote extensively about his association with Riley. The two were studies in contrast. Riley, who was lithe, athletic, nervous, and ever on the go, projected the image of a flamboyant artist-naturalist, conspicuous by his stylish dress, prominent sombrero, and luxurious hair and mustache. Howard, by contrast, sported a balding head, round stomach, and wore modest clothing. Riley, intent on improving the world of farmers and horticulturists, frequently clashed with those who moved too slowly for his timetable. Howard, the plodding, reliable organization man, fulfilled the role of first assistant, dutifully balancing Riley's checkbook and performing editorial and administrative tasks to Riley's exacting standards. Whereas Riley often improvised with ad hoc administrative constructions, Howard bided his time, waiting for the right political moment to promote his programs.

Howard's interpretation of Riley, expressed most clearly in his *History of Applied Entomology,* set the dominant tone for subsequent evaluations of Riley, mostly by entomologists, for a century. Riley's primary achievements, according to Howard, included his Missouri Reports, which were based on "study and observation" and illustrated by "the most admirable woodcuts of insects . . . ever . . . published in this country; his founding of the US Entomological Commission, which undertook "the first broad national investigation of . . . insect plagues;" his founding and staffing of the Division of Entomology; his Grape Phylloxera investigations; his initiation of the first successful international experiment in classical biological control (the Vedalia Beetle); and his investigation of "non-economic" species, especially the Yucca Moth and hypermetamorphic blister beetles.[22]

While Howard's assessment of Riley's achievements was positive, his assessment of Riley's personality was definitely negative: "He was a restless, ambitious man, a great schemer, and striving constantly to make his work appear more important. He was ambitious to build up a large organization. Unfortunately, he made many enemies. . . . [T]here was [much] dissatisfaction [among staff members] and . . . unfair treatment of subordinates."[23] Howard's negative description of Riley's personality became the standard interpretation. The biographical sketch by Arnold Mallis in *American Entomologists*, for example, relies primarily on Howard, including his account of Riley's insomnia, for which he sought relief by sleeping in a barber's chair.[24]

Howard's statement that Riley made enemies is correct; these included Fredrick Watts, William G. LeDuc, and John H. Comstock, to name the most conspicuous. There is no general consensus among Riley's contemporaries that he alienated colleagues inside and outside his division. Riley was clearly ambitious, but so were John Wesley Powell, Beverly T. Galloway, and other bureau chiefs. They strove to build effective scientific agencies to carry out their mandate to improve society, not to exert personal power. Riley was often dictatorial and unfair in his treatment of subordinates, which was a regrettable failing; however, following his death, former employees (with the exception of Howard, much later) expressed their satisfaction at having worked under his supervision rather than resentment toward their former chief.

Howard's assertion that Riley appropriated his work and the work of other assistants is not substantiated by the historical record. His claim that he, not Riley, wrote the silkworm bulletin, for example, seems unlikely, since Riley had written extensively on silk culture in his Missouri Reports.[25] Howard correctly notes that appropriation of the work of students and subordinates was generally practiced during Riley's era. Riley, however, regularly credited the assistance of others in his Missouri Reports, and in division reports. Among the more than seven hundred publications credited to him for the years 1878–87, there is no indication of his appropriating the work of others.[26] He had no need to do so. He repeatedly demonstrated his vast knowledge of insect form and habits, he had superb resources in the collection at the National Museum and the entomological literature in department and Smithsonian libraries, and he wrote rapidly and effortlessly.

While Howard's negative portrayal of Riley's personality became widely accepted, Riley's family quietly erected an anti-Howard, anti-USDA interpretation that remained largely unknown to the public. From the time of Riley's retirement, the family was convinced that the Department of Agriculture, with orchestration from Howard, engaged in a systematic program to belittle Riley. In 1897, two years after Riley's death, Nina, in a letter to Emilie, noted that, whereas Riley had expressed his debt to Benjamin D. Walsh as mentor, Howard had never expressed his indebtedness to Riley.[27] Cathryn Vedalia disputed Howard's unflattering description of Riley in marginal notes she wrote in Howard's *History of Applied Entomology*: "Howard owed his appointment to and his training in Entomology to Riley, yet his jealousy made him lose no opportunity to belittle him . . . as witness Howard's own autobiography [*Fighting the Insects*, 1933] (written when he must have been senile) . . . He carefully waited to write his slanderous publicity until after most of the men who had known and admired Riley were dead."[28]

Beginning with A. Hunter Dupree, who in *Science in the Federal Government* (1957) credits Riley with initiating the problem-centered approach that transformed agricultural science in the late nineteenth century, historians have presented a more balanced assessment. Carol Anelli has examined Riley's relationship to Walsh and she and her coworker Richard A. Oliver have discussed his discovery (with George Engelmann) of yucca pollination. Edward H. Smith and Janet R. Smith, in collaboration with Carol Anelli, Conner Sorensen, and Yves Carton, have investigated Riley's leadership in applied entomology, his pioneering of entomological illustration, and his Grape Phylloxera campaign. Gene Kritsky has treated the relationship of Riley and to Darwin and Darwinism, and Conner Sorensen in *Brethren of the Net* underscores Riley's leadership in the American entomological community. Smith and Kritsky have edited Riley's Kankakee diary.[29] Of special importance for an evaluation of Riley's scientific abilities and accomplishments is the account of Riley's contributions as a taxonomist by Alfred G. Wheeler Jr., Richard Hoebecke, and Edward H. Smith.[30]

Wheeler et al. maintain that Riley's virtuosity in insect taxonomy and insect biology was the indispensable basis for his accomplishments as a scientific organizer and administrator. They point out that Riley's estimate of the total number of insect species—10 million—more than five times the estimates of his contemporaries, coincides remarkably with modern estimates ranging between 5 million and 30 million.[31] They note that the proposals and decisions of the committee on nomenclature of the Entomological Club, which Riley chaired, set taxonomic standards for entomologists worldwide. They confirm that his collection formed the basis of the National Insect Collection that today is ranked as the world's second largest. Although Riley named relatively few new species compared to his contemporaries (compare Riley's 203 new species to LeConte's 10,000 Coleoptera), he proposed the scientific and popular names for many of the most economically important species, like the Rocky Mountain Locust and Grape Phylloxera. His study of yucca pollination was one of the first investigations of plant-insect coevolution.[32] Wheeler et al. credit him with revealing the complex life history of the Grape Phylloxera and other species and the oviposition and voltinism of the Army Worm. They credit him with solving intricate problems like parasites whose instars exhibit different larval types, including hypermetamorphic larvae, and the alternation of generation in cynipid wasps on different parts of the hackberry representing different species.[33]

Wheeler et al. conclude that, had biosystematics been recognized as a field in the nineteenth century, Riley would have been perhaps its leading practitioner and that his insights have pointed the way to fruitful investigations. For example, his observations concerning the changes in characteristics of herbivores as they switch from native to cultivated plants presaged recent interest in rapid evolution.[34]

Riley was a key player in every aspect of American entomology in its formative period. As noted above, he was a central figure in insect taxonomy (in its largest sense). He embodied and led the transition from an entomological community consisting mainly of self-trained collectors primarily concerned with taxonomy to a community of university-trained scientists employed in public agencies with primary mandate of controlling insect pests. In the context of this transition, Riley played the key role in the organization of entomologists, the popularization of entomology, entomological illustration, and chemical and biological control of insect pests.

Unlike most of his contemporaries who specialized in particular groups or orders, Riley ranged over the entire insect fauna (including the arachnids). Only Walsh, and to some extent Packard, approximated Riley's catholic approach. In his first Missouri report (1868), when he was twenty-five, he addressed the question of why and how the Oyster Shell Bark Louse, a pest of the apple tree, produced a particular kind of scale:

> We might with equal reason try to learn why and how the thousand different excrescences and galls caused by insects are formed! Why is it that the larva hatching from an egg deposited on a rose leaf by a little four-winged fly . . . causes a peculiar growth . . . in the form of a mangel-wurzel, or beet seed, to surround it, while that of a similar fly, belonging to the very same genus . . . hatched from eggs deposited in the root of the same plant, causes an entirely different gall? Why is it that the puncture of a little yellow louse . . . causes an unnatural growth to surround and entomb it in the shape of the little green globular galls of different sizes . . . while the same sized puncture of another louse . . . produces no such effect? Why, again does a little Lepidopterous larva, often found in the golden rod . . . produce an elongated hollow gall, while a Dipterous larva . . . in a neighboring stalk produces one that is round and solid? Or, lastly, why should the suction of different species of Dipterous larvae . . . produce the wonderful galls found on our willows, causing in many instances not only a total change in the texture of the leaf, but also in its mode of growth?[35]

One might ask which entomologist in that era, or any era, has had such command of the biology and taxonomy of insects in various orders. Riley freely challenged the findings of lepidopterists like Joseph A. Lintner, Samuel H. Scudder, Alpheus Spring Packard, William Henry Edwards, and Augustus R. Grote; coleopterists like John LeConte; and hymenopterists like Ezra T. Cresson, regardless of their standing, academic or otherwise. In his sixth Missouri report, Riley pointed out that two Emperor Butterflies of the genus *Apatura,* much prized by collectors, had been misrepresented by succeeding generations of lepidopterists beginning with incorrect illustrations by Jean Baptiste Alphonse Chauffour de Boisduval and John Eaton LeConte [father of John L. LeConte] (1833), and continuing with John G. Morris, Townend Glover, and Asa Fitch, who copied the figures, thereby perpetuating the errors. Riley credited Edwards and Scudder with clearing up part of the confusion, and he congratulated Scudder for retaining names that had been familiar for forty years. He graciously concluded that the generic relationship of the two species "must be left to the future . . . Meanwhile Mr. Edwards's opinion is, in one sense, as rightfully held as Mr. Scudder's or mine."[36]

In 1885, Riley confided to Stephen Forbes that, although he encouraged his assistants to pursue independent investigations, he reserved for himself "special questions," not only because of their economic importance but for the sake of science: "Did I abandon all interest in such studies, my life as Entomologist here [in the Division of Entomology] would become one of mere drudgery and routine, devoted to the education and training of others."[37] Riley occupied the nerve center of economic entomology. When in the course of the Cottony Cushion Scale crisis, Albert Koebele sent specimens of parasitic insects to Riley for identification, Riley personally identified each species on the basis of specimens in the National Museum. On trips to Europe, he took along breeding cages with insects so he could observe their development. Riley continued his voluminous reading of entomological literature, though he admitted to Lintner that he found it increasingly difficult to stay abreast.[38]

Riley was the primary transmitter of learning from early American entomologists, including John Abbott, John Eaton LeConte, William Dandridge Peck, Thaddeus William Harris, Frederick Ernst Melsheimer, John Morris, and Asa Fitch, to his generation that included Walsh, Cyrus Thomas, Packard, Scudder, and Lintner, and to the succeeding generation of college- and university-trained entomologists, including Howard, Charles L. Marlatt, and Herbert Osborn. Although he did not personally know Harris (who died in 1856), he corresponded with Thomas Affleck, his Texas colleague in Cotton Worm matters, who had corresponded with Harris. William LeBaron, whom Riley knew well, considered himself a student of Fitch (whom Riley also knew from at least one visit). What Riley lacked in direct personal contact with early American entomologists, he compensated for by his familiarity with their writings. In connection with his discovery that the "Canker Worm" represented two distinct species, the Spring Canker Worm (*Anisopteryx vernata*) and the Fall Canker Worm (*Anisopteryx pometaria*), Riley reproduced Peck's 1795 essay on the Canker Worm to identify the source of confusion and its resolution.[39] Riley cited the history of this taxonomic problem because, he insisted, this would help prevent repeating such errors and thus provide a solid basis for effective control measures. For the Spring Canker Worm, he prescribed fall plowing under the trees and removing loose bark in the spring to destroy eggs; for the Fall Canker Worm, he prescribed traps and barriers applied in the fall to prevent the worms from climbing up the trees.[40]

Riley also forged Americans' primary link to the international entomological community. Although many Americans, like Packard and Scudder, made sojourns in Europe, Riley had the most extensive experience with European cultures, and he maintained the most frequent and sustained contact with European entomologists. He participated frequently in English and French scientific meetings, and he met and corresponded regularly with entomologists in Europe, Australia, New Zealand, the Cape Colony (South Africa), and elsewhere. When investigating the Colorado Potato Beetle and allied species, Riley examined Latin American specimens in the London collection of Henry Walter Bates, concluding that the tibial groove that Carl Stål had employed to erect a new genus was of no generic value.[41] This seemingly technical distinction, Riley pointed out, cleared up confusion that originated with members of the Belgian Entomological Society, who had lumped Stål's Latin American species together with the true (North American) Colorado Potato Beetle. He pointed out further that European efforts to prevent the pest's importation into Europe depended on correct identification of the true Colorado Beetle.[42]

Riley had formidable competitors. John Henry Comstock, for example, outshone Riley as a theoretician in evolutionary biology and as an organizer of academic entomological institutions. Forbes was Riley's most direct rival as an economic entomologist. Born in 1844, Forbes fought as a cavalry officer in more than twenty Civil War battles (compared with Riley's brief and uneventful volunteer service). Forbes studied medicine and taught in public schools before his appointment, in 1872, as director and curator of the Illinois Natural History Society and Museum at Illinois State Normal University (later Southern Illinois University) at Carbondale.[43] Concentrating first on the botany of Illinois, then on fish and other aquatic life in Illinois waters, then on the feeding habits of birds and other vertebrates and their role in the control of injurious insects, and finally on the insect pests of crops, Forbes emerged around 1890 as the leading exponent of ecology in the United States.[44] As Illinois state entomologist and director of the Illinois Laboratory of Natural History, Forbes presided over entomological agencies that rivaled the Division of Entomology.[45] In 1893, Forbes succeeded Riley as the second president of the American Association of Economic Entomology.[46] Following Riley's death, Howard declared that Forbes was better qualified than he to succeed Riley as chief of the Division of Entomology, and he accepted the appointment only because Forbes preferred to remain in Illinois.[47]

Had Forbes, rather than Riley, been confronted with the Rocky Mountain Locust emergency, their career paths may have been quite different. When Riley was named Missouri state entomologist in 1868, Forbes was teaching school and pursuing botanical investigations in Illinois. In 1874, when the locusts devastated western Missouri, Forbes was curator of the Natural History

Museum at Illinois State Normal University near Bloomington, Illinois. Even more problematic is the question of what Riley might have done if he had not met an untimely end. Forbes, Comstock, Howard, and Marlatt each enjoyed thirty years of productive activity after Riley's death.

Though it was cut short at fifty-two years, Riley led a remarkable life. As illegitimate son of an Anglican clergyman and a mother from a family with literary talents, he learned classics, language, and natural history from masters in England, France, and Germany. In Kankakee, Chicago, St. Louis, and finally Washington, DC, he contributed his unique talents as naturalist, artist, author, and organizer to the emerging discipline of applied entomology. The assessment of his friend and colleague James Fletcher is still the most appropriate: "[A]s an economic entomologist, take him all in all, he was far and away the most eminent the world has ever seen."[48]

Appendix

List of Insects and Other Arthropods Mentioned in the Text

This list is alphabetical by the common name (or, if common name is not mentioned, the scientific name) used in the text. Many insect names, common and scientific (Latin), have changed since Riley's time. However, many of the insects he studied were economically important and therefore have common names approved by the Entomological Society of America, as noted by an asterisk.* Each scientific name is a binomial; the first part is the genus (generic name) which is always capitalized and italicized, and the second part is the species (specific) name, which is also italicized but never capitalized. The author is the person who named the species; if the author name is in parentheses, the name (typically, the generic name) has been changed since originally named. A generic name followed by "sp." means that the species is unspecified or unknown. Orders (of insects and other organisms) consist of multiple families, which in turn consist of multiple genera (plural of genus); a genus consists of one or more species. Arthropods (Phylum Arthropoda) include hexapods (Subphylum Hexapoda) which includes the Insects (Class Insecta), which includes thirty Orders. Of the approximately 1 million insect species described (an estimated 10–20 percent of those actually in existence), the most diverse Orders are Coleoptera (beetles), Lepidoptera (butterflies and moths), Diptera (flies), Hymenoptera (ants, bees, and wasps), Hemiptera (true bugs, aphids, etc.), and Orthoptera (grasshoppers, katydids, and crickets).

* *Common Names of Insects Database*, Entomological Society of America, Annapolis, Maryland, www.entsoc.org/common-names (accessed February 27, 2018).

Insect common name in text	Insect scientific name in text	Current common name *
Ailanthus Silkworm	—	cynthia moth *
—	*Amara* (sp.)	[a ground beetle]
American Oak Phylloxera	*Phylloxera rileyi*	—
American Silk Worm	*Telea polyphemus*	polyphemus moth *
—	*Anthomyia*	[many species]
—	*Apatura*	[several species, Emperor butterflies]
Apple Curculio	—	apple curculio *
Apple Maggot	—	apple maggot *
Apple Root Louse	—	woolly apple aphid *
Apple Tree Bark Louse	—	woolly apple aphid *
Apple Tree Borer	—	apple bark borer *
Apple Tree Plant Louse	—	apple aphid *
Apple Worm: *see* Codling Moth		
Apple-Tree Borer	—	apple bark borer *
Army Worm	*Pseudaletia unipuncta*	armyworm *
Ash Gray Blister-Beetle	—	ashgray blister beetle *
Asparagus Beetle	—	asparagus beetle *
—	*Attacus cecropia*	cecropia moth *
Bean Weevil	—	bean weevil *
Bee-fly larvae	—	bee fly
Black Scale [of olive and citrus]	—	black scale *

Current scientific name with author	Order: Family
Samia cynthia (Drury)	Lepidoptera: Saturniidae
Amara sp.	Coleoptera: Carabidae
Phylloxera riley Riley	Hemiptera: Phylloxeridae
Antheraea polyphemus (Cramer)	Lepidoptera: Saturniidae
Anthomyia sp.	Diptera: Anthomyiidae
Apatura sp.	Lepidoptera: Nymphalidae
Anthonomus quadrigibbus Say	Coleoptera: Curculionidae
Rhagoletis pomonella (Walsh)	Diptera: Tephritidae
Eriosoma lanigerum (Hausmann)	Hemiptera: Aphididae
Eriosoma lanigerum (Hausmann) (most likely)	Hemiptera: Aphididae
Synanthedon pyri (Harris)	Lepidoptera: Sesiidae
Aphis pomi De Geer	Hemiptera: Aphididae
Synanthedon pyri (Harris)	Lepidoptera: Sesiidae
Mythimna unipuncta (Haworth)	Lepidoptera: Noctuidae
Epicauta fabricii (LeConte)	Coleoptera: Meloidae
Crioceris asparagi (L.)	Coleoptera: Chrysomelidae
Hyalophora cecropia (L.)	Lepidoptera: Saturniidae
Acanthoscelides obtectus (Say)	Coleoptera: Chrysomelidae
[many species]	Coleoptera: Bombyliidae
Saissetia oleae (Olivier)	Hemiptera: Coccidae

Insect common name in text	Insect scientific name in text	Current common name *
Blister beetle	—	blister beetle
Blue-Spangled Peach Worm	—	colona moth
Bogus Yucca Moth	*Prodoxus decipiens*	bogus yucca moth
Buck Moth	—	buck moth *
Borer of Red Currant	—	currant borer *
bumble bee	—	bumble bee [many species]
cabbage butterfly	—	cabbage butterfly
Cabbage Plant Louse	—	cabbage aphid *
Cabbage Plusia	—	cabbage looper *
Cabbage Plutella	—	diamondback moth *
Cabbage Tinea	—	leafminer
—	*Caloptenus atlanis*	migratory grasshopper * (in part)
Canker Worm	—	
—	*Carabus truncaticolis*	[a ground beetle]
Carpet-Clothes-and-Fur Moths	—	
chalcidid parasite	—	
Cheese Maggot	—	cheese skipper *
Chinch Bug	—	chinch bug *
—	*Chrysobothris* sp.	
Clover Worm	—	green cloverworm *
Cockroach	—	

Current scientific name with author	Order: Family
[several species]	Coleoptera: Meloidae
Haploa colona (Hübner)	Lepidoptera: Arctiidae
Prodoxus decipiens Riley	Lepidoptera: Prodoxidae
Hemileuca maia (Drury)	Lepidoptera: Saturniidae
Synanthedon tipuliformis (Clerck)	Lepidoptera: Sesiidae
Bombus sp. (many species)	Hymenoptera: Apidae
Pieris sp.	Lepidoptera: Pieridae
Brevicoryne brassicae (L.)	Hemiptera: Aphididae
Trichoplusia ni (Hübner)	Lepidoptera: Noctuidae
Plutella xylostella (L.)	Lepidoptera: Plutellidae
Phytomyza sp. (probably)	Diptera: Agromyzidae
Melanoplus sanguinipes subsp. *atlanis* (Riley)	Orthoptera: Acrididae
[many species]	Lepidoptera: Geometridae
Carabus truncaticolis Eschscholtz	Coleoptera: Carabidae
[various species]	Lepidoptera: Tineidae
[many species]	Hymenoptera: Chalcidoidea
Piophila casei (L.)	Diptera: Piophilidae
Blissus leucopterus (Say)	Hemiptera: Blissidae
[many species]	Coleoptera: Buprestidae
Hypena scabra (F.)	Lepidoptera: Erebidae
[many species]	Blattodea: several families

Insect common name in text	Insect scientific name in text	Current common name *
Codling Moth	*Carpocapsa [pomonella]*	codling moth *
Colorado Potato Beetle	*Doryphora 10-lineata*	Colorado potato beetle *
Common Bagworm	—	bagworm *
Common Flesh-Fly	—	
Common Meal Worm	—	yellow mealworm *
Convergent Ladybird	—	convergent lady beetle *
Cotton Boll Weevil	—	cotton boll weevil *
Cotton Stainer	—	cotton stainer *
Cotton Worm	*Aletia argillacea*	cotton leafworm *
Cottony Cushion Scale	*Dortheasia characias*	cottony cushion scale *
Crickets, Mormon	—	Mormon Cricket *
Croton Bug	—	German cockroach *
Curculio: *see* Plum Curculio		
Currant Plant Louse	—	currant aphid *
Cutworms	—	
Cylindrical Tiger Beetle	—	tiger beetle
cynipid wasps	—	gall wasps
—	*Cyrtoneura stabulans* (Fallén)	false stable fly *
deaths-head moth	—	death's head hawk moth
—	*Doryphora juncta*	false potato beetle *
earwig predator	—	

Current scientific name with author	Order: Family
Cydia pomonella (L.)	Lepidoptera: Tortricidae
Leptinotarsa decemlineata (Say)	Coleoptera: Chrysomelidae
Thyridopteryx ephemeraeformis (Haworth)	Lepidoptera: Psychidae
[many species]	Diptera: Sarcophagidae
Tenebrio molitor L.	Coleoptera: Tenebrionidae
Hippodamia convergens Guérin-Méneville	Coleoptera: Coccinellidae
Anthonomus grandis subsp. *grandis* Boheman	Coleoptera: Curculionidae
Dysdercus suturellus (Herrich-Schäffer)	Hemiptera: Pyrrhocoridae
Alabama argillacea (Hübner)	Lepidoptera: Noctuidae
Icerya purchasi Maskell	Hemiptera: Margarodidae
Anabrus simplex Haldeman	Orthoptera: Tettigoniidae
Blattella germanica (L.)	Blattodea: Blattellidae
Cryptomyzus ribis (L.)	Hemiptera: Aphididae
[many species]	Hymenoptera: Noctuidae
Cicindela sp.	Coleoptera: Carabidae
[many species]	Hymenoptera: Cynipidae
Muscina stabulans (Fallén)	Diptera: Muscidae
Acherontia atropos (L.)	Lepidoptera: Sphingidae
Leptinotarsa juncta (Germar)	Coleoptera: Chrysomelidae
[unknown species]	Dermaptera: several families

Insect common name in text	Insect scientific name in text	Current common name *
Emperor Butterfly	—	hackberry emperor
Emperor caterpillar	—	small emperor moth
—	*Euplectrus comstockii*	—
European Cabbage Butterfly: *see* Imported Cabbage Worm		
Fall Army Worm	—	fall armyworm *
Fall Canker Worm	*Anisopteryx pometaria*	fall cankerworm*
Fluted Scale: *see* Cottony Cushion Scale		
Forest Tent Caterpillar	—	forest tent caterpillar *
Glassy-Winged Soldier Bug	—	glassy-winged soldier bug
Goat Weed Butterfly	—	goatweed leafwing
Gooseberry Span-Worm	—	currant spanworm *
Grain Plant Louse	—	western wheat aphid *
Grain Weevil	—	granary weevil *
Grape Leaf Folder	—	grape leaffolder *
Grape Phylloxera	*Phylloxera vastatrix*	grape phylloxera *
Grape Vine Plume Moth	—	grape plume moth *
Grapevine Aphis	—	grapevine aphid *
Grasshoppers	—	
Gypsy Moth	—	gypsy moth *
Harlequin Cabbage Bug	—	harlequin bug *
Hessian Fly	—	Hessian fly *

Current scientific name with author	Order: Family
Asterocampa celtis (Boisduval & Leconte)	Lepidoptera: Nymphalidae
Saturnia pavonia (L.)	Lepidoptera: Saturniidae
Euplectrus comstockii Howard	Hymenoptera: Eulophidae
Spodoptera frugiperda (J.E. Smith)	Lepidoptera: Noctuidae
Alsophila pometaria (Harris)	Lepidoptera: Geometridae
Malacosoma disstria Hübner	Lepidoptera: Lasiocampidae
Hyaliodes vitripennis (Say)	Hemiptera: Miridae
Anaea andria Scudder	Lepidoptera: Nymphalidae
Itame ribearia (Fitch)	Lepidoptera: Geometridae
Diuraphis tritici (Gillette) (probably)	Hemiptera: Aphididae
Sitophilus granarius (L.)	Coleoptera: Curculionidae
Desmia funeralis (Hübner)	Lepidoptera: Crambidae
Daktulosphaira vitifoliae (Fitch)	Hemiptera: Phylloxeridae
Pterophorus periscelidactylus Fitch	Lepidoptera: Pterophoridae
Aphis illinoisensis Shimer	Hemiptera: Aphididae
[many species]	Orthoptera: Acrididae
Lymantria dispar (L.)	Lepidoptera: Erebidae
Murgantia histrionica (Hahn)	Hemiptera: Pentatomidae
Mayetiola destructor (Say)	Diptera: Cecidomyiidae

Insect common name in text	Insect scientific name in text	Current common name *
Hessian Fly parasite	*Entedon epigonus*	[a hymenopteran parasitoid]
Hickory Bark-Borer	—	hickory bark beetle *
honey bee	—	honey bee *
Hop Vine Hypena	—	hop looper *
House Fly	—	house fly *
Ichneumon flies	—	
Imported Cabbage Worm	—	imported cabbageworm *
Katydid	—	
large scale parasite	*Lestophonus iceryae*	[a dipteran parasitoid]
Leaf Beetle of Elm	—	elm leaf beetle *
lucky moth	—	
—	*Mantis carolina*	Carolina mantid *
May Beetle	—	May or June beetle
meal moth	—	meal moth *
mealy bugs	—	
Monarch Butterfly	*Danaus archippus*	monarch butterfly *
mud wasp	—	
Myriapods		millipedes, centipedes, etc.
Olive tree scale: *see* Black Scale		
Orange-Leaf Notcher	—	little leaf notcher
Orange-Leaf Nothris	—	orange webworm
—	*Orgyia pudibunda*	Pale Tussock Moth

Current scientific name with author	**Order: Family**
Pediobius epigonus (Walker)	Hymenoptera: Eulophidae
Scolytus quadrispinosus Say	Lepidoptera: Curculionidae
Apis mellifera L.	Hymenoptera: Apidae
Hypena humuli Harris	Lepidoptera: Erebidae
Musca domestica L.	Diptera: Muscidae
[many species]	Hymenoptera: Ichneumonidae
Pieris rapae (L.)	Lepidoptera: Pieridae
[many species]	Orthoptera: Tettigoniidae
Cryptochetum iceryae Williston (=*Cryptochaetum*)	Diptera: Cryptochetidae
Xanthogaleruca luteola (Müller)	Coleoptera: Chrysomelidae
[unknown]	Lepidoptera
Stagmomantis carolina (Johansson)	Mantodea: Mantidae
Phyllophaga sp.	Coleoptera: Scarabaeidae
Pyralis farinalis L.	Lepidoptera: Pyralidae
[many species]	Hemiptera: several families
Danaus plexippus (L.)	Lepidoptera: Danaidae
[many species]	Hymenoptera: Sphecidae or Crabronidae
[subphylum of arthropods]	[several classes]
Artipus floridanus Horn	Coleoptera: Curculionidae
Dichomeris citrifoliella (Chambers)	Lepidoptera: Gelechiidae
Calliteara pudibunda (L.)	Lepidoptera: Lymantriidae

Insect common name in text	Insect scientific name in text	Current common name *
Oyster Shell Bark Louse	—	oystershell scale *
Pear Tree Flea Louse	—	pear psylla *
Periodical Cicada	—	periodical cicada *
Pernyi	—	Chinese tussar moth
Phlox Worm	—	Darker-spotted straw moth
—	*Phora aletiae*	—
—	*Phylloxera quercus*	[species complex]
Phylloxera: *see* Grape Phylloxera		
Pickle Worm	—	pickleworm *
Pine-leaf Scale	—	pine needle scale *
Plum Curculio	—	plum curculio *
Pou blanc	*Icerya sacchari*	—
Queen [butterfly]	*Danaus gilippus*	queen
Rape Butterfly: *see* Imported Cabbage Worm		
Rascal Leaf-Crumpler (of fruit trees)	—	leaf crumpler *
Red Rust (of citrus)	—	citrus rust mite *
Red Scale (of citrus)	*Aonidia aurantii*	California red scale *
Red-legged Locust	*Caloptenus femur-rubrum*	redlegged grasshopper *
Red-spotted Purple [butterfly]	*Nymphalis ursula*	red-spotted purple
Rocky Mountain Locust	*Caloptenus spretus*	Rocky Mountain grasshopper *

Current scientific name with author	Order: Family
Lepidosaphes ulmi (L.)	Hemiptera: Diaspididae
Cacopsylla pyricola Foerster (probably)	Hemiptera: Psyllidae
Magicicada septendecim (L.) (species complex)	Hemiptera: Cicadidae
Antheraea pernyi (Guérin-Méneville)	Lepidoptera: Saturniidae
Heliothis phloxiphaga Grote and Robinson	Lepidoptera: Noctuidae
Megaselia aletiae (Comstock)	Diptera: Phoridae
Phylloxera quercus Boyer de Fonscolombe	Hemiptera: Phylloxeridae
Diaphania nitidalis Cramer	Lepidoptera: Crambidae
Chionaspis pinifoliae (Fitch)	Hemiptera: Diaspididae
Conotrachelus nenuphar (Herbst)	Coleoptera: Curculionidae
Icerya seychellarum (Westwood)	Hemiptera: Margarodidae
Danaus gilippus (Cramer)	Lepidoptera: Danaidae
Acrobasis indigenella (Zeller)	Lepidoptera: Pyralidae
Phyllocoptruta oleivora (Ashmead)	Acari: Eriophyidae
Aonidiella aurantii (Maskell)	Hemiptera: Diaspididae
Melanoplus femurrubrum (De Geer)	Orthoptera: Acrididae
Limenitis arthemis (Drury) subsp. *astyanax*	Lepidoptera: Nymphalidae
Melanoplus spretus (Walsh)	Orthoptera: Acrididae

Insect common name in text	Insect scientific name in text	Current common name *
Rose Chafer	—	rose chafer *
San Jose Scale	—	San Jose scale *
—	*Saperda* sp.	
Satellite Sphinx	—	satellite sphinx
Scales	—	
—	*Scenopinus*	window fly
Seventeen-year locust: *see* Periodical Cicada		
Silkworm	—	silkworm *
Snout Beetles	—	
Soldier [butterfly]	*Danaus eresimus*	soldier
Southern Cabbage Butterfly	—	southern cabbageworm *
Spanworm	—	
Spring Canker Worm	*Anisopteryx vernata*	spring cankerworm*
Stalk Borer	—	stalk borer *
Strawberry Crown-borer	—	strawberry crown borer *
Strawberry Leaf Beetle	—	strawberry leaf beetle
Strawberry Leaf-Roller	—	strawberry leafroller *
Streaked Cottonwood Leaf Beetle	—	cottonwood leaf beetle *
Striped Blister-Beetles	—	striped blister beetle *
Sweet Potato Beetle	—	sweet potato leaf beetle *
Tachinid (or Tachina) flies	—	

Current scientific name with author	Order: Family
Macrodactylus subspinosus (F.)	Coleoptera: Scarabaeidae
Quadraspidiotus perniciosus (Comstock)	Hemiptera: Diaspididae
[many species]	Coleoptera: Cerambycidae
Eumorpha satellitia (L.)	Lepidoptera: Sphingidae
[many species]	Hemiptera: several families
[many species]	Diptera: Scenopinidae
Bombyx mori (L.)	Lepidoptera: Bombycidae
[many species]	Coleoptera: Curculionidae
Danaus eresimus Cramer	Lepidoptera: Danaidae
Pontia protodice (Boisduval & LeConte)	Lepidoptera: Pieridae
[many species]	Lepidoptera: Geometridae
Paleacrita vernata (Peck)	Lepidoptera: Geometridae
Papaipema nebris (Guenée)	Lepidoptera: Noctuidae
Tyloderma fragariae (Riley)	Coleoptera: Curculionidae
Galerucella tenella (L.)	Coleoptera: Chrysomelidae
Ancylis comptana (Frölich)	Lepidoptera: Tortricidae
Chrysomela scripta F.	Coleoptera: Chrysomelidae
Epicauta vittata (F.)	Coleoptera: Meloidae
Typophorus nigritus (Crotch)	Coleoptera: Chrysomelidae
[many species of parasitoid flies]	Diptera: Tachinidae

Insect common name in text	Insect scientific name in text	Current common name *
Thirteen-year Cicada	—	periodical cicada
Thurberia Weevil	—	thurberia weevil
Tobacco Worm	—	tobacco hornworm *
Tomato Fruit Worm	—	tomato fruitworm * (= corn earworm* = bollworm*)
Twice-Stabbed Lady Bird	—	twicestabbed lady beetle *
Vedalia beetle	—	vedalia *
Viceroy [butterfly]	*Nymphalis disippus*	viceroy *
Walnut Tortrix Moth	—	southern ugly-nest caterpillar moth
Wheat Midge	—	wheat midge *
Wheat-Head Army Worm	*Leucania albilinea*	—
White Pine Scale: *see* Pine-leaf Scale		
White-Lined Sphinx Moth	—	whitelined sphinx *
White Winged Morning Sphinx Moth: *see* White-Lined Sphinx Moth		
yamamai silkworm	—	Japanese oak silk moth
Yucca Moth	*Pronuba yuccasella*	yucca moth *

Current scientific name with author	Order: Family
Magicicada tredecim (L.) (species complex)	Hemiptera: Cicadidae
Anthonomus grandis subsp. *thurberia* (or "variant")	Coleoptera: Curculionidae
Manduca sexta (L.)	Lepidoptera: Sphingidae
Helicoverpa zea (Boddie)	Lepidoptera: Noctuidae
Chilocorus stigma (Say)	Coleoptera: Coccinellidae
Rodolia cardinalis (Mulsant)	Coleoptera: Coccinellidae
Limenitis archippus (Cramer)	Lepidoptera: Nymphalidae
Archips rileyanus (Grote)	Lepidoptera: Tortricidae
Sitodiplosis mosellana (Géhin)	Diptera: Cecidomyiidae
Faronta albilinea (Hübner)	Lepidoptera: Noctuidae
Hyles lineata (F.)	Lepidoptera: Sphingidae
Antheraea yamamai Guérin-Méneville	Lepidoptera: Saturniidae
Tegeticula yuccasella (Riley)	Lepidoptera: Prodoxidae

Notes

Preface

1. "Entomology," no source (December 1875), box 1, scrapbook vol. 9, p. 113, Charles Valentine Riley Papers, Record Unit 7076, Smithsonian Institution Archives, Washington, DC (hereafter cited as SIA, CVR). Riley and his assistants created these scrapbooks by using outdated USDA annual reports and pasting articles from newspapers, agricultural publications, and other publications over the printed pages of the reports .

Chapter 1

1. Birth certificates, Charles Valentine Riley and George Riley, St. Luke, South Chelsea, Middlesex County, Riley Family Records (hereafter cited as RFR), Special Collections, USDA, National Agricultural Library, Beltsville, MD (hereafter cited as NAL). Previously held as a private collection, the Riley Family Records were officially moved to the National Agricultural Library in April 2018. As these records are processed by the NAL, some recataloging of material may take place, resulting in changes to folder numbers and other organizational systems. Citations in this book refer to the original folder numbers as they appeared in the privately held collection. Certificates of baptism for [Charles Valentine Riley], Parish of Saint Luke, Chelsea, July 7, [1844]: name (blank), parents names (blank), notation "illegitimate;" and George, October 2, 1845: name of parents (blank), notation "illegitimate." Folder 7D, RFR, NAL; Thomas Carlyle, who leased a house in Chelsea in the 1830s, described it as "genteel," noting that his neighbors on both sides had pianos. *Froude's Life of Carlyle*, ed. John Clubbe (Columbus: Ohio State University Press, 1979), 307–8. The 1851 Census for Chelsea lists Robert Pritchard (Wylde's father-in-law) and family as occupants of Caroline Cottage. Stanton Folder, RFR, NAL.

2. Birth certificate of unnamed female, July 11, 1837, Folder 2C, RFR, NAL; *Clerical Directory*, London, 1842–47, Folder 7D, RFR, NAL.

3. Ian Dyck, "From 'Rabble' to 'Chopsticks': The Radicalism of William Cobbett," *Albion* 21 (Spring 1989): 56–87; Martin J. Wiener, "The Changing Image of William Cobbett," *Journal of British Studies* 13 (May 1974): 135–54; H. D. Traill, ed., *Social England: A Record of the Progress of the People* (London: Cassell and Company, 1894–98) 6: 120.

4. Shelley detested Cannon personally, but he nevertheless agreed to correct the proofs for publication. Louise Schutz Boas, "'Erasmus Perkins' and Shelley," *Modern Language Notes* 70 (June 1955): 408–9.

5. Photocopy of handwritten court record, November 10, 1815, Folder 7C, RFR, NAL.

6. Mira Cannon Young to C. V. Riley, April 26, 1874, Folder 7C, RFR, NAL.

7. Mira Cannon Young to C. V. Riley, April 26, 1874, Folder 7C, RFR, NAL.

8. Laura, Mira, and Amelia stayed at home and perfected domestic skills. In 1851, Amelia and Laura immigrated to Australia. Amelia married, became a lady of fashion, and gradually broke off all contact with her family. Laura died in Australia in 1855, reportedly suffering from depression ("broken heart") at the news of her father's death. Mira remained in London where, at age twenty-eight, she married Timothy Young, a teacher and tutor. Mira had no children but devoted herself to Mary's children. Mira Cannon Young to C. V. Riley, April 20, 1877, Folder 7C; and Journal, December 24, 1851 and October 20, 1854, both in RFR, NAL.

9. Mary's sister Mira later explained to Riley that a gold locket Mary wore with the inscription "Garde d'amor, 1843" [beware of love] referred to the "loss of the dream of her girlhood and the one man she had really been dominated by." Mira to Charlie, April 20, 1877, Folder 7C, RFR, NAL.

10. Jane Darby Knox Wylde died of "consumption," following a long illness, on May 25, 1844. Death certificate for Jane Wylde, Folder 2C, RFR, NAL; Marriage license for Antonio Hipolito Lafargue, bachelor, and Mary St. Valentine Cannon, spinster, January 15, 1846, Parish Church, St. Giles-without-Cripplegate, London, Register of Marriages, East London Registrations District, General Register Office, London, no. 409, Folder 7D, RFR, NAL.

11. *Census 1851*, Chelmarsh, Shropshire, Upland House. The notation lists the following individuals: Head–John. F. Wylde, 74 widower, Halfpay surgeon, 'Army' member of the Royal College of Surgeons, London, not practicing. Farmer of 282 acres, employs 8; Charles E., nephew, widower, 42, Clergyman without care of souls. Stanton Folder, RFR, NAL. In 1972, Riley's daughter Helen and granddaughter, Emilie Wenban-Smith Brash, visited the Wylde estate, and Emilie recorded her mother's reminiscences of the Wylde and Riley families. See "Visit to Bridgenorth 1972," Folder 7D, RFR, NAL.

12. Marriage license for Charles Edmund Fewtrell Wylde (bachelor, clerk in orders) and Cecilia Elizabeth Bell (spinster), parish church, Walmer, County of Kent, August 17, 1852, Folder 7D, RFR, NAL.

13. During one parole in 1856, Wylde and his wife Cecilia produced a son, Tracy Fewtrell Wylde, who died at age six of "croup" [probably diphtheria]. Death certificate, Tracy Fewtrell Wylde, January 21, 1862, "six years of age," Lewisham Village, Kent County, Folder 7D, RFR, NAL; Inquisition of Coroner and Jurors and Certificate of death for Charles Edmund Fewtrell Wylde, age 51, Queen's Prison, December 17, 1859, St. George Southwark Registration District, General Registration Office, London. Wylde was committed to King's [later Queen's] Bench Prison May 21, 1855, and "discharged" at the time of his death, December 15, 1859. Public Record Office,

RC 6875, Pris 10/19, King's Bench, Fleet and Marshalsea Prisons Committal Book and Discharge Records, Folder 7D, RFR, NAL. Charles Dickens, whose father was incarcerated for debt in Marshalsea Prison, describes prison conditions in his novel, *Little Dorrit*.

14. Mira Cannon Young to C. V. Riley, April 20, 1877, Folder 7C, RFR, NAL.

15. Birth certificate for Louise Josephine Lafargue, Trinity Cottage, Trinity Street, Rotherhithe, County Surrey, July 17, 1848, Register of Births, Rotherhithe Registration District, County Surrey, Folder 7C, RFR, NAL. Louise Josephine's father is listed as Antonio Hipolito Lafargue, and her mother as Mary Saint Valentine Lafargue, formerly Cannon.

16. The *London Census*, 1851, lists the following inhabitants for Walton-on-Thames, Back Street: Andrew Bissett, gardener, 53 yr. born in Scotland; Annie, wife, 54 yr. born in Surrey; son George, groom, 20 yrs.; Annie daughter, [17?] yrs.; Charles Lafargue, 7, nurse child, scholar; George Lafargue, 5, nurse child, scholar. Riley Family Tree, Folder 7D, RFR, NAL. Riley's first Journal entry (1850) refers to two families living with the Bissetts: "Mr. Groom" (perhaps George Bissett, who worked as a groom) and "Mrs. Greathead," who might have been the daughter, Annie.

17. Journal, June 10, 1851 and Mira to Charlie, July 3, 1877, Folder 7C, both in RFR, NAL. Wylde's move to Bridgenorth in 1851 may have prompted Mary to inform her husband of her sons. It is likely that Charles and George lived with the Lafargue family for a time, although Riley makes no mention of this in his journal. Charles and George are listed as "nurse children" in the census for 1851 at both the Bissett residence in Walton and the Lafargue residence in East London. In 1848, and perhaps at other times, Charles and George stayed at their grandparents' home, where Mary's sisters still lived. Handwritten note (probably Mira's): "[Dec.] 25th, 1848. G & Ch a week with us. They are dear little pets." Folder 7C, RFR, NAL.

18. Journal, May 10, 1853, RFR, NAL.

19. Journal, June 10, 1851, RFR, NAL.

20. Journal, October 10, 1851, RFR, NAL.

21. Journal, March 14, 1851, and April 2, 1852, RFR, NAL.

22. David Elliston Allen, *The Naturalist in Britain: A Social History*. Princeton Science Library (Princeton, NJ: Princeton University Press, 1994), 90–91.

23. Norman J. Colman, a close friend of Riley's, cited Hewitson's influence in "Prof. C. V. Riley, U.S. Entomologist," *Colman's Rural World* (May 12, 1892): 4 (separate reprint). Two years after Riley's death, his half-sister Nina explained Riley's success without benefit of university training by citing Hewitson's influence as being equivalent to formal training. Nina Lafargue to Mrs. C. V. Riley, February 4, 1897, Folder 7C, RFR, NAL; Robert McLachlin, a British entomologist who knew both Riley and Hewitson, states that Riley was acquainted with Hewitson. R[obert] McLachlin, "[Riley] Obituary," *Entomologist's Monthly Magazine* 31 (1895): 269–70.

24. William C. Hewitson and John O. Westwood, *The Genera of Diurnal*

Lepidoptera: Their Generic Characters, a Notice of their Habits and Transformations, and a Catalogue of the Species of Each Genus, 2 vols. (London: Longman, Brown, Green, and Longmans, 1846–1852); William C. Hewitson and W. Wilson Saunders, *Illustrations of New Species of Exotic Butterflies, Selected Chiefly from the Collections of W. Wilson Saunders and William C. Hewitson*, 5 vols. (London: Van Voorst, 1851–1878).

25. William Kirby and William Spence, *An Introduction to Entomology; or, Elements of the Natural History of Insects*, 4 vols. (London: Longman, Hurst, Rees, Orme, and Brown, 1816–1826).

26. Allen, *Naturalist in Britain*, 50, 74–99, 101–5; Harriett Ritvo, "Animal Pleasures: Popular Zoology in Eighteenth and Nineteenth-Century England," *Harvard Library Bulletin* 33 (1985): 240–41; Adrian Desmond, "The Making of Institutional Zoology in London, 1822–1836, Part 1," *History of Science* 23 (1985): 156–58, 167–68, 176–77; and Adrian Desmond, "The Making of Institutional Zoology in London, 1822–1836, Part 2," *History of Science* 23 (1985): 234, 240–43.

27. J. F. M. Clark, *Bugs and the Victorians* (New Haven and London: Yale University Press, 2009), 9–32.

28. C. V. Riley, "The Gooseberry Span-Worm," *Ninth Annual Report on the Noxious, Beneficial, and Other Insects of the State of Missouri* (Jefferson City: State Printer, 1877), 5. Riley's Missouri Reports are hereafter cited as *First* [*–Ninth*] *Annual Report.*

29. Journal, September 20, 1853, and July 20, 1854, RFR, NAL.

30. Charles recorded that a fellow student who in rage threw an inkwell at their teacher, Hubbard, received a severe thrashing. Journal, April 15, 1854, RFR, NAL.

31. Journal, May 20, 1854, RFR, NAL; Francis Watson, *The Year of the Wombat: England, 1857* (London: Harper & Row, 1974), 32.

32. Journal, August 20, 1854, RFR, NAL. The date in Riley's journal is apparently inaccurate. British and allied troops landed in the Crimea in mid-September 1854, a month after Riley's journal entry. The futile charge of the British Light Brigade on the fortification at Sebastopol, immortalized by Alfred Lord Tennyson, occurred on October 25, 1854.

33. Simona Pakenham, *Sixty Miles from England: The English at Dieppe, 1840–1914* (London: Macmillan, 1967).

34. Journal, November 28, 1855, RFR, NAL. Details about the school were provided by Claude Feron, secretary of the Historical Society in Dieppe. Feron to Edward Smith, November 13, 1985, Folder 7C, RFR, NAL. In 1855, the boys' father was committed to prison in London, but there is no indication that they were aware of this.

35. Journal, May 20 and June 6, 1856, RFR, NAL.

36. Journal, February 28 and June 29, 1857, RFR, NAL.

37. The art book of sixty-six pages includes illustrated histories of the silkworm, the honey bee, the bumble bee, emperor caterpillar, deaths-head moth, and lucky moth. USDA, National Agricultural Library, Special Collections, Manuscript Collection (MC) 143, Beltsville, Maryland.

38. Nina [Lafargue] to Emilie [Riley], February 4, 1897, Folder 7C, and Journal, July 2, 1857, RFR.

39. Journal, July 14, 1857, RFR, NAL.

40. Mira to Charlie, April 20, 1877, Folder 7C, RFR, NAL.

41. The last entry in Riley's Journal, July 25, 1857, recounts the two brothers swimming. RFR, NAL.

42. George's whereabouts during Riley's Bonn years (1858–1860) are not known. For Riley's years in Bonn, see W. Conner Sorensen and Edward H. Smith, "Charles Valentine Riley: Art Training at Bonn, 1858–1860," *American Entomologist* 43 (Summer 1997): 92–104 (hereafter cited as *AE*); and W. Conner Sorensen and Edward H. Smith (with assistance of Norbert Schloßmacher), "Charles V. Riley in Bonn 1858–1860," *Bonner Geschichtsblätter* 43/44 (1996): 301–39.

43. Karl Heinz Stader, "Bonn und der Rhine in der englishchen Reiseliteratur," in *Aus Geschichte und Volkskunde von Stadt und Raum Bonn: Festschrift Joseph Dietz zum 80. Geburtstag am 8. April 1973. Veröffentlichungen des Stadtarchiv Bonn*, Band 10 (Bonn: Ludwig Rohrscheid Verlag, 1973), 117–53.

44. Klaus Peter Sauer, "Hermann Schaaffhausens Beitrag zur Entwicklung des Evolutionsgedanken und des Artbegriffs vor Darwin," *Mitteilungen der Deutschen Zoologischen Gesellschaft* (2002): 39-51.

45. Riley's only reference to his teacher by name appears indirectly in an article by Norman Colman, "Prof. C. V. Riley, U.S. Entomologist," *Colman's Rural World* (1892): 4. There the name is given as "Prof. A. Hoe," a misspelling of the name of the well-known artist, Nicolaus Christian Hohe. Riley's awareness of Bonn's university professors is alluded to in W. B. Davis and D. S. Durrie, *An Illustrated History of Missouri, Comprising Its Early Records, and Civil, Political, and Military History from the First Exploration to the Present Time, Including . . . Biographical Sketches of Prominent Citizens* (St. Louis: A. J. Hall & Co., 1876), 575. This publication appeared while Riley was Missouri state entomologist and was based on personal interview.

46. Sabine Gertrud Cremer, "Nicolaus Christian Hohe (1798–1868): Universitätszeichenlehrer in Bonn," PhD diss., Rheinischen Friedrich-Wilhelms-Universität zu Bonn, 2001 (hereafter cited as Cremer, "Hohe"). See especially pp. 82–85 and 100–106 (art instruction books), 94–95 (tourism and tourist art), 89–94 (portraits), and 95–99 (Hohe's landscape paintings). See also Heinrich Gerhartz, "Christian Hohe: Ein Beitrag zur Geschichte der rheinischen Malerei im 19. Jahundert," *Annalen des historischen Vereins für den Niederrhein* 128 (Dusseldorf: Von L. Schwann, 1936): 90–120.

47. Cremer compares, for example, Riley's depiction of Mehlem and Königswinter with Hohe's lithograph of the same scene. Cremer, "Hohe," 82–83.

48. The original album, with the fly leaf, "Bonn: Charles V. Riley, Sketched in Prussia, Dec. 1, 1858," is in the library of Riley's recently deceased granddaughter, Emilie Wenban-Smith Brash, Headley Down, Hampshire, England. The word "*nette*" is German for "neat" or "pleasing."

49. The crisis is mentioned in Colman, "Prof. C. V. Riley, U.S. Entomologist," (1892): 3, and R. McLachlan, "Obituary," *Entomologist's Monthly Magazine* (May 31, 1895): 269–70. His father's death, in December 1859, probably did not precipitate the crisis. It is not known how much he knew about his father at that time.

50. Riley, Bonn art book, cover page, and nos. 12, 46, and 47, RFR, NAL. Riley named the butterfly on the cover page "Origia pudibunda." Professor C. M. Naumann of the Zoological Research Institute, Bonn, identifies it as the *Calliteara pudibunda* (the Pale Tussock Moth), of the family Lymantriidae, a species still common there.

51. C. V. Riley, "How to Counterwork Noxious Insects," *Fifth Annual Report* (1873), 26. Riley no doubt revised his recollections in light of later knowledge.

52. W. Conner Sorensen, *Brethren of the Net: American Entomology, 1840–1880* (Tuscaloosa: University of Alabama Press, 1995), 155–56.

Chapter 2

1. *London Times*, "Weather," August 27, 1860; "Shipping Intelligence," *New York Times*, October 3, 1860; "Packet Ships," *The World Book Encyclopedia*, vol. 17 (Chicago: World Book, Inc., 1995).

2. Carlton J. Corliss, *Main Line of Mid America: The Story of the Illinois Central* (New York: Creative Age Press, 1950), 65; Riley to Grandmamma and Aunt, December 12, 1860, Folder 7C, RFR, NAL.

3. John Klasey and Mary Jean Houde, *Of the People: A Popular History of Kankakee County* (Chicago, IL: General Printing Co., 1968), 43–53; additional information supplied by Connie Licon and John Klasey of the Kankakee Historical Society. Exactly how Riley knew the Edwards is not known.

4. Vern LaGesse, a resident of Kankakee County, supplied logistics of the Kankakee vicinity, including historical maps. LaGesse/Kankakee Folder, RFR, NAL. Additional information on local geography supplied by John Klasey.

5. Vic Johnson, "Up 'til Now," *The Kankakee Sunday Journal*, March 18, 1990; Burt E. Burroughs and Vic Johnson, *The Story of Kankakee's Earliest Pioneer Settlers* (Bradley, IL: Lindsay Publications, Inc., 1986); J. Loreena Ivens et al., *The Kankakee River Yesterday and Today*, Miscellaneous Publication 60 (Champaign: Illinois Department of Energy and Natural Resources, Illinois State Water Survey, 1981).

6. Riley to Grandmamma and Aunt, December 12, 1860, Folder 7C, RFR, NAL.

7. In 1860, Riley reported that Charles Edwards owned one hundred head of cattle. Riley to Grandmamma and Aunt, December 12, 1860, Folder 7C, RFR, NAL. According to the records of the Kankakee County Tax Assessor for 1860, George owned 160 acres of land, 136 cattle, and draft animals plus tools and buildings valued at $50.

8. Riley to Grandmamma and Aunt, December 12, 1861, Folder 7C, RFR, NAL.

9. Riley mentions attending church only once, when he heard Mr. "Cleniguy" [*sic* Chiniquy] preach at St. Anne. Diary, August 17, 1862, RFR, NAL. At that time Charles Chiniquy, the charismatic, French-Canadian Catholic priest turned

Presbyterian, was pastor of the First Presbyterian Church of St. Anne. John Klasey, personal communication to Conner Sorensen, October 13, 2012.

10. Margaret B. Bogue, *Patterns from the Sod: Land Use and Tenure in the Grand Prairie, 1850–1900* (Springfield: Illinois State Historical Society, 1959), 254; Allen G. Bogue, *From Prairie to Corn Belt: Farming on the Illinois and Iowa Prairies in the Nineteenth Century* (Chicago: University of Chicago Press, 1963), 185.

11. Riley to Grandmamma and Aunt, December 12, 1860, Folder 7C, RFR, NAL.

12. Riley to Grandmamma and Aunt, December 12, 1860, Folder 7C, RFR, NAL.

13. Riley to Grandmamma and Aunt, December 12, 1860, Folder 7C, RFR, NAL.

14. Riley to Grandmamma and Aunt, December 12, 1860; and Riley to Aunt [Mira Cannon Young] and Uncle [Timothy Young], March 5, 1861, both in Folder 7C, RFR, NAL.

15. Riley to Grandmamma and Aunt, December 12, 1860; and Riley to Aunt [Mira] and Uncle [Timothy], March 5, 1861, both in Folder 7C, RFR, NAL.

16. Diary, January 12, 16, 17, 19, 24, 25, and 31, 1861; February 4, 15, and 18, 1861, RFR, NAL.

17. Diary, February 7 and 8, 1861, RFR, NAL; and Bogue, *From Prairie to Corn Belt*, 100. Portions of Riley's diary for the Kankakee and Chicago years have been published in Edward Smith and Gene Kritsky, "Charles Valentine Riley: His Formative Years," *AE* 57 (Summer 2011): 74–80.

18. Landholding patterns in Great Britain changed radically during the Napoleonic Wars, when many British yeomen sold their holdings at high prices. As a result, by the 1830s, large landholders had largely replaced individual farmsteads. Traill, *Social England*, 6: 102–14 and 468–82.

19. Diary, February 2, 1861, RFR, NAL; Recorder of Deeds and Tax Assessor, Kankakee County. On January 24, 1864, George Edwards sold his three parcels of land to Laura H. Mafit and was not on the tax rolls thereafter. He apparently returned to England. Charles Edwards was on the tax rolls in 1868 but not in 1869. He also presumably returned to England. Vic Johnson to Edward Smith, July 1, 1994, RFR, NAL.

20. Riley noted in his diary, "Mr. E. bought 40 acres for $5 an acre, I bought 40 acres for $6 per acre. Paid interest for 6 months on my 40 acres." Diary, February 2 and 9, 1861, RFR, NAL.

21. Diary, September 23, 1861, RFR, NAL. The Coyer-Riley transaction recorded in the diary is not mentioned in the Recorder of Deeds and the Tax Assessor for Kankakee County. The relevant documents on file in the Recorder's Office are: (1) Trust Deed: C. V. Riley to Thomas Bonfield; (2) Quit Claim Deed: C. V. Riley to Trustees of Schools T30N R12W, of 2nd PM; and (3) Release: Thomas Bonfield to Charles Riley. These records suggest that Riley's purchase was based on a loan from Bonfield or the township school district. The trustee deed was made in Bonfield's name; in effect, he owned the land until Riley paid off his debt. On June 6, 1863, (by which time

Riley had moved to Chicago) Riley gave the "Trustees of Schools of Township Number Thirty (30) North Range Number, Twelve (12) West of the Second Principal Meridian" a Quit Claim Deed and asked to be released from the note of February 2, 1861. On October 24, 1863, Thomas Bonfield released Riley from the note.

22. Diary, May 18, June 1 and 2, 1861, RFR, NAL.

23. Diary, April 1, 1861, RFR, NAL; Bogue, *From Prairie to Corn Belt*, 84.

24. Diary, April 3 and April 13, 1861, RFR, NAL. Vern LaGesse has identified marshy areas along Spring Creek, just south of the Edwards farm, that still pose hazards to livestock. Vern LaGesse to Edward H. Smith, December 28, 1993, LaGesse Folder, RFR, NAL.

25. Diary, April 22 and 24, 1861, RFR, NAL.

26. Diary, March 4 and April 12, 1861, RFR, NAL.

27. Diary, March 13 and 16, April 3, 16, 23, 27, and 30, 1861, RFR, NAL.

28. Diary, June 25 and July 16, 1861; April 28 and July 21, 1862, RFR, NAL.

29. Diary, April 18, 1861, RFR, NAL.

30. Bogue, *From Prairie to Corn Belt*, 132.

31. Diary, April 25, 1861, RFR, NAL.

32. Diary, May 1 and 11, 1861, RFR, NAL.

33. Diary, June 6 and 7, 1861, RFR, NAL; Bogue, *From Prairie to Corn Belt*, 144.

34. Diary, June 6 and 18, 1861, RFR, NAL.

35. Diary, June 13, 1861, and September 28, 1862, RFR, NAL; Bogue, *From Prairie to Corn Belt*, 120–22.

36. Paul W. Gates, *The Farmer's Age: Agriculture, 1815–1860* (New York: Holt, Reinhardt and Winston, 1960), 162; Bogue, *From Prairie to Corn Belt*, 82.

37. Diary, July 9, 1861, RFR, NAL. Riley noted many similar occurrences. Diary, October 12 and November 18, 1861; August 15 and September 17, 1862, RFR, NAL.

38. Diary, July 24, 1861, RFR, NAL. George resided elsewhere in the neighborhood but visited regularly.

39. Diary, August 14 and 30, 1861, RFR, NAL.

40. Diary, September 25 and 26, 1861, RFR, NAL.

41. Diary, September 25 and October 9, 1861, RFR, NAL. "The Turk" was either the original or a copy of painting no. 17, "The Turk," in Riley's Bonn art book, RFR, NAL.

42. Diary, September 18, 1861, RFR, NAL.

43. Diary, October 9, 1861, RFR, NAL.

44. Diary, October 9, 1861, RFR, NAL. A month earlier, in September 1861, Riley sold his forty acres (see above). He did not record his pay or his earnings from the sale of honey.

45. Diary, October 28, November 15, 17, 1861; January 1, 10, February 1, May 28, 1862, RFR, NAL.

46. Riley to Grandmamma and Aunt, December 12, 1860, Folder 7C, RFR, NAL.

47. Diary, October 3, 1861, RFR, NAL.

48. Diary, December 27, 1862, RFR, NAL.

49. Diary, October 6, 11, December 27, 1862, RFR, NAL.

50. Diary, February 2, December 19, 1862, RFR, NAL.

51. Diary, May 28, November 14–15, December 18, 1861; January 1, March 2, October 28, 1862, RFR, NAL.

52. Diary, April 15–16, May 3, 21, 1862, RFR, NAL.

53. Riley to Grandmamma and Aunt, December 12, 1860, Folder 7C, RFR, NAL.

54. Diary, June 17, 1862, RFR, NAL.

55. H. W. Beckwith, *History of Iroquois County, Historical Notes on the Northwest* (Chicago: H. H. Hill and Co., 1880), 284–85.

56. Diary, August 21, 1862, RFR, NAL.

57. Ironically, the unit's heaviest losses were suffered in the assault on Fort Blakely, Mississippi, shortly after General Lee's surrender to General Grant on April 9, 1865. A. L. Whitehall, "The Seventy-Sixth Regiment Illinois," in Beckwith, *History of Iroquois County* (1880), 284–92; Vic Johnson, "Up 'til Now," *The Kankakee Sunday Journal*, November 22 and 29, 1992; December 6, 13, 20, and 27, 1992; January 17, 24, and 31, 1993; and February 7 and 14, 1993; *Report of the Adjutant General of the State of Illinois, 1886, containing Reports for the Years 1861–1866*, 8 vols. (Springfield, IL: H. W. Rokker, 1886), 4: 622–47; William Kenaga and G. R. Letourneau, *History of Kankakee County, Seventy-Sixth Illinois Infantry* (Chicago: Middle West Publishing Co., 1906), 706–9.

58. On the impact of the Civil War on agriculture, see Paul Wallace Gates, *Agriculture and the Civil War* (New York: Alfred A. Knopf, 1965), 157–250; William Cronen, *Nature's Metropolis* (New York: W. W. Norton & Co., 1991), 218–30; and Wayne Rasmussen, "The Civil War: A Catalyst of Agricultural Revolution," *Agricultural History* 39 (October 1965): 187–95 (hereafter cited as *AgH*).

59. Shortly after his arrival on the farm, Riley ordered paint brushes and canvas from a school mate in New York. Riley to Grandmamma and Aunt, December 12, 1860, Folder 7C, RFR, NAL. Riley's art work during his Kankakee years has disappeared; however, he recorded painting and sketching frequently in his diary: January 12, 1862, "painted"; May 9, 1862, "bought picture frame"; July 30, 1862, "made sketch of the farm"; January 8, 1863, "painted Mrs. Edwards' brother"; January 11, 1863, "finished painting Mr. Johnson." All in Diary, RFR, NAL.

60. Riley to Aunt Mira and Uncle Timothy Young, March 5, 1861, Folder 7C, RFR, NAL. L. O. Howard later related attending a séance with Riley in Washington, DC, where both he and Riley were skeptical. L. O. Howard, *Fighting the Insects: The Story of an Entomologist* (New York: The Macmillan Company, 1933), 302–4.

61. Diary, Sept 8–9, 15–20, 22, 26; October 13, 17–18; November 5, 9, 14, 21; December 6, 14, 1862; January 6, 1863, RFR, NAL.

62. Diary, January 2, 5, 1863, RFR, NAL.

63. Diary, September 11, 1861, RFR, NAL.

64. Quoted in Mallis, *American Entomologists* (New Brunswick, NJ: Rutgers University Press, 1971), 71. *Colman's Rural World* (May 12, 1892) also stated that Riley "secured a good quarter section." Riley purchased forty acres, not a quarter section (160 acres). He lived on the farm approximately two and a half years, not three years.

65. Diary, January 10, 1863, RFR, NAL.

66. Diary, October 3, 9, 24, 1861; May 13, 1863, RFR, NAL.

67. C. V. Riley, "The Buck Moth or Maia Moth," *Fifth Annual Report* (1873), 127.

Chapter 3

1. Diary, January 12–22, 1863, RFR, NAL.

2. Diary, January 28, 1863, RFR, NAL.

3. Diary, January 15, 17, 19, 22, 28, 29, 30, 31; February 14, 17, 19, 1863, RFR, NAL.

4. Diary, March 24, 1863, RFR, NAL; *Chicago Tribune*, March 25, 26, 27, 1863.

5. *Chicago Tribune*, March 23, 1863; Diary, March 21, 1863, RFR, NAL.

6. Later, Emery did experiment with a German edition. C. V. Riley, "Der neue Kartoffel-Käfer [The New Potato Beetle]," *Deutsche Prairie Farmer* (August 1866): 14.

7. Diary, April 1, 15, 17, 18, 1863, RFR, NAL.

8. United States Census, 1860; "Unfortunate St. Louis," *St. Louis Post Dispatch* (July 8 [1880]), SIA, CVR, box 4, scrapbook vol. 27, p. 154; Robert Cromie, *Short History of Chicago* (San Francisco: Lexikos, 1984), 75–76; "Chicago," *Encyclopedia Britannica*, 9th ed. (1878).

9. "Kinsbury Hall" is possibly Kingsbury Hall, still open in Chicago. Diary, February 11, April 13, 1863, RFR, NAL.

10. Diary, January 18; February 18, 19, 21, 22; March 2, 20–24, 28, 31; April 1, 12, 13, 19, 24; May 20, 1863, RFR, NAL.

11. Diary, February 1; March 19, 23, 27; April 5, 23, 27, 1863, RFR, NAL.

12. "The Prairie Farmer: What it Now is and What it has Been," *Prairie Farmer* 55 (September 29, 1883), box 7, scrapbook vol. 54, p. 143, SIA, CVR (hereafter cited as *PF*); Diary, April 7, 18, 20, 23, 24, 25, 27; May 11, 12, 15, 17, 19, 26, 28, 29, 1863, RFR, NAL.

13. Diary, June 3; October 23; November 1, 19, 1863, RFR, NAL.

14. Diary, May 28, 1863, RFR, NAL.

15. Diary, July 1, 14, 15, 1863, RFR, NAL.

16. Upon Riley's request, Charles Edwards brought the collection from Kankakee when he came to Chicago on business, and Pool paid Riley two dollars for it. Diary, April 9, 15; May 13, 15; June 14, 21, 28; July 5, 16, 1863, RFR, NAL.

17. Diary, June 1, 10, 11, 13, 17, 22, 24, 26; July 25, 1863, RFR, NAL. The journal is in the National Archives (hereafter cited as NA), Record Group (hereafter cited as RG) 7, Records of the Bureau of Entomology and Plant Quarantine, E 24, Notes, Division of Entomology 1863–1905, Insect Rearing Notebooks, "History and Description of

Insects Raised by C. V. Riley (1863–1866)." This is the first of many notebooks by Riley and his assistants. The series was continued by others after his death through 1903.

18. Diary, February 10, 1864, RFR, NAL.

19. Diary February 12, 1864, RFR, NAL.

20. Diary, June 3; August 3, 7, 1873, RFR, NAL; Carol A. Sheppard, "Benjamin Dann Walsh: Pioneer Entomologist and Proponent of Darwinian Theory," *Ann. Rev. Ent.* 49 (2004): 1–4; Walsh to Baird, September 17, 1859, Office of the Assistant Secretary, Incoming Correspondence, 18:402, SIA; Walsh to Baron Osten Sacken, May 12, 1862, Museum of Comparative Zoology Archives, Ernst Mayr Library, Harvard University, Cambridge, Massachusetts (herafter cited as MCZ); Walsh to Scudder, October 20, 1862, Scudder Corr., Boston Society of Natural History; Charles V. Riley, "In Memoriam," *AE* 2 (Dec.–Jan. 1869–70): 67–68; Benjamin D. Walsh, "Insects Injurious to Vegetation in Illinois," *Transactions of the Illinois State Agricultural Society* 4 for 1859–60 (1861): 335–72 (hereafter cited as *Trans. ISAS*).

21. Both Walsh and Thomas were awarded $25.00 for their essays. *Trans. ISAS* 5 (1861–1864), 63, 67; Carol A. Sheppard and Richard A Weinzierl, "Entomological Lucubrations: The 19th Century Spirited Conflict Concerning the Natural History of the Armyworm, *Pseudaletia unipuncta* (Haworth) (Lepidoptera: Noctuidae)," *AE* 48 (Summer 2002): 108–13; Mallis, *American Entomologists*, 49–50 (for LeBaron).

22. Diary, June 15, 1863, RFR, NAL.

23. Riley to Fitch, November 7, 13, 17, 1863, New York State Museum (hereafter cited as NYSM), copies in Asa Fitch Correspondence, RFR, NAL; Diary, November 11, 13, 19, 1863, RFR, NAL.

24. Riley to Uncle Tim, April 10, 1864, Folder 7C, RFR, NAL.

25. "Independent Order of Good Templars," *Brooklyn Daily Standard Union* (February 17, 1871).

26. On Templar rituals practiced during Riley's time in Chicago, see *Ritual of the Independent Order of Good Templars, for Subordinate Lodges under the Jurisdiction of the Right Worthy Grand Lodge of North America: Adopted at Cleveland Session, May 24, 1864* (Chicago: Right Worthy Grand Lodge, 1864), preface, 5–10, 27–29.

27. In Kankakee, Riley attended at least one temperance meeting with Charles and Mary Edwards, but they apparently went more out of curiosity than conviction, as Charles Edwards continued to brew his own beer. Diary, January 8, 1861; June 19, 1862, RFR, NAL.

28. Diary, April 6, 1863, RFR, NAL. The news was premature. Confederate forces in Charleston held out until February 1865. https://en.wikipedia.org/wiki/Charleston, South Carolina in the American Civil War.

29. Diary, July 6, 1863, RFR, NAL.

30. Victor Hicken, *Illinois in the Civil War*, 2nd ed. (Urbana and Chicago: University of Illinois Press, 1991), 240–42; T. M. Eddy, *The Patriotism of Illinois: A Record of*

the Civil and Military History of the State in the War for the Union, with a History of the Campaigns in which Illinois Soldiers have been Conspicuous, 2 vols. (Chicago: Clarke and Co., 1865), 1:148–49; Bruce Catton, *A Stillness at Appomattox* (1953; repr., New York: Pocket Books, Inc., 1958), 329.

31. Daniel E. Sutherland, *A Savage Conflict: The Decisive Role of Guerrillas in the Civil War* (Chapel Hill: University of North Carolina Press, 2009), 38–39.

32. Diary, April 30, 1864, RFR, NAL; Kankakee County, with a population of 15,393 in 1860, contributed a total of 2,575 enlisted men by 1864. Eddy, *Patriotism of Illinois*, 1:607.

33. Diary, May 8, 24, 1864, RFR, NAL; for a capsule history of the 134th Infantry Regiment, see Frederick H. Dyer, *A Compendium of the War of the Rebellion*, 3 vols. (New York and London: Thomas Yosaloff, 1959), 3:1101.

34. Diary, May 16, 1864, RFR, NAL.

35. Diary, May 19, 22, 31, 1864, RFR, NAL.

36. Diary, May 8, 11–13, 18–19, 22, 25, 1864, RFR, NAL.

37. Diary, June 4, 5, 1864, RFR, NAL.

38. Diary June 5, 7, 10, 1864, RFR, NAL.

39. Diary, June 13, 1864, RFR, NAL; Eddy, *Patriotism of Illinois*, 1:216.

40. Diary, June 27, 1864, RFR, NAL.

41. Diary, June 13, 1864, RFR, NAL.

42. Diary, June 8, 11, 1864, RFR, NAL.

43. Diary, May 13, 14, 29; June 1, 25; August 25, 29; September 11; November 10, 1864, RFR, NAL.

44. Diary, June 22, 25, 29, 30; August 1–2, 6; October 29; November 1, 10, 18, 1864, RFR, NAL.

45. Diary, June 25, 29, 30; October 29; November 1, 10, 1864, RFR, NAL.

46. C. V. Riley, "Report of Committee on Entomology," *Transactions of the Illinois State Horticultural Society* for 1867, n.s. 1 (1868): 105 (hereafter cited as *Trans. ISHS*). He also captured other creatures, including a beautiful little snake, a large snake measuring over five feet long, and a blue-tailed lizard. Diary, May 25; June 9, 11, 31; July 2, 1864, RFR, NAL.

47. Diary, July 4–7, 1864, RFR, NAL.

48. Diary, July 7, 1864, RFR, NAL.

49. Diary, July 8; August 4–8, 1864, RFR, NAL.

50. Diary, August 14, 21–25, 1864, RFR, NAL. A few months previously, Paducah was captured by Confederate guerrillas and then retaken by Union troops. *PF* 29 (April 2, 1864): 240.

51. Such executions, while not official policy, were nevertheless common on both sides. Diary, August 26, 30, 1864; William C. Winter, *The Civil War in St. Louis: A Guided Tour* (St. Louis: Missouri Historical Society Press, 1994), 96–97; Sutherland, *A Savage Conflict*, 38–39.

52. Diary, September 12, 1864, RFR, NAL.

53. Diary, August 23–24, 26, 28, 30–31, 1864, and sketch of fortifications in pocket of Civil War diary, RFR, NAL. The fort was not completed.

54. Diary, September 6, 9, 1864, RFR, NAL.

55. Diary, September 2, 4, 13, 1864, RFR, NAL.

56. Diary, September 19–23, 1864, RFR, NAL.

57. Diary, September 26–27, 29, 1864, RFR, NAL.

58. Diary, October 4, 9, 17, 1864, RFR, NAL.

59. Diary, October 4, 9, 1864, RFR, NAL.

60. Diary, October 1–2, 21–22, 25, 28, 1864, RFR, NAL.

61. According to Dyer, *A Compendium of the War of the Rebellion*, 3:1101, the 134th volunteers were mustered out October 5, 1864. The Illinois adjutant general gives the date as October 25. Riley's diary entries indicate a later date.

62. Diary, November 7, 1864, RFR, NAL; "Rebels in Chicago," *PF* 30 (October 2, 1864): 320; and Catton, *A Stillness at Appomattox*, 328–30.

63. A biographical sketch in 1877, written with Riley's collaboration, indicates his Union sympathies. See C. R. Barnes, ed., *The Commonwealth of Missouri, a Centennial Record* (St. Louis: Bryan, Brand and Co., 1877), 673.

64. C. V. Riley, "Letter from the Hundred Days Boys," *PF* 14 (July 9, 1864): 26.

65. Riley to Aunt Mira and Uncle Tim, January 22, 1865, Folder 7C, RFR, NAL.

66. C. V. Riley, "Queries Answered," *PF* 36 (September 7, 1867): 148.

67. Asa Fitch, "Address on our most Pernicious Insects," *Transactions of the New York State Agricultural Society* 19 (1860): 588–98.

68. Walsh, "Insects Injurious to Vegetation," *Trans. ISAS* 4 (1861): 338–39.

69. C. V. Riley, "Smith's Patent Curculio Trap," *PF* 36 (July 13, 1867): 21. For his exposé of false remedies for the Chinch Bug, see C. V. Riley, "The Chinch Bug," *PF* 32 (September 9, 1865): 190.

70. C. V. Riley, "Entomology," *The Prairie Farmer Annual, Agricultural and Horticultural Advertiser* 1 (Chicago: Prairie Farmer Co., 1868): 53–59.

71. Riley, "Report on Entomology," *Trans. ISHS* for 1864, n.s. 1 (1868): 114.

72. Samuel Henshaw, ed., *Bibliography of the More Important Contributions to American Economic Entomology: Parts 1, 2, and 3, the more important writings of Benjamin Dann Walsh and Charles Valentine Riley* (Washington, DC: Government Printing Office, 1890) (hereafter cited as Henshaw, *Bibliography*). The count is based on article summaries that are listed chronologically.

73. James Whorton, *Before Silent Spring: Pesticides and Public Health in Pre-DDT America* (Princeton, NJ: Princeton University Press, 1974), title, chapter 1.

74. *Trans. ISHS* (1862), included in *Trans. ISAS* 5 (1862): 740.

75. *Trans. ISHS* (1868), included in *Trans. ISAS* 7 (1867–1868): 547.

76. *Trans. ISHS* (1865), included in *Trans. ISAS* 6 (1865–66): 365; *Trans. ISAS* 6 (1865–66): 18.

77. Sheppard, "Benjamin Dann Walsh," *Ann. Rev. Ent.* 49 (2004): 4–5; *Trans. ISHS* (1868), included in *Trans. ISAS* 7 (1867–68): 494, 499; Walsh to Cresson, April 12, 1869, American Entomological Society Correspondence, Collection 150, Academy of Natural Sciences of Philadelphia (hereafter cited as ANSP).

78. "The Illinois State Fair," *PF* 34 (October 1866): 218.

79. The Illinois Horticultural Society was founded in 1856, absorbing the Northwest Fruit Growers Association that dated back to 1845. Three years later, in 1859, the Missouri Horticultural Society was organized. W. C. Flagg, "History of the Illinois State Horticultural Society," *Trans. ISHS* (1868), included in *Trans. ISAS* (1867–68): 486–90. The Missouri society was organized as the Missouri Fruit Growers Association but changed its name in 1862. *Proceedings of the Missouri State Horticultural Society* (hereafter cited as *Proc. MSHS*), included in *Second Annual Report of the Missouri State Board of Agriculture* (Jefferson City: Missouri State Board of Agriculture, 1866), 339–40 (hereafter cited as *Second [–Fourth] Annual Report, MSBA*).

80. Edgar Saunders, "Chicago and St. Louis Horticulturally Considered," *PF* 36 (November 16, 1867): 310.

81. See Riley's recollection of meeting with the Southern Illinois Fruit Growers Association in November 1867 in C. V. Riley, "The Strawberry Crown-borer," *Third Annual Report* (1871), 42, and his recollection of his visiting the vineyard of George Hussman in Hermann, Missouri, in 1867 in C. V. Riley, "The Satellite Sphinx," *Second Annual Report* (1870), 77. After his appointment as Missouri state entomologist in April 1868, Riley continued to meet with and advise fruit growers in Illinois. See his recollections in C. V. Riley, "The Strawberry Leaf-Roller," *First Annual Report* (1869), 142.

82. Report of the meeting in *Colman's Rural World* (February 1, 1867): 19 (hereafter cited as *CRW*); and C. V. Riley, "Remarks on Saperda, Chrysobothris, Carpocapsa, and Conotrachelus," *PF* 35 (January 12, 1867): 23.

83. *Proc. MSHS*, included in *Third Annual Report, MSBA* (1867), 415–17.

84. *Fourth Annual Report, MSBA* (1868), iii; George Francis Lemmer, *Norman J. Colman and Colman's Rural World: A Study in Agricultural Leadership* (Columbus: University of Missouri Press, 1953), 44; and C. V. Riley, "Progress of Economic Entomology," in *Fifth Annual Report* (1873), 22.

85. Anonymous, "A State Entomologist," *CRW* 19 (December 1, 1867): 359. Colman's support of Riley is cited in Lemmer, *Norman J. Colman and Colman's Rural World*, 44, and C. V. Riley, "Progress of Economic Entomology," in *Fifth Annual Report* (1873), 22.

Chapter 4

1. C. R. Barns, ed., *The Commonwealth of Missouri, A Centennial Record* (St. Louis: Bryan, Brand and Co., 1877), 652; "Unfortunate St. Louis," *St. Louis Post Dispatch* (July 8, 1880), SIA, CVR, box 4, scrapbook vol. 27, p. 154.

2. Darryl Pinckney, "Elite Black and Quite Different," review of *Negroland: A*

Memoir, by Margo Jefferson, *New York Review of Books* 63 (May 26–June 8, 2016): 49 (hereafter cited as *NYRB*).

3. C. V. Riley, Introduction, *First Annual Report* (1869), 3. Jesse James and his band continued to fight for the rebel cause until the leader was killed in 1882. T. J. Stiles, *Jesse James: Last Rebel of the Civil War* (New York: Alfred A. Knopf, 2002), prologue, 5–6.

4. Laws of the State of Missouri passed at the Adjourned Session of the 24th General Assembly (Jefferson City: Elwood Kirby, Public Printer, 1868), 299.

5. Riley, "Entomology," *Fifth Annual Report* (1873), 22.

6. L. O. Howard, "Charles V. Riley, Ph.D.," *Proceedings of the Entomological Society of Washington* 3 (October 6, 1896): 295 (hereafter cited as *Proc. ESW*).

7. Quoted in Carol A. Sheppard and Richard A. Weinzerl, "Entomological Lucubrations: The 19th Century Spirited Conflict Concerning the Natural History of the Armyworm, *Pseudaletia unipuncta* (Haworth) (Lepidoptera: Noctuidae)," *AE* 48 (Summer 2002): 114.

8. Riley, *First Annual Report* (1869), 176. The author and title of the poem have not been identified. On Andrew Bolter, see biographical sketch, *National Cyclopaedia of American Biography* 24 (New York: James T. White and Company, 1935), 186, and William F. Rapp Jr., "The Andrew Bolter Insect Collection," *Entomological News* 56 (October 1945): 209 (hereafter cited as *Ent. News*).

9. C. V. Riley, "Insects Infesting the Potato," *First Annual Report* (1869), 115.

10. Riley, "Insects Infesting the Potato," *First Annual Report* (1869), 115, 112.

11. C. V. Riley, "The White-Winged Morning Sphinx," *Third Annual Report* (1871), 140–41.

12. C. V. Riley, "The Periodical Cicada," *First Annual Report* (1869), 18–19.

13. C. V. Riley, "Katydids," *Sixth Annual Report* (1874), 150.

14. Riley, "Katydids," *Sixth Annual Report* (1874), 150–51.

15. Riley, "Katydids," *Sixth Annual Report* (1874), 153.

16. Riley, "Katydids," *Sixth Annual Report* (1874), 152.

17. Riley, Introduction, *First Annual Report* (1869), 5.

18. Riley, "Entomology," *Fifth Annual Report* (1873), 27–28, and C. V. Riley, "Hackberry Butterflies," *Sixth Annual Report* (1874), 137.

19. The National Agricultural Library Special Collections contains several hundred linoleum-wood blocks that Riley's used in the Missouri Reports. Around 1880, photoengraving began replacing hand carved wood blocks. Erving A. Denis and John D. Jenkins, *Comprehensive Graphic Arts* (Indianapolis, IN: Howard W. Sams & Co., Inc., 1974), 195.

20. C. V. Riley, "The Chinch Bug," *Seventh Annual Report* (1875), 19. Cyrus Thomas, Illinois state entomologist, complained that his reports were likewise little known within Illinois. Cyrus Thomas, "Entomological," *PF* 46 (July 17, 1875): 226.

21. Riley, "The Chinch Bug," 56, 60, 69.

22. Leland O. Howard to M. V. Slingerland, October 15, 1904, Entomology Library, Cornell University.

23. *Prairie Farmer* listed Riley and Walsh, in that order, as editors of the new publication. *PF* 36 (August 1, 1868): 36. Walsh, by age and experience, was "senior editor," but as Riley pointed out to John L. LeConte, he (Riley) initiated the project, oversaw its publication in St. Louis, and continued it after Walsh's death. Riley to LeConte, October 8, 1873, LeConte Corr., American Philosophical Society, Philadelphia, PA (hereafter cited as APS).

24. Preface, *Third Annual Report* (1871), 3–4; "Entomology," St. Louis *Globe-Democrat*, April 1, 1876; and Barns, *Commonwealth*, 674. Examples of Riley's Insect Wall Charts are shown for several Featured Insects (pp. 37, 41, etc.), and others may be viewed at the Kansas State University Library Special Collections website, http://www.lib.k-state.edu/depts/spec/archives.html. Compare, for example, the Colorado Potato Beetle drawing in Figure 4.1 with his Wall Chart on p. 37.

25. G. A. Dean, "The Contribution of Kansas Academy of Science to Entomology," *Transactions of the Kansas Academy of Science* 41 (1938): 62–63. Riley taught courses for Benjamin F. Mudge, who was away on geological fieldwork.

26. J. T. Willard, *History of the Kansas State College of Agriculture and Applied Science* (Manhattan: Kansas State College Press, 1940), 467. A grateful correspondent in Emporia, Kansas, wrote "the people of Kansas feel as if you belonged to us." C. V. Riley, "The Rocky Mountain Locust," *Seventh Annual Report* (1875), 50.

27. L. O. Howard, *A History of Applied Entomology* (*Somewhat Anecdotal*) (Washington, DC: Smithsonian Institution Press, 1930), 71.

28. "Lecture Notes," Charles V. Riley Collection, Field Museum of Natural History, Chicago. This collection contains notes Riley used at Cornell, Manhattan, and elsewhere. Riley used the same set of notes throughout his career, adding and expanding with additional examples, from his Missouri Reports and other publications.

29. From its founding in 1839, the official name was University of the State of Missouri. In 1901, the name was shortened to University of Missouri. In Riley's time, it was commonly referred to, unofficially, as Missouri State University. Gary Cox, University of Missouri Archives, to Conner Sorensen, May 11, 2012, Sorensen files, Eschbach, Germany.

30. "An Agricultural College for the State of Missouri," *St. Louis Democrat*, August 20, 1870, box 1, scrapbook vol. 5, p. 110, SIA, CVR; and "State University," n.s., March 25, 1873, in box 1, scrapbook vol. 7, pp. 145–46, SIA, CVR.

31. Donna A. Brunette, "Charles Valentine Riley and the Roots of Modern Insect Control," *Missouri Historical Review* 86 (April 1992): 232.

32. C. V. Riley, "Agricultural: State University–Agricultural College," n.d., n.s., box 1, scrapbook vol. 9, pp. 70–71, SIA, CVR. Swallow complained that the state of Missouri appropriated only $500 per year for college buildings and furnishings, whereas other agricultural colleges had budgets of $25,000 to $100,000. G. C. Swallow, "The Agricultural College," *Daily Tribune* (April 5, 1877), in box 2, scrapbook vol.

13, pp. 169–70, SIA, CVR. For background on conflicts in college agricultural programs, especially the emphasis on classics in that era, see Alan I. Marcus, "The Ivory Silo: Farmer-Agricultural Tensions in the 1870s and 1880s," *AgH* 60 (Spring 1986): 22–36.

33. "Education in St. Louis," *St. Louis Globe Democrat*, September 5, 1876.

34. Riley, Preface, *Seventh Annual Report* (1875), v; and Barns, *Commonwealth*, 677.

35. C. V. Riley, "Thos. Wier's Apple Worm and Curculio Trap," *Fourth Annual Report* (1872), 23–25.

36. Riley, "Thos. Wier's Apple Worm and Curculio Trap," 25.

37. "Horticultural Session in the Hall of the House," 27th General Assembly, and *Globe* (St. Louis, MO), February 22, 1873, box 1, scrapbook vol. 6, pp. 145–47, SIA, CVR.

38. *Globe* (St. Louis, MO), February 22, 1873, box 1, scrapbook vol. 6, p. 145, SIA, CVR.

39. *Journal of the Missouri State Senate at the Adjourned Session of the 27th General Assembly, Commencing on Wednesday, January 7, 1874* (Jefferson City: Regan and Carter, Public Printers, 1875), 16–17.

40. *St. Louis Dispatch*, January 24, 1874.

41. "State Capital: Friday's Fight in the House to Abolish State Entomologist: Uproarious Scenes—Bushwhacking the Bugs," n.d., n.s., clipping from House Journal, n.d., box 1, scrapbook vol. 9, pp. 14–18, SIA, CVR; and article from *Colman's Rural World* (January 24, 1874) by Charles R. Allen, master of the Missouri Grange, defending Riley, clipping in MC 143, Series IV, Box 4, Folder 111a, NAL.

42. Clippings [from *Colman's Rural World*?], 1874, NAL, MC 143, Series IV, Box 4, Folder 115; and "Entomology," n.d., n.s., box 1, scrapbook vol. 6, p. 145, SIA, CVR.

43. "Bugs and Worms: Shall Missouri Abolish the Office of the State Entomologist?" *St. Louis Dispatch*, February 4, 1874.

44. Quoted in "Retrenchment," *Kansas City Journal of Commerce*, January 12, 1874. See also communication from the St. Louis Academy of Science, "Governor Woodson vs. Prof. Riley," *St. Louis Democrat*, January 9, 1874; and "Riley Abolished," *Times* [Kansas City, MO?], January 24, 1874. Only one paper in this admittedly selective collection favored abolishing Riley's position. Box 1, scrapbook vol. 9, pp. 1–7, SIA, CVR. See also resolution of Missouri State Horticultural Society to maintain the office in NAL, MC 143, Series IV, Box 4, Folder 112.

45. See discussion of post of state entomologist by the Missouri State Senate on February 4, 1874, *Journal of the Missouri State Senate*, 255.

46. "Questioning the Utility of the Agricultural Board and Entomologist," *St. Louis Evening Dispatch*, February 4, 1875; "Board of Agriculture and Entomologist," n.s., February 5, 1875, box 1, scrapbook vol. 9, p. 99, SIA, CVR.

47. Brunette, "Charles Valentine Riley," 233.

48. The connection between the locust threat and the office of entomologist was made explicit in "Our State Entomologist," *Journal of Agriculture* (July 1875), box 1, scrapbook vol. 9, p. 80, SIA, CVR.

49. Mira to Charlie, April 26, 1874, Folder 7C, RFR, NAL.

50. Mira to Charlie, April 20, 1877, and Mira to Charlie, July 3, 1877, both in Folder 7C, RFR, NAL.

51. Plans for Nina's move to St. Louis were made in 1876, prior to Mary Cannon's death. See chap. 10.

52. Riley to Osten Sacken, September 6, 1869, MCZ.

53. Riley to Lintner, January 28, 1872, Lintner Corr., NYSM.

54. Quoted in Donald W. Meyer, Superintendent, Bellefontaine Cemetery Association, St. Louis, to Janet R. Smith, March 19 and April 3, 1985, regarding Gottlieb Conzelman Lot 42 and George Byron Riley, Folder 7E, RFR, NAL. This correspondence includes a copy of a letter from Riley to Ward Burlingame, January 29, 1875, Governor Thomas Osborne, Received Correspondence, 1875, Subject File, Box 2, Folder 2, Kansas State Historical Society, Topeka, KS.

55. "He Once Lived in St. Louis" [referring to J. M. Tracy], *Republic* (New York, NY), February 1895, box 13, scrapbook vol. 95, p. 30, SIA, CVR. Information on Tracy (1843–1893) and his dog paintings is available on many sites on the Internet.

56. Riley to Dr. J. S. Billings, US Army, December 26, 1894, Box 27, SIA, CVR. Copy in Misc. Corr., RFR, NAL. See chap. 10.

57. C. V. Riley, "On the Causes of Variation in Organic Forms," *Proceedings of the American Association for the Advancement of Science* 37 (1888): 268 (hereafter cited as *Proc. AAAS*); James Good, "The Value of Thomas Davidson," *Transactions of the Charles S. Peirce Society* 40 (Spring 2004): 291–94. The Philosophical Society even had its own publication, the *Journal of Speculative Philosophy.*

58. N.d., n.s., box 1, scrapbook vol. 6, p. 48, SIA, CVR.

59. Under Harris, St. Louis schools pioneered the addition of the kindergarten in their system. Harris served as Federal Commissioner of Education from 1889 to 1906. See Reese, William J. "The Philosopher-King of St. Louis," *Curriculum & Consequence*, ed. by Barry M. Franklin, 155–77. New York: Teachers College, 2000.

60. Charles V. Riley, "Educational Interests of Farmers," (1872), address to the St. Louis School Board, box 1, scrapbook vol. 1, p. 5, SIA, CVR. See also his talk on "Natural Science in Our Agricultural Colleges and Schools," delivered to the Kansas State Horticultural Society in 1873. Box 3, scrapbook vol. 19, pp. 228–34, SIA, CVR.

61. "Prof. Turner's Speech on Education," box 1, scrapbook vol. 8, pp. 171–73, SIA, CVR.

62. *Campbell's New Atlas of Missouri with Descriptions Historical, Scientific, and Statistical* (St. Louis, MO: R. A. Campbell, 1873), 93–100; Riley, "Entomology," *Fifth Annual Report* (1873), 5; dispatch from Topeka, KS, to Missouri *Democrat*, December 12, [1872], box 1, scrapbook vol. 6, p. 143, SIA, CVR.

63. See Riley's collection of articles on industrial and scientific education in box 1, SIA, CVR; Conzelman's views and activities in "Notes on the Gottlieb Conzelman Family," Folder 7A, RFR, NAL; and National Register of Historic Places Registration

Form, Crunden-Martin Manufacturing Company, alias Conzelman-Crunden Realty Co., Bowman Stamping Co., Swayzee Glass Co., St. Louis, Missouri, Section 8, p. 12, including biographical sketch of Theophilus Conzelman from the Missouri Historical Society Collections 5:3 (1928), https://dnr.mo.gov/shpo/nps-nr/05000013.pdf (accessed March 13, 2018).

64. Riley, "Entomology," *Fifth Annual Report* (1873), 28–29; and C. V. Riley, Preface, *Sixth Annual Report* (1874), 6.

65. C. R. Riley, "Silkworms," *Fourth Annual Report* (1872), 111.

66. In 1885, he rejoined Riley in the Division of Entomology, remaining there until 1888, when he was appointed state entomologist of Minnesota, a post he held until his death in 1901. Mallis, *American Entomologists*, 196.

67. See chap. 8.

68. During the years 1877–1888, Missouri had no state entomologist. See Anon., "Mary Esther Murtfeldt," *Journ. Econ. Ent.* 6 (1913): 288–89; and Hermann Schwarz, "Miss Mary E. Murtfeldt," *Ent. News* 24 (June 1913): 241–42.

69. Moritz Schuster obituary, *Ent. News* 5 (March 1894): 96; and A. S. Packard, ed., *Record of American Entomology for the Year 1877* (Salem, MA: Naturalist's Book Agency, Essex Institute Press, 1878).

70. J. J. Davis, "Joseph Tarrigan Monell," *Journ. Econ. Ent.* 8 (1915): 503; J. J. Davis, J. T. Monell obituary, *Ent. News* 26 (October 1915): 380–83; Henshaw, *Bibliography*, 234.

71. See discussion of Riley's finances in chap. 10.

72. Confirmation of this comes from an unnamed supporter of Riley, who, in the legislative hearings regarding the state entomologist in 1874, pointed out that Riley's salary for the years 1868–74 totaled $16,500, of which Riley paid out $10,210.66 for expenses. If averaged over six years, his expenses amounted to roughly $1,700 per year, significantly less than the $2,350 cited by Riley but still of comparable magnitude. "Public Opinion," *Globe* (St. Louis, MO), January 9, 1874, box 1, scrapbook vol. 9, p. 15, SIA, CVR.

73. "A New Enterprise," n.s., n.d. [1874], in box 1, scrapbook vol. 9, p. 105, SIA, CVR.

74. "The City Council," *Leader* (Springfield, MO), February 5, 1874, in box 1, scrapbook vol. 9, p. 74, SIA, CVR.

75. See chaps. 6 and 9.

76. See chaps. 6 and 7.

77. Benjamin D. Walsh, "The New Potato Bug and its Natural History," *PE* 1 (October 1865): 1–4.

78. C. V. Riley, "The Colorado Potato Beetle," *PF* 21 (December 1867): 389. In 1868, Riley noted Cyrus Thomas's doubts that the beetle originated in the foothills of the Rocky Mountains, but Riley concluded that Walsh's theory of its origin was essentially correct. C. V. Riley "The Colorado Potato Beetle, *Eighth Annual Report* (1868), 8–10. Later investigations have established that the beetle was present along the

Iowa-Nebraska border as early as 1811. It apparently added domestic potatoes to its diet around 1859, becoming a pest in Nebraska and spreading eastward. R. A. Casagrande, "The 'Iowa' Potato Beetle, its Discovery and Spread to Potatoes," *Bulletin of the Entomological Society of America* (Summer 1985): 27–29.

79. C. V. Riley, "The Colorado Potato Beetle," *Seventh Annual Report* (1875), 1.

80. C. V. Riley, "Notes of the Year: The Colorado Potato Beetle," *Fourth Annual Report* (1872), 5.

81. Riley, "The Colorado Potato Beetle," *Seventh Annual Report* (1875), 1.

82. Riley, "The Colorado Potato Beetle," *Seventh Annual Report* (1875), 3–4; and J. F. M. Clark, "The eyes of our Potatoes are weeping: The Rise of the Colorado Beetle as an Insect Pest," *Archives of Natural History* 34, no. 1 (April 2007): 113–18.

83. C. V. Riley, *Potato Pests: Being an illustrated account of the Colorado Potato-Beetle and the other insect foes of the potato in North America, with suggestions for their repression and methods for their destruction* (New York: Orange Judd Company, 1876).

84. Riley, "Periodic Cicada," *First Annual Report* (1869), 19–31.

85. Riley, "Periodic Cicada," *First Annual Report* (1869), 29.

86. C. V. Riley, "The Bark-Lice of the Apple Tree," *First Annual Report* (1869), 8.

87. In fact, Riley reported, the adult beetles that hatched out in 1870 were still alive in February 1871. C. V Riley, "Snout Beetles," *Third Annual Report* (1871), 11–13.

88. X. Zhang, Z. Tu, S. Luckhart, and D. G. Pfeiffer, "Genetic diversity of plum curculio (Coleoptera: Curculionidae) among geographical populations in the eastern United States," *Annals of the Entomological Society of America* 101, no. 5 (2008), 824–32.

89. C. V. Riley, "The Army Worm," *Eighth Annual Report* (1876), 32; Sheppard and Weinzierl, "Entomological Lucubrations," 108–13.

90. C. V. Riley, "The Army Worm," *Eighth Annual Report* (1876), 56.

91. Report of AAAS meeting in *Canadian Entomologist* 4 (October 1872): 24–27 (hereafter cited as *CE*).

92. See chap. 14.

93. See chap. 5.

94. C. V. Riley, "The Cotton Worm," *Sixth Annual Report* (1874), 17. See also chaps. 8 and 9.

95. C. V. Riley, "The Critic Criticized," *PF* 35 (March 16, 1867): 169.

96. C. V. Riley, "The Blue-Spangled Peach Worm," *Third Annual Report* (1871), 133n.

97. Riley to Lintner, March 7, 1872, Lintner Corr., NYSM.

98. C. V. Riley, "The Colorado Potato Beetle," *Seventh Annual Report* (1875), 5–6.

99. Riley to Lintner, July 15, 1881, Lintner Corr., NYSM.

100. *Rural New Yorker*, n.d., box 1, scrapbook vol. 5, p. 237, SIA, CVR.

101. See chap. 5.

102. Excerpt from the London society proceedings, n. d., box 1, scrapbook vol. 9, p. 79, SIA, CVR. See chap. 6 for Riley's travels in France.

103. *Western Ruralist* (November 1871), box 1, scrapbook vol. 5, p. 247, SIA, CVR.

104. *The Nationalist*, September 20, 1872, History Index, Kansas State University Library, Manhattan, KS.

105. A writer for the *St. Louis Dispatch* declared, "Professor Riley knows what he is talking about [regarding locusts]. There are prophets in science as well as there are in religion." "The Grasshoppers," *St. Louis Dispatch*, May 15, 1875.

106. Mira to Charlie, July 3, 1877, Folder 7C, RFR, NAL.

107. The *Fourth Annual Report* was reviewed in Germany and extracts of his reports were regularly translated into French and German. Box 1, scrapbook vol. 5, pp. 72, 241–42, SIA, CVR; "The Return of the State Entomologist," *St. Louis Dispatch*, August 25, 1875.

108. Barns, *Commonwealth*, 679.

Chapter 5

1. Walsh to Darwin, April 24, 1864, *The Correspondence of Charles Darwin* (Cambridge: Cambridge University Press), 12: 161–62 (hereafter cited as *Darwin Correspondence*); Carol A. Sheppard, "Benjamin Dann Walsh: Pioneer Entomologist and Proponent of Darwinian Theory," *Ann. Rev. Ent.* 49 (2004): 13.

2. Benjamin D. Walsh, "On Certain Entomological Speculations of the New England School of Naturalists," *Proceedings of the Entomological Society of Pennsylvania* 3 (August 1864): 207–49 (hereafter cited as *Proc. ESP*).

3. Benjamin D. Walsh, "On Certain Entomological Speculations of the New England School of Naturalists," *Proc. ESP* 3 (August 1864): 207–11, 223–29, 236–41.

4. Benjamin D. Walsh, "On Certain Entomological Speculations of the New England School of Naturalists," *Proc. ESP* 3 (August 1864): 223.

5. Darwin to Walsh, October 21, 1864, *Darwin Correspondence* 12: 374–75. On the Walsh-Darwin relationship, see Carol M. Anelli, "Darwinian Theory in Historical Context and its Defense by B. D. Walsh: What is Past is Prologue," *AE* 52 (Spring 2006): 11–18; Gene Kritsky, "Entomological Reactions to Darwin's Theory in the Nineteenth Century," *Ann. Rev. Ent.* 53 (2008): 356; and Kritsky, "Darwin, Walsh, and Riley: The Entomological Link," *AE* (Summer 1995): 89–92. See also Edward J. Pfeiffer, "United States," in Thomas F. Glick, ed., *The Comparative Reception of Darwinism* (Austin: University of Texas Press, 1974), 184–85.

6. Benjamin D. Walsh, "On Phytophagic Varieties and Phytophagic Species," *Proc. ESP* 3 (November 1864): 403–30; and "On the Insects, Coleopterous, Hymenopterans, and Dipterous, Inhabiting the Galls of Certain Species of Willow, Part 1, Diptera," *Proc. ESP* 3 (December 1864): 534–644.

7. Diary, February 13, 1864, RFR, NAL.

8. C. V. Riley, "Review of Charles Darwin, *Animals and Plants under Domestication*," *PF* 37 (June 6, 1868): 364–65. Regarding opposition to Darwinism on moral grounds, see Ronald Numbers, *Darwinism Comes to America* (Cambridge, MA: Harvard University Press, 1998), 2–3, 138.

9. C. V. Riley, "Mimicry as Illustrated by these two Butterflies, with Some Remarks on the Theory of Natural Selection," *Third Annual Report* (1871), 172. Like other annual reports, the third report was submitted in December 1870, prior to its publication the following year (1871).

10. In Riley's scrapbook is a twelve-page summary of Darwin's *Origin of Species*, referring to page numbers in the book; however, there is no date indicating when this was written. Riley Scrapbook No. 2, page 2, Field Museum of Natural History, Chicago.

11. Riley, "Review of Charles Darwin, *Animals and Plants*," (1868), 365.

12. C. V. Riley, "The Walnut Tortrix," *First Annual Report* (1869), 153–54. In his third report, Riley cited additional examples of phytophagic variation, commenting that "The study of these variations—of phytophagic varieties and phytophagic species—must ever prove interesting as well as important, by throwing light on the question of the origin of species." C. V. Riley, "The White-Lined Sphinx," *Third Annual Report* (1871), 141.

13. Walsh to Darwin, August 29, 1868 (original letter), box 1, scrapbook vol. 9, between pp. 66–67, SIA, CVR. See copy in *Darwin Correspondence* 16:2, p. 698.

14. Darwin to Walsh, August 29, 1868, *Darwin Correspondence* 16:2, pp. 758–59.

15. The discussion of mimicry in American butterflies is drawn in part from Sorensen, *Brethren of the Net*, 202–4. On Bates's relationship to Darwin and the reception of his findings on mimicry in butterflies among entomologists, see Kritsky, "Entomological Reactions," 354–55.

16. Mary Alice Evans, "Mimicry and the Darwinian Heritage," *Journal of the History of Ideas* 26 (1965): 213 (hereafter cited as *JHI*); Barbara Bedall, ed., *Wallace and Bates in the Tropics: An Introduction to the Theory of Natural Selection. Based on the Writings of Alfred Russel Wallace and Henry Walter Bates* (London: Macmillan, 1969), 104–5. See the appendix for modern nomenclature.

17. Bates to Darwin, March 28, 1861, *Darwin Correspondence*, 9:71.

18. Darwin to Bates, November 20, 1862, *Darwin Correspondence*, 10:539–40.

19. Evans, "Mimicry," 213; and Murial Louise Blaisdell, "Darwinism and its Data, The Adaptive Coloration of Animals" (PhD diss., Harvard University, 1976), 142. Also available under the same title in "Harvard Dissertations in the History of Science" (New York: Garland, 1992).

20. Darwin to Walsh, March 1, 1865, *Darwin Correspondence* 13:71 and note 30.

21. William Leach, *Butterfly People: An American Encounter with the Beauty of the World* (New York: Pantheon Books, 2013), 167, 173; In the nineteenth century, the nomenclature on these species varied, some authors preferring generic names, others common names. See Lincoln P. Brower, "Understanding and Misunderstanding the Migration of the Monarch Butterfly (Nymphalidae) in North America: 1857–1995," *Journal of the Lepidopterists' Society* 49 (1995), 313–14. Letter of Walsh to Darwin, May 1, 1868, reproduced in the Darwin Correspondence Project, https://www.darwinproject.ac.uk/letters. See the appendix for modern nomenclature.

22. S. B. Malcolm and L. P. Brower, "Evolutionary and ecological implications of cardenolide sequestration in the monarch butterfly." *Experientia* 45.3 (1989): 284–95.

23. Benjamin D. Walsh and Charles V. Riley, "Imitative Butterflies," *AE* 1 (1869): 191.

24. This is not actually true; Nyphalidae is a diverse and abundant family.

25. Walsh and Riley, "Imitative Butterflies," 192. Until the 1990s, scientists interpreted the relationship between the Monarch and Viceroy in terms of classical Batesian mimicry, where a palatable (non-noxious) species resembles an unpalatable (noxious) species. Recent research has demonstrated that this mimicry is actually "Müllerian," in which both species are unpalatable to predators, though the Monarch's toxins have a different chemical composition and are much more powerful than those of the Viceroy. In Müllerian mimicry, both species benefit from reduced predation, though to different degrees. David B. Ritland and Lincoln P. Brower, "The Viceroy Butterfly is not a Batesian Mimic," *Nature* 350 (1991): 497–98.

26. C. V. Riley, "Mimicry as Illustrated by these two Butterflies, with some Remarks on the Theory of Natural Selection," *Third Annual Report* (1871), 164–65.

27. Riley, "Mimicry," (1871), 159–60. Modern writers on the Monarch Butterfly consider Riley to be the leading early authority on this species, having investigated its range, food sources, mimicry by other species, and its migration and hibernation. Brower, "Understanding . . . the Monarch Butterfly," 305; and Leach, *Butterfly People*, 168.

28. Riley, "Mimicry," (1871), 169. Riley cited Lugger, who reported observing a bird seizing a Monarch on the wing and then suddenly dropping it. In his first report (with Walsh), Riley argued that the larvae of the two species would be preyed upon at the same rate, indicating his belief that they occurred in the same habitat (see discussion above). He apparently concluded later that Monarch and Viceroy larvae occurred in separate habitats.

29. Riley, "Darwin's Work in Entomology," (1882), 77–80.

30. C. V. Riley, "The Goat Weed Butterfly," *Second Annual Report* (1870), 127.

31. C. V. Riley, "Two of Our Common Butterflies," *Third Annual Report* (1871), 148–49.

32. C. V. Riley, "The Grape-Vine Plume," *Third Annual Report* (1871), 67.

33. Riley, "The Grape-Vine Plume," (1871), 66.

34. This discussion of the Yucca Moth is based in part on Sorensen, *Brethren of the Net*, chap. 11, "The Yucca Moth."

35. Joseph Ewan, "George Engelmann," *Dictionary of Scientific Biography*, 130–31 (hereafter cited as *DSB*); and Carol A. Sheppard and Richard A. Oliver, "Yucca Moths and Yucca Plants: Discovery of 'the most wonderful Case of Fertilization,'" *AE* 50 (Spring 2004): 33–34.

36. H. Lewis McKinney, "Fritz Müller," *DSB* (1974), 559–600; and William Trelease, review of *The Mutual Relations Between Flowers and the Insects Which Serve to Cross Them* by Hermann Müller, *American Naturalist* 13 (July 1879): 451–52 (hereafter cited as *AN*).

37. Sheppard and Oliver, "Yucca Moths and Yucca Plants," 35.

38. C. V. Riley, "On a New Genus in the Lepidopterous Family *Tineidae*: with Remarks on the Fertilization of Yucca," *Transactions of the Academy of Science of St. Louis* 3 (June 1873): 55–64 (hereafter cited as *Trans. SLAS*).

39. Reported in *CE* 4 (October 1872): 182. Riley's findings, printed the next year as an abstract of his paper at the AAAS meeting in August 1872, were widely reported in the United States and in Europe. On September 2, 1872, he reported on the Yucca Moth at the St. Louis Academy of Science, and his report was published in the society's *Transactions* the following year. The same article, with additional notes and correspondence, was published in Riley's *Fifth Annual Report* (1873), 150–60.

40. Müller to Riley, December 15, 1873, box 2, scrapbook vol. 17, p. 76. Riley quoted this letter in Charles V. Riley, "Further Notes on the Pollination of Yucca and on Pronuba and Prodoxus," *Proc. AAAS* 29 (1881): 623n4.

41. C. V. Riley, "Some Interrelations of Plants and Insects," *Proceedings of the Biological Society of Washington, DC* 7 (1892): 93 (hereafter cited as *Proc. BSW*); and Sheppard and Oliver, "Yucca Moths and Yucca Plants," 41–42, 44.

42. C. V. Riley, "On a New Genus," *Fifth Annual Report* (1873), 150–60; C. V. Riley, "The Yucca Moth—*Pronuba yuccasella* Riley," *Sixth Annual Report* (1874), 131–35; C. V. Riley, "Supplementary Notes on *Pronuba Yuccasella*," in *Trans. SLAS* 3 (1874): 178–80; C. V. Riley, "On the Oviposition of the Yucca Moth," *AN* 7 (October 1873): 619–23.

43. George Engelmann, "Notes on the genus *Yucca*, no. 2," *Trans. SLAS* 3 (1873): 210–14; and C. V. Riley, "The Yucca Moth and Yucca Pollination," *Report of the Missouri Botanical Garden* 3 (1892): 109.

44. Riley, "Further Notes on the Pollination of Yucca," (1881), 617–39. Riley's most succinct account is *The Yucca Moth and Yucca Pollination* (1892), 99–158, cited above. For a modern review, see O. Pellmyr, "Yuccas, Yucca Moths, and Coevolution: a Review," *Annual Report of the Missouri Botanical Garden* 90 (2003): 35–55 (hereafter cited as *Ann. Rept. Mo. Bot. Garden*); and Sheppard and Oliver, "Yucca Moths and Yucca Plants."

45. Riley, "On the Oviposition," *AN* 7 (1873): 619–23. Extract reprinted in *Trans. SLAS* (1873): 208–10.

46. Riley, "On a New Genus," *Fifth Annual Report* (1873), 156–57.

47. Sheppard and Oliver, "Yucca Moths and Yucca Plants," 44–45.

48. J. W. Hall, "In Memoriam–Vactor T. Chambers," *Journal of the Cincinnati Society of Natural History* 6 (1883): 239–44.

49. Sheppard and Oliver, "Yucca Moths and Yucca Plants," 39.

50. C. V. Riley, "The True and Bogus Yucca Moth; with Remarks on the Pollination of Yucca," *AE* 3 (1880): 142–45.

51. Riley, "On a New Genus," *Fifth Annual Report* (1873), n59. For an extended discussion, including recent discoveries and taxonomic revisions in this group of insects, see Sheppard and Oliver, "Yucca Moths and Yucca Plants," 39–40.

52. Darwin to Hooker, April 7, 1874, cited in Sheppard and Oliver, "Yucca Moths

and Yucca Plants," 42. A similar statement by Hermann Müller is noted above.

53. Darwin to Riley, September 28, 1881, quoted in Sheppard and Oliver, "Yucca Moths and Yucca Plants," 42.

54. For example, in a talk in New York City in 1872, Riley cited the Yucca Moth to illustrate Darwinian adaptation by slow degrees. *New York World*, box 1, scrapbook vol. 5, p. 259, SIA, CVR. Yucca pollination figured prominently in Riley's Lowell lectures in Boston in 1892, discussed later.

55. Edward S. Morse, "What American Zoologists Have Done for Evolution," *Proc. AAAS* 25 (1877): 154.

56. B. D. Walsh, "Imported Insects; The Gooseberry Sawfly," *PE* 1 (September 29, 1866): 117–18.

57. C. V. Riley, "Imported Insects and Native American Insects," *Second Annual Report* (1870), 8–13.

58. Walsh, "Imported Insects," 117–19.

59. C. V. Riley, "The Hickory Bark-Borer," *Fifth Annual Report* (1873), 104. See similar observations in "The Rocky Mountain Locust," *Seventh Annual Report* (1875), 180, 191.

60. He filled several volumes of his scrapbooks with clippings on forests and the presumed climatic consequences of deforestation.

61. C. V. Riley "The Influence of Climate on Trees, Plants, and Animals," box 1, scrapbook vol. 6, p. 54, SIA, CVR; and *Western Christian Advocate* 49 (May 3, 1882): 141.

62. Report of St. Louis Academy of Science proceedings, *PF* 43 (May 25, 1872): 62.

63. C. V. Riley, "Cabbage Worms," *Second Annual Report* (1870), 107.

64. Riley, "Cabbage Worms," (1870), 107.

65. C. V. Riley, "Apple-Tree Borers," *First Annual Report* (1869), 42–3.

66. C. V. Riley, "The Wheat-Head Army Worm," *Ninth Annual Report* (1877), 50.

67. Riley, "Wheat-Head Army Worm," (1877), 52.

68. C. V. Riley, "The Pickle Worm," *Second Annual Report* (1870), 70.

69. Riley, "Wheat-Head Army Worm," (1877), 50–51.

70. C. V. Riley, "The Codling Moth or Apple Worm," *First Annual Report* (1869), 65.

71. Riley, "Codling Moth or Apple Worm," (1869), 65.

72. C. V. Riley, "The Codling Moth, Again," *Fourth Annual Report* (1872), 22–23.

73. C. V. Riley, "The Grape-Leaf Gall-Louse," *Third Annual Report* (1871), 91–92. The latter possibility related to Packard's evolutionary theory discussed below. Riley also cited a new race of the White Pine Scale around St. Louis that now infested Scotch Pines.

74. C. V. Riley, "The Glassy-Winged Soldier Bug," *Third Annual Report* (1871), 137.

75. C. V. Riley, "The American Bean-Weevil," *Third Annual Report* (1871), 52–53.

76. C. V. Riley, "The Apple Curculio," *Third Annual Report* (1871), 30.

77. C. V. Riley, "Snout Beetles," *Third Annual Report* (1871), 28.

78. Riley, "Snout Beetles," (1871), 28–29.

79. Riley, "The Glassy-Winged Soldier-Bug," 137–38.

80. Riley, "The Glassy-Winged Soldier-Bug," 137–38.

81. C. V. Riley, "The Pine-Leaf Scale Insect," *Fifth Annual Report* (1873), 100.

82. C. V. Riley, "The Oyster Shell Bark Louse of the Apple," *Fifth Annual Report* (1873), 90.

83. C. V. Riley, "Darwin's Work in Entomology," *Proc. BSW* (1882): 70.

84. Hae-Gyung Geong, "Exerting Control: Biology and Bureaucracy in the Development of American Entomology, 1870–1930" (PhD diss., University of Wisconsin-Madison, 1999), UMI no. 9937208, pp. 103–4.

85. Geong, "Exerting Control," 179.

86. George Gale, "Saving the Vine from Phylloxera: a Never-Ending Battle," in *Wine: A Scientific Exploration*, ed. M. Sandler and R. Pinder (London: Taylor & Francis, 2003), 87–88. See the discussion of Grape Phylloxera in the next chapter.

87. A. S. Packard. Jr., "On Certain Entomological Speculations: a Review," *Proc. ESP* 6 (November 1866): 218.

88. T. D. A. Cockerell, "Alpheus Spring Packard, 1839–1905," in *Biographical Memoirs of the National Academy of Sciences*, vol. 9 (Washington, DC: National Academy of Sciences, 1920), 181–236; Theodore John Greenfield, "Variation, Heredity, and Scientific Explanation in the Evolutionary Theories of Four American Neo-Lamarckians, 1867–1897" (PhD diss., University of Wisconsin, Madison, 1986), UMI no. 8614370, pp. 64–68, 95.

89. Samuel Henshaw, ed., "The Entomological Writings of Dr. Alpheus Spring Packard," US Department of Agriculture, Division of Entomology, Bulletin No. 16 (Washington, DC: GPO, 1887).

90. Greenfield, "Variation, Heredity, and Scientific Explanation," 69–74; Morse, "What American Zoologists have Done for Evolution," 158–59; A. S. Packard Jr., *Lamarck, the Founder of Evolution: His Life and Work* (New York: Longmans, Green, 1901), 402; Stephen Bocking, "Alpheus Spring Packard and Cave Fauna in the Evolution Debate," *Journal of the History of Biology* 21, no. 3 (Fall 1988): 439–48 (hereafter cited as *JHB*). Packard later discussed cave fauna in "The Cave Fauna of North America, with Remarks on the Anatomy of the Brain, and Origin of the Blind Species," *Biographical Memoirs of the National Academy of Sciences* 4 (1888). An abstract under the same title appeared in *AN* 21 (1887): 82–83.

91. A. S. Packard, "A Monograph of the Geometrid Moths or Phalaenidae of the United States." F. V. Hayden, *Report of the United States Geological and Geographical Survey of the Territories*, vol. 10 (Washington, DC: GPO, 1876); Greenfield, "Variation, Heredity, and Scientific Explanation," 68. Packard had been investigating the effect of past climate on moths and other animals since 1862 or earlier. Packard to Scudder, September 18, 1862, Scudder Corr., Boston Museum of Science.

92. Greenfield, "Variation, Heredity, and Scientific Explanation," 66, 74.

93. Ralph W. Dexter, "The Impact of Evolutionary Theories on the Salem Group of

Agassiz Zoologists (Morse, Hyatt, Packard, Putnam)," *Historical Collections of the Essex Institute* 115 (1979): 163–64.

94. As far as is known, Packard and Riley never debated their differences in person or in their correspondence, but their different viewpoints are evident in their writings. A. S. Packard, "Rapid as well as Slower Evolution," *Independent* 29 (August 23, 1877): 6–7. Packard's ideas were later developed in his address as vice president for zoology at the AAAS in 1898. A. S. Packard, "A Half Century of Evolution, with Special Reference to the Effects of Geological Change on Animal Life," *AN* 32 (September 1898): 623–74.

95. Many contemporary scientists, Darwin among them, confessed they could not understand the Cope-Hyatt theory. Numbers, *Darwinism Comes to America*, 34–35.

96. Greenfield, "Variation, Heredity, and Scientific Explanation," 250–91, 261, 281–82.

97. Riley, "A New Genus," *Fifth Annual Report* (1873), 157.

98. On Weismann, see the following by Frederick B. Churchill, "August Weismann and a Break from Tradition," *JHB* 1 (1968): 91–112; "Hertwig, Weismann, and the Meaning of Reduction Division Circa 1890," *Isis* 61 (1970): 429–57; and "The Weismann-Spencer Controversy over the Inheritance of Acquired Characters," in *Human Implications of Scientific Advance: Proceedings of the XVth International Congress of the History of Science, Edinburgh, 10–15 August, 1977*, ed. Eric G. Forbes (Edinburgh: Edinburgh University Press, 1978), 451–68. See also Frederick B. Churchill, *August Weismann: Development, Heredity, and Evolution* (Cambridge, MA: Harvard University Press, 2015).

99. Numbers, *Darwinism Comes to America*, 36.

100. Numbers, *Darwinism Comes to America*, 35–37 and note 27, where Riley is cited in a contemporary list of neo-Lamarckians.

101. Riley to Varigny, the exact date of the letter is uncertain, but it was written before 1887, as Riley refers in it to his 1887 AAAS address. NAL, MS 143, Series 1, Box 1, Folder 11. See also Yves Carton, *Henry de Varigny, Darwinien convaincu, médecin, chercheur et journaliste (1855–1934)* (Paris: Hermann, 2008).

102. Riley, "Interrelations of Plants and Insects," *Proc. BSW* 7 (May 1892): 102.

103. C. V. Riley, "Darwin's Work in Entomology," *Proc. BSW* 1 (1882): 70–80; C. V. Riley, "On the Causes of Variation in Organic Forms," *Proc. AAAS* 37 (1889): 225–73; Riley, "Interrelations of Plants and Insects," 81–104.

104. Riley, "Interrelations of Plants and Insects," 103.

105. Riley, "Causes of Variation," 255.

106. Riley, "Causes of Variation," 234.

107. Lester F. Ward, "Neo-Darwinism and Neo-Lamarckism," Presidential Address, January 24, 1891, *Proc. BSW* 6 (January 1891): 11–71. Cited by Riley in "Interrelations of Plants and Insects," 101.

108. Ward, "Neo-Darwinism and Neo-Lamarckism," 54–55.

109. Ward, "Neo-Darwinism and Neo-Lamarckism," 13–14.

110. Ward, "Neo-Darwinism and Neo-Lamarckism," 22–23.

111. Riley, "Causes of Variation," 239, 235–36, 239–40.

112. Riley, "Causes of Variation," 248.

113. Riley, "Causes of Variation," 254–55.

114. Riley, "Causes of Variation," 251–52.

115. Riley, "Causes of Variation," 252, 262.

116. Oliver Sacks, "The Mental Life of Plants and Worms, Among Others," *NYRB* 61 (April 24–May 7, 2014): 4–8.

117. Ward, "Neo-Darwinism and Neo-Lamarckism," 68. On American naturalists' concern that natural selection undermined morality see Numbers, *Darwinism Comes to America*, 37–38.

118. Riley, "Causes of Variation," 266.

119. Riley, "Causes of Variation," 269.

120. Riley, "Causes of Variation," 273.

121. Riley, "Mimicry as Illustrated by these two Butterflies," *Third Annual Report* (1871), 174–75.

122. In the mimicry article, Riley discussed Darwin's theory extensively and refuted various objections.

123. Numbers, *Darwinism Comes to America*, 73–74, 109.

124. Riley, "Causes of Variation," 271.

125. Riley, "Causes of Variation," 235. See also chap. 6.

126. Drawing on a personal example, one of the authors (W. Conner Sorensen), while serving as a seasonal park ranger in Joshua Tree National Monument in Southern California in 1969–1970, often cited Riley's Yucca Moth studies when interpreting the Joshua Tree (Yucca) to park visitors.

127. For his role in converting Engelmann to evolution, see W. B. Hendrickson, "An Illinois Scientist Defends Darwinism: A Case Study in the Diffusion of Scientific Theory," *Transactions of the Illinois Academy of Science* 65 (1972): 25–26.

128. Richard Billon, "Inspiration in the Harness of Daily Labor: Darwin, Botany, and the Triumph of Evolution 1859–1868," *Isis* 102 (September 2011): 410. See also Kritsky, "Entomological Reactions," 354–55. Ronald Numbers points out that the majority of naturalists agreed with Darwin that new species evolved but remained skeptical that natural selection served as the prime agency in evolution. Numbers, *Darwinism Comes to America*, 33–34, 44.

129. Billon, "Darwin, Botany, and the Triumph of Evolution 1859–1868," 394–96, 407–12, 415.

130. Riley's reports on the Colorado Potato Beetle in the *Chicago Tribune* were cited in foregoing chapters. The following chapters on phylloxera and the Rocky Mountain Locust contain citations in other newspapers.

Chapter 6

This chapter appeared in an extended form in W. Conner Sorensen, Edward H. Smith, Janet Smith, and Yves Carton, "Charles V. Riley, France, and *Phylloxera*," *AE* 54 (Fall 2008): 134–49.

1. On wine culture in Missouri, see C. V. Riley, "Insects Injurious to the Grape-Vine," *First Annual Report* (1869), 124. The first mention of phylloxera as a pest of grape vines in Riley's annual reports is in the third (1871), submitted in December 1870.

2. C. V. Riley, "The Grape-Leaf Gall-Louse—*Phylloxera vitifoliae*, Fitch," *Third Annual Report* (1871), 84.

3. The term "plant louse" is an older term for aphid, which are insects related to phylloxera and previously grouped with them.

4. Frédéric Cazalis, publisher of the *Messager agricole du Midi*, and Louis Vialla were also members of the commission but did not participate in the field investigations.

5. G. Bazille, J. É. Planchon, and F. Sahut, "Sur une maladie de la vigne actuellement régnante en Provence," *Compte rendus de l'Académie des Sciences, Paris* 67 (August 3, 1868): 333–36. The question of who "discovered" the root form became a matter of controversy. It now seems clear that Sahut first observed the root form. Yves Carton, "La découvert du Phylloxera en France: Un sujet de polémique: les archives parlent (Hemiptera, Chermesidae)," *Bulletin de la Société entomologique de France* 111, no. 3 (2006): 305–16.

6. Signoret was president of the Entomological Society of France in 1861 and 1883.

7. L. J. H. Boyer de Fonscolombe, "Description des Kermès qu'on trouve aux environs d'Aix," *Annales de la Société entomologique de France* 3 (1834): 201–24.

8. J. É. Planchon, "Nouvelles observations sur le Puceron de la vigne (*Phylloxera vastatrix*) (*nuper Rhizaphis*, Planch.)," *Compte rendus de l'Académie des Sciences, Paris* 67 (September 14, 1868): 588–94.

9. Riley's citations indicate his familiarity with the French literature. For example, he cited Planchon and Lichtenstein's 140-page bibliography containing 481 items for the years 1868–1871. C. V. Riley, "Insects Injurious to the Grape-Vine: The Grape Phylloxera-*Phylloxera vastatrix* Planchon," *Sixth Annual Report* (1874), 32.

10. [J. O. Westwood], "New Vine Diseases," *The Gardeners' Chronicle and Agricultural Gazette* 45 (January 30, 1869): 109. See Riley's comments on Westwood's article in Riley, "Grape-Vine Leaf-Gall," *AE 1* (August 1869): 248.

11. Interestingly, the only American Westwood mentioned was Henry Shimer, an Illinois physician and naturalist. Westwood, "New Vine Diseases," 109.

12. C. V. Riley, "Answers to Correspondents: Grape leaf louse," *PF* 18 (August 4, 1866): 73.

13. Riley, "Grape-Vine Leaf-Gall," 248.

14. C. V. Riley, "Grape-Leaf Gall-Louse," *Third Annual Report* (1871), 86.

15. Paul A. Gagnon, *France Since 1789* (New York: Harper and Row, 1964), 196–212.

16. J. Granett, A. Walker, J. De Benedictis, G. Fong, and E. Weber, "California Grape Phylloxera More Variable than Expected," *California Agriculture* 50, no. 4 (July–August, 1996): 391, 395. It is not clear why the vines sent to Westwood in 1867–68 contained both root and leaf infestations.

17. J. É. Planchon and Jules Lichtenstein, "Des modes d'invasion des vignobles par le Phylloxera," *Messager du Midi* 25 (August 25, 1869); J. É. Planchon and Jules Lichtenstein, "De l'identité spécifique du phylloxera des feuilles et du phylloxera des racines," *Compte rendus de l' Académie des Sciences, Paris* 69 (1870): 298–300; Minutes of meeting, August 11, 1869, *Bulletin de la Société entomologique de France*, 4th ser., 9 (1869):41–44.

18. J. É. Planchon and Jules Lichtenstein, "Notes entomologiques sur le *Phylloxera*," *Insectologie Agricole* 12 (1869): 315–24.

19. Jules Lichtenstein and J. É. Planchon, "De l'identité spécifique du phylloxera des feuilles et du phylloxera des raciness de la vigne," *Journal d'Agriculture Pratique* 34 (August 11, 1870): 181–82.

20. Christy Campbell quotes Signoret as saying that, at the time he received Riley's letter, he was preparing a report on phylloxera for the Entomological Society of France and that he declined to examine Riley's findings until he was finished; however, the source for this information is not given. See Christy Campbell, *The Botanist and the Vintner: How Wine was Saved for the World* (Chapel Hill: Algonquin Books, 2005), 68.

21. Later that year, Riley received word from Signoret, who wrote him by means of a balloon from besieged Paris, that, although he himself was reduced to eating "cats, dogs, and horse-flesh," his phylloxera cultures were in good health. Riley, "Grape-Leaf Gall-Louse," *Third Annual Report* (1871), 86 and footnote.

22. Riley, "Grape-Leaf Gall-Louse," *Third Annual Report* (1871), 86 (report submitted December 2, 1870).

23. Riley, "Grape-Leaf Gall-Louse," *Third Annual Report* (1871), 95–96.

24. C. V. Riley, "Grape-leaf Gall-louse (*Phylloxera vitifoliae*, Fitch.)," *Am. Ent. and Bot.* 2 (December 1870): 354–56; Riley, "Grape-Leaf Gall-Louse," *Third Annual Report* (1871), 88–89 (report submitted December 2, 1870).

25. F. Cazalis, "Maladie de la vigne," minutes of Scientific Congress of France, Montpellier, *Bulletin de la Société centrale d'agriculture de l'Hérault* 56 (1869): 196–214; Gale, "Saving the vine from Phylloxera," 71–72.

26. Minutes of meeting, December 13, 1871, *Bulletin de la Société entomologique de France*, 6th ser., 9 (1871): 80.

27. Comtesse de Fitz-James, *La viticulture franco-américaine (1869–1889)* (Montpellier: Coulet; Paris: Masson, 1889), 11.

28. Victor Signoret, "*Phylloxera vastatrix*. Hémiptère-Homoptère de la familie des Aphidiens, cause prétendue de la maladie actuelle de la vigne," *Annales de la Société entomologique de France* 9 (1869): 548–88.

29. Campbell, *Botanist and Vintner*, 69–70.

30. Campbell, *Botanist and Vintner*, 69–70; Minutes of the Entomological Society of France, May 25, 1870, *Bulletin de la Société entomologique de France*, 5th ser., 10 (1870), 45.

31. Minutes of the Entomological Society of France, July 10, 1870, *Bulletin de la Société entomologique de France*, 5th ser., 10 (1870), 64; Riley, "Grape-leaf Gall-louse," *Am. Ent. and Bot.* 2 (1870): 354; and Riley, "Grape-Leaf Gall-Louse," *Third Annual Report* (1871), 86.

32. Minutes of meeting, June 22, 1870, *Bulletin de la Société entomologique de France*, 5th ser., 10 (1870): 60.

33. Riley, "Grape-Leaf Gall-Louse," *Third Annual Report* (1871), 95.

34. Louis Vialla, reminiscences of Riley's 1871 visit in speech welcoming Riley, *Bulletin de la Société centrale d'agriculture de l'Hérault* 71 (1884): 416.

35. C. V. Riley, "Grape Disease: On the Cause of Deterioration in some of our Native Grape-vines, and one of the probable reasons why European Vines have so generally failed with us," *Fourth Annual Report* (1872), 69; C. V. Riley, "Grape Deterioration: On the Cause of Deterioration in some of our Native Grape-Vines, and the probable reason why European Vines have so generally failed in the Eastern Half of the United States," *Moore's Rural New Yorker* 24 (November 4, 1871): 283.

36. C. V. Riley, "Grape Phylloxera," *Fifth Annual Report* (1873), 67.

37. Jules Lichtenstein, report, minutes of meeting, August 23, 1871, *Bulletin de la Société entomologique de France*, 6th ser., 9 (1871): 46–47.

38. Campbell, *Botanist and Vintner*, 61–62; Gale, "Saving the vine," 73.

39. C. V. Riley, "Grape-leaf Gall-louse," *Am. Ent. and Bot.* 2 (1870), 354–59. Lichtenstein arranged for publication of the article as "Insects nuisibles à la vigne, le puceron de la vigne," in *Bulletin de la Société centrale d'Agriculture de l' Hérault* 58 (February 20, 1871): 172–80; and as "Insects nuisibles à la vigne, le puceron de la vigne," in *Le Messager Agricole du Midi* 2 (April 10, 1871): 84–89.

40. Riley, "Grape Disease," *Fourth Annual Report* (1872), 62.

41. Léo Laliman, *Etudes sur les divers travaux phylloxériques et les vignes américaines, notamment sur les études faites en Espagne par Don Miret y Tarral, Vocal de la commission à Madrid* (Paris: Librairie Agricole, 1879), 169; Campbell, *Botanist and Vintner*, 70–71.

42. Campbell, *Botanist and Vintner*, 143, 194, 214.

43. For example, Bazille, president of the agricultural society in Montpellier, recommended grafting French vines onto American rootstock. Minutes of meeting, August 2, 1869, *Bulletin de la Société centrale d'agriculture de l' Hérault* 58 (1869): 289.

44. Riley, "Grape Disease," *Fourth Annual Report* (1872), 62.

45. Signoret's comments in minutes of meeting, December 13, 1871, *Bulletin de la Société entomologique de France*, 6th ser., 11 (1871): 80; Gale, "Saving the vine," 74, 77–80.

46. Isidor Bush & Son & Meissner, *Illustrated Descriptive Catalogue of American Grape Vines, with Brief Directions for their Culture* (St. Louis, MO: R. P. Studley & Co., 1875), 3. Campbell places Riley's visit to the Bush nursery in 1870, prior to his trip to France in summer 1871. This chronology assumes that Riley's "discovery" of the

American root form, which Riley places in the fall of 1870, refers to his excavation of root phylloxera in the Bush nursery. Campbell, *Botanist and Vintner*, 86n31. This seems unlikely, as Bush dates Riley's visit to 1871, and more importantly, Riley's "discovery" in 1870 referred to the identity of the root and leaf forms of phylloxera on American, not European, vines. See Riley, "Grape-Leaf Gall-Louse," *Third Annual Report* (1871), 86 (report submitted December 2, 1870); Riley referred to his "discovery" again in C. V. Riley, "The Grape Phylloxera," *Seventh Annual Report* (1875), 95.

47. Riley, "Grape Disease," *Fourth Annual Report* (1872), 60–64.

48. C. V. Riley, "Grape Deterioration: On the Cause of Deterioration in some of our Native Grape-vines, and the Probable Reason Why European Vines Have so Generally Failed in the Eastern Half of the United States," *Moore's Rural New Yorker* 24 (October 21, 1871): 251, (October 28, 1871): 268, and (November 4, 1871): 283, translated in *Le Messager du Midi* (December 3, 1871); in *Le Messager agricole du Midi* (December 10, 1871); and as "Le Phylloxera," *Bulletin de la Société centrale d'agriculture de l' Hérault* 58 (1871): 246–54. See also Isidor Bush & Son, *Illustrated Descriptive Catalogue of Grape Vines, Potatoes, Cultivated and For Sale at the Bushberg Vineyards & Orchards, Jefferson Co., Mo., with Brief Directions for Planting and Cultivating* (St. Louis, MO: R. P. Studley Co., 1869); Riley, "Grape Disease," *Fourth Annual Report* (1872), 60–63; and Campbell, *Botanist and Vintner*, 110.

49. Gale, "Saving the vine," 75; Louis Vialla and J. É. Planchon, *Etat des vignes américaines dans le Départment de l'Herault pendant l'année 1875. Rapport adressé au nom d'une commission spéciale à la Société centrale d'agriculture du départment de l'Hérault* (Montpellier: Grollier, 1875), 42; Siegmar Muehl, "Isidor Bush and the Bushberg Vineyards of Jefferson County," *Missouri Historical Review* 94 (October 1999), 53.

50. Jules Lichtenstein, *Histoire du Phylloxera* (Montpellier: Coulet; Paris: Bailliére, 1879), 19.

51. Riley, "Grape Disease," *Fourth Annual Report* (1872), 62.

52. J. É. Planchon, *Les vignes américaines, leur culture, leur résistance au Phylloxera et leur avenir en Europe* (Montpellier: Coulet; Paris: Delahaye, 1875), 18; Dwight W. Morrow Jr., "The American Impressions of a French Botanist, 1873," *AgH* 34 (1960): 71.

53. After leaving St. Louis, Planchon completed his three-month tour with visits to vineyards in western New York and eastern Massachusetts. Morrow, "American Impressions," 72.

54. Riley, "Grape Disease," *Fourth Annual Report* (1872), 66.

55. Bush & Son & Meissner, *American Grape Vines* (1875), 28n53. The 1875 edition was translated the next year by Louis Bazille, with annotations by Planchon, as *Les Vignes Américaines: Catalogue Illustré et descriptif avec de bréves indications sur leur culture* (Montpellier: C. Coulet; Paris: V.-A. Delahaye, 1876). In the French version, the citation from Riley appears on p. 49.

56. On French resistance to Darwinism, see Robert E. Stebbins, "France," in Glick, ed., *The Comparative Reception of Darwinism*, 117–39; and Carton, *Henry de Varigny*.

57. J. É. Planchon, "Le *Phylloxera* en Europe et en Amérique," *Revue des deux Mondes* (February 1–15, 1874): 51–52.

58. Riley, "Darwin's Work in Entomology," 80.

59. C. V. Riley, "The Grape Phylloxera," *Eighth Annual Report* (1876), 167. Bazille had written to Riley in 1874 that the vineyards in the Midi were practically destroyed. Riley, "Grape Phylloxera," *Seventh Annual Report* (1875), 104.

60. Riley, "Insects Injurious to the Grape-Vine," *Sixth Annual Report* (1874), 65.

61. Riley's itinerary in France is based on Riley's recollections recorded in minutes of the Hérault Agricultural Society, June 30, 1884, *Bulletin de la Société centrale d'agriculture de l' Hérault* 71 (1884): 397–416; and Minutes of the Entomological Society of France, July 14, 1875, *Bulletin de la Société entomologique de France*, [10th ser. ?] 55 (July 14, 1875): 151–53; *Le Journal illustré* (Paris), September 22, 1878. At the entomological society meeting in Paris on July 14, 1875, Riley gave a talk on phylloxera and served the members Rocky Mountain Locust delicacies.

62. Louis Vialla and J. É. Planchon, "Etat des vignes américaines," *Messager Agricole du Midi*, December 10, 1874; Riley, "Grape Phylloxera," *Eighth Annual Report* (1876), 167.

63. Campbell, *Botanist and Vintner*, 147–48. The Concord variety of *Vitis labrusca*, though it evolved in North America, was vulnerable to Grape Phylloxera.

64. Pierre Viala, *Une mission viticole en Amérique* (Montpellier: C. Coulet; Paris: G. Masson, 1889), 303–75.

65. Dr. Crolas and V. Vermorel, *Manuel pratique des sulfurages. Guide du vigneron pour l'emploi du sulfure de carbone contre le Phylloxera* (Villefranche, Montpellier, and Lyon: Bibliothèque du progrès agricole et viticole, 1886), 106; Campbell, *Botanist and Vintner*, 109–11, 120, 174–76.

66. At the Phylloxera Congress in Bordeaux in 1881, vintners agreed that grafting French vines on American stock was the best remedy. "Farmer's Review," December 10, 1881, box 6, scrapbook vol. 47, pp. 40–41, SIA, CVR; Campbell, *Botanist and Vintner*, 153–54, 190, 204–15.

67. Riley, "Insects Injurious to the Grape-Vine," *Sixth Annual Report* (1874), 44.

68. Riley, "Insects Injurious to the Grape-Vine," *Sixth Annual Report* (1874), 39.

69. Riley, "Grape Deterioration" (October 21, 1871), 251; Riley, "Grape-Leaf Gall-Louse," *Third Annual Report* (1871), 86n14.

70. Signoret considered these to be females that had already deposited their eggs; Riley considered this unlikely, as they were not yet mature. Riley, "Insects Injurious to the Grape-Vine," *Sixth Annual Report* (1874), 38–39.

71. Planchon, *Les vignes américains*, 62–65.

72. Riley, "Grape Phylloxera," *Seventh Annual Report* (1875), 118–21. The Oak-feeding phylloxera of Europe are now considered a species complex including *Phylloxera quercus* Boyer de Fonscolombe (Hemiptera: Phylloxeridae).

73. Riley, "Insects Injurious to the Grape-Vine," *Sixth Annual Report* (1874), 41–42.

74. Riley, "Grape Phylloxera," *Seventh Annual Report* (1875), 90–91.

75. Riley "Insects Injurious to the Grape-Vine," *Sixth Annual Report* (1874), 41–42; and Riley, "Grape Phylloxera," *Eighth Annual Report* (1876), 160n78.

76. Riley, "Insects Injurious to the Grape-Vine," *Sixth Annual Report* (1874), 64–65; and Riley, "Grape Phylloxera," *Seventh Annual Report* (1875), 90–91.

77. Riley, "Grape Phylloxera," *Eighth Annual Report* (1876), 163.

78. Riley, "Grape Phylloxera," *Eighth Annual Report* (1876), 163.

79. Riley, "Grape Phylloxera," *Eighth Annual Report* (1876), 160.

80. Campbell, *Botanist and Vintner*, 163–66.

81. Riley, "Insects Injurious to the Grape-Vine," *Sixth Annual Report* (1874), 55.

82. Riley, "Grape-Leaf Gall-Louse," *Third Annual Report* (1871), 85, and notes on pp. 93–95.

83. Riley, "Grape-Leaf Gall-louse," *American Entomologist and Botanist* 2 (1870): 353–54 (hereafter cited as *Am. Ent. and Bot.*); Riley, "Grape Disease," *Fourth Annual Report* (1872), 55n.

84. Riley, "Grape-leaf Gall-louse," *Am. Ent. and Bot.* 2 (1870): 354; Riley, "Grape-Leaf Gall-Louse," *Third Annual Report* (1871), 94–95, 95n1.

85. Riley, "Grape Disease," *Fourth Annual Report* (1872), 55n.

86. For background on this controversy, see Sorensen, *Brethren of the Net*, chap. 12, "The Debate over Entomological Nomenclature."

87. Riley, "Grape-Leaf Gall-Louse," *Third Annual Report* (1871), 95n.

88. Riley, "Grape Disease," *Fourth Annual Report* (1872), 55n.

89. Riley, "Grape Phylloxera," *Fifth Annual Report* (1873), 57n. Riley defended his decision in French publications as well. C. V. Riley, "Les espèces américaines du genre *Phylloxera*," *Compte rendus de l'Académie des Sciences, Paris* 79 (1874): 1384–88. Signoret preferred Fitch's name, perhaps because he felt that this would weaken Planchon's claim to a new discovery. See V. Signoret, "Observations sur les points qui paraissent acquis à la science, au sujet des espèces connues du genre Phylloxera," *Compte rendus de l' Académie des Sciences, Paris* 79 (1874): 778.

90. Taxonomists continue to struggle with the issues faced by Riley and his contemporaries. In 1958, in a concession to "general usage," zoologists agreed that names that had been in usage for fifty years could not be changed, even if a previously unknown describer was identified. Kjell B. Sandved and Michael G. Emsley, *Insect Magic* (New York: Viking Press, 1978), 28.

91. *Common Names of Insects Database*, Entomological Society of America, Annapolis, Maryland, www.entsoc.org/common-names (accessed October 27, 2016).

92. Campbell, *Botanist and Vintner*, 204–15, 191n.

93. Riley recounted this meeting in C. V. Riley, ed., *Agriculture*, Reports of the United States Commissioners to the Universal Exposition of 1889 at Paris, vol. 5 (Washington, DC: GPO, 1891), 282, 362 (hereafter cited as Riley, ed., *Agriculture*).

94. Lichtenstein quoted this statement in an article featuring Riley as Chief of the

Division of Entomology. See Jules Lichtenstein, *Riley et l'Entomologie aux Etats Unis* (Montpellier: Imprimerie Centrale du Midi, Hamelin frères, 1883), 5; Minutes of Meeting, June 16, 1884, *Bulletin de la Société centrale d'agriculture de l' Hérault* 71 (1884): 390.

95. C. V. Riley, "Quelques mots sur les insecticides aux Etats Unis et proposition d'un nouveau remede contre le Phylloxera," *Bulletin de la Société centrale d'agriculture de l' Hérault* 71 (1884): 397–416.

96. Yves Carton has found many reprints of Riley's 1884 Montpelier address in French libraries and archives. See also Bush and Meissner, *Les Vignes Américaines.*

97. Riley, "Quelques mots sur les insecticides aux Etats Unis," 397–416 (esp. n80); Gustave Foëx, *Manuel pratique de viticulture pour la reconstitution des vignobles méridionaux-vignes américaines, submersion, plantation dans les sables* (Montpellier: Coulet; Paris: Delahaye et Lecrosnier, 1887), 248.

98. G. Claudey, "Victor Vermorel, connu et méconnu," *Bulletin de l'Académie de Villefranche sur Saône* (1991): 65–76.

99. Riley, ed., *Agriculture*, 282, 362.

100. Viala, *Une mission viticole.*

101. Viala, *Une mission viticole*; Campbell, *Botanist and Vintner*, 224–26; D. Hawthorne, ed., "France Will Honor Early Texas Pioneer: Phylloxera Fighter Munson Finally Recognized," *Vinifera Wine Growers Journal* (Spring 1988): 1–6; S. S. McLeRoy and R. E. Renfro, *Grape Man of Texas: The Life of T. V. Munson* (Austin, TX: Eakin Press, 2004).

102. Riley, ed., *Agriculture* (1891). In March 1889, the Agricultural Department was elevated to cabinet status. Thereafter, the commissioner of agriculture was the secretary of agriculture, who presided over the Department of Agriculture. Howard, *History of Applied Entomology*, 92.

103. Minutes of meeting, May 22, 1889, *Bulletin de la Société entomologique de France*, 24th ser. (1889): 47; Riley to L. O. Howard, June 30, 1889, box 27, SIA, CVR; C. V. Riley, "The Outlook for Applied Entomology," *Insect Life* 3 (1891): 181–90 (hereafter cited as *IL*).

104. V. Vermorel, "La station viticole de Villefranche," *Revue trimestrielle de la Station Viticole de Villefranche (Rhône)* 1–2 (1890), vii–x. In a manuscript copybook with names of visitors to the laboratory appears the notation, "station honorée en 1889 d'une visite de M. Ryley [*sic*] entomologiste à Washington (Etats-Unis)," Library of Villefranche sur Saône, 'Fond ancien Vermorel.'

105. Riley, ed., *Agriculture*, 534–35.

106. Riley to Howard, June 30, 1889, box 27, SIA, CVR; *London Morning Post*, September 3, 1889.

107. Planchon, *Les vignes américaines*, 18; "Complementary," *St. Louis Democrat*, February 25, 1874, box 1, scrapbook vol. 9, p. 100, SIA, CVR.

108. Riley, ed., *Agriculture*, vol. 1, pp. N. and O. Riley's granddaughter, Emilie Wenban-Smith Brash, presented the Legion of Honor medal to the Entomological Society of Washington on the occasion of its centennial in 1984.

109. Riley's daughters, Thora M. and Cathryn V. Riley, presented the statuette to the Entomological Society of America in 1962. R. H. Nelson, ed., "Charles Valentine Riley," *Bulletin of the Entomological Society of America* 9, no. 3 (1963): 183; E. H. Smith, "The Grape Phylloxera: A Celebration of Its Own," *AE* 38 (Winter 1992): 213, 219n2.

110. C. V. Riley and L. O. Howard, "The *Phylloxera* Problem Abroad as it Appears to-day," *IL* 2 (April 1890): 310.

111. Planchon, *Les vignes américaines*, 61. Planchon also considered Agassiz's laboratories, classrooms, and collections superior to those in France. See Morrow, "American Impressions," 74.

112. Riley, "Grape Phylloxera," *Eighth Annual Report* (1876), 161.

113. J. Granett et al., "California grape phylloxera," 9–13; Gale, "Saving the vine," 86–87.

Chapter 7

1. B. D. Walsh, "Grasshoppers and Locusts," *PE* 2 (1866): 1–5; C. V. Riley, "Locusts," *PF* 34 (November 3, 1866): 290; and "Grasshoppers and Locusts," *PF* 34 (November 24, 1866): 333. Riley recalled his reporting on the 1860s grasshopper and locust outbreaks in "The Rocky Mountain Locust–*Caloptenus spretus* Thomas," *Seventh Annual Report* (1875), 171–72, 187–88. Riley used the same title for all his articles in the seventh, eighth, and ninth annual reports. Walsh and Riley's distinction between grasshoppers and locusts was accurate as far as it went, but later investigators developed the more complex "phase theory" discussed later.

2. Craig Miner, *West of Wichita: Settling the High Plains of Kansas, 1865–1890* (Lawrence: University Press of Kansas, 1986), 52; and Patricia Wickham Mills, "The Bug Invasion" (September 14, 2010), a memoir in manuscript that tells the experience of her grandparents who, as settlers in central Kansas in 1874, experienced the locust invasion firsthand. Manuscript in the files of Conner Sorensen, Eschbach, Germany.

3. "Bucolic Observations," [noting prairie schooners with dispirited Kansans heading east] *St. Louis Dispatch*, August 5, 1875. Riley, "Rocky Mountain Locust," *Seventh Annual Report* (1875), 149–50.

4. C. V. Riley, "Rocky Mountain Locust, *Eighth Annual Report* (1876), 60–61, 91. Entomologists now use the term "nymph" when referring to the immature stage of grasshoppers and locusts; however, Riley used the term "hopper," and we will follow his usage.

5. Riley, "Rocky Mountain Locust," *Seventh Annual Report* (1875), Preface, 64–66, 144–50; Riley, "Rocky Mountain Locust," *Eighth Annual Report* (1876), 60–66, 75–76.

6. Riley, "The Rocky Mountain Locust," *Seventh Annual Report* (1875), 121. For Riley's practice of collecting insects while riding on the front of the locomotive, see Riley, "The Rocky Mountain Locust," *Eighth Annual Report* (1876), 111.

7. Riley used the information to give a detailed account, by state and territory, of the 1874 invasion. Riley, "Rocky Mountain Locust," *Seventh Annual Report* (1875),

144–55. In 1876, he distributed a follow-up questionnaire. Riley, "Rocky Mountain Locust," *Ninth Annual Report* (1877), 57, 67, 76.

8. Riley, "Rocky Mountain Locust," *Seventh Annual Report* (1875), 124–32. For a comparison of the size of the Rocky Mountain Locust with normal grasshoppers, see the illustration in Miner, *West of Wichita*, 57.

9. C. V. Riley, "Chronological History of Locust Injuries," US Entomological Commission, *First Annual Report for the Year 1877 relating to the Rocky Mountain Locust and the best means of preventing its injuries and of guarding against its invasions* (Washington, DC: GPO, 1878), 53–113. (herafter cited as USEC, *First Annual Report*). This history was compiled from Riley's *Seventh* through *Ninth Annual Reports* (1875-77).

10. Riley, "The Rocky Mountain Locust," *Seventh Annual Report* (1875), 156–57. Riley admitted some poetic license in sketching the scene.

11. Riley, "The Rocky Mountain Locust," *Seventh Annual Report* (1875), 122, 142. A map appeared also in C. V. Riley, "The Rocky Mountain Locust," *Ninth Annual Report* (1877), between pp. 56 and 57.

12. Riley, "The Rocky Mountain Locust," *Eighth Annual Report* (1876), 91. Compare with Riley's quotation of a newspaper report in the Kansas City *Journal of Commerce* (June–July 1867), in Riley, "The Rocky Mountain Locust," *Eighth Annual Report* (1876), 59: "A Farmer near the Platte [R]iver informs us that a morning or two ago he went out to plow his corn, which was about four inches high the day before, and found it all gone."

13. Joel 2:5. Quoted in "The Rocky Mountain Locust," *Seventh Annual Report* (1875), 158. Jeffrey A. Lockwood (University of Wyoming) furnished the name of the desert locust described by Joel.

14. The incident is related in C. V. Riley, *The Locust Plague in the United States: Being More Particularly a Treatise on the Rocky Mountain Locust or so-called Grasshopper, as it occurs East of the Rocky Mountains, with Practical Recommendations for its Destruction* (Chicago: Rand McNally, 1877), 215. Since early times, colonial governors and other officials had proclaimed days of prayer in response to grasshopper plagues and other natural calamities. Conference of Governors, *The Rocky Mountain Locust, or Grasshopper, Being the Report of Proceedings of a Conference of the Governors of Several Western States and Territories . . . held at Omaha, Nebraska, on the Twenty-fifth and Twenty-sixth Days of October, 1876, to consider the Locust Problem; also a Summary of the Best Means now known for Counteracting the Evil* (St. Louis, MO: R. R. Studley Company, 1876), 28, 31–32 (hereafter cited as *Conference of Governors*).

15. Riley, *The Locust Plague*, 214; and Riley, "The Rocky Mountain Locust," *Eighth Annual Report* (1876), 96–97.

16. *St. Louis Globe,* May 19, 1875, quoted in Riley, "The Rocky Mountain Locust," *Eighth Annual Report* (1876), 96. See also p. 92.

17. Riley, "The Rocky Mountain Locust," *Eighth Annual Report* (1876), 72–73, 76.

18. Riley, "The Rocky Mountain Locust," *Eighth Annual Report* (1876), 127.

19. *St. Louis Post Dispatch,* June 21, 1875; Riley to LeConte, on board the *Abyssinia,* June 18, 1875, LeConte Coll., APS.

20. Riley, "The Rocky Mountain Locust," *Eighth Annual Report* (1876), 66–67.

21. Riley, "Rocky Mountain Locust," *Seventh Annual Report* (1875), 186.

22. "Very Latest: Grasshopper Grub," *St. Louis Dispatch,* May 29, 1875.

23. Riley, "The Rocky Mountain Locust," *Eighth Annual Report* (1876), 146.

24. "Hungry Hoppers," *Kansas City Times* [June 3, 1875], box 1, scrapbook vol. 9, p. 134, SIA, CVR.

25. "Grasshopper Soup," *St. Louis Globe-Democrat,* n.d., box 1, scrapbook vol. 9, p. 137, SIA, CVR.

26. Riley, "Rocky Mountain Locust," *Seventh Annual Report* (1875), 147. Riley served locust delicacies to the members of the Entomological Society of France in Paris at their meeting of July 14, 1875 (Bastille Day). Minutes of the Entomological Society of France, July 14, 1875, *Bulletin de la Société entomologique de France,* [10th ser.?] 55 (July 14, 1875): 53.

27. Riley, "Rocky Mountain Locust," *Seventh Annual Report* (1875), 144–47. Riley's AAAS address is reprinted here and also in USEC, *First Annual Report* (1878), 438–41.

28. Quoted in Riley, "Rocky Mountain Locust," *Eighth Annual Report* (1876), 82–83.

29. Riley, "Rocky Mountain Locust," *Seventh Annual Report* (1875), 151.

30. For Riley's views on public and private relief, see Riley, "Rocky Mountain Locust," *Seventh Annual Report* (1875), 152–53; *Eighth Annual Report* (1876), 67, 71–72; and *Ninth Annual Report* (1877), 111–16.

31. Riley, "Rocky Mountain Locust," *Seventh Annual Report* (1875), 149–50. Riley's attitude regarding relief was somewhat contradictory. While he castigated the Kansas legislature for withholding relief funds, elsewhere he advised Kansans not to request "outside aid," meaning, presumably, federal aid. Address to Kansas Academy of Science in the *Kansas Farmer* (November [1875]), box 1, scrapbook vol. 10, p. 235, SIA, CVR.

32. Riley, "The Rocky Mountain Locust," *Eighth Annual Report* (1876), 120.

33. Dickens distrusted the House of Commons and felt he could be most effective by writing about social ills; however, he later appealed to London's Metropolitan Sanitary Association to correct deplorable housing conditions. Claire Tomalin, *Charles Dickens: A Life* (London: Viking, 2011), 44, 227–31.

34. Riley, *Eighth Annual Report* (1876), 126.

35. Riley, "Rocky Mountain Locust," *Ninth Annual Report* (1877), 86–90, 99–106, and chart of daily temperatures at St. Louis, November to January 1876–77, p. 120.

36. Riley, "Rocky Mountain Locust," *Ninth Annual Report* (1877), 181–84.

37. Earle D. Ross, "The U. S. Department of Agriculture in the Commissionership," *AgH* 20 (July 1946): 134–36.

38. Dupree, *Science in the Federal Government: A History of Policies and Activities to 1940* (1957; reprint, New York: Arno Press, 1980), 155–56.

39. Riley to LeConte, December 3, 1873, LeConte Coll., APS.

40. John L. LeConte, "Hints for the Promotion of Economic Entomology," *Proc. AAAS* 22 (1874): 20 (also published with extended title in *AN* 7 [November 1873]: 710–22).

41. Riley to LeConte, December 3, 1873, LeConte Coll., APS.

42. Riley to LeConte, December 23, 1874, LeConte Coll., APS.

43. Riley to LeConte, December 23, 1874, LeConte Coll., APS.

44. LeConte, "Promotion of Economic Entomology," 20.

45. John L. LeConte, "Methods of Subduing Insects Injurious to Agriculture," *CE* 7 (September 1875): 168–69, 171.

46. Riley, "Rocky Mountain Locust," *Seventh Annual Report* (1875), preface, v.

47. Charles V. Riley, "The Locust Plague; How to Avert it," *Proc. AAAS* 24 (1876): 216.

48. Riley, "The Rocky Mountain Locust," *Eighth Annual Report* (1876), 150.

49. Thomas to Hayden, March 18, 1876, NA, RG 48, Records of the Secretary of the Interior, Hayden Survey, Letters Received (hereafter cited as Hayden Survey, Letters Received).

50. Thomas to Scudder, February 18, 1874, Scudder Corr., Museum of Science, Boston (hereafter cited as MSB).

51. Senator H. B. Anthony to LeConte, February 7, 1875, LeConte Coll., APS; John L. LeConte, "On the Method of Subduing Insects Injurious to Agriculture," *Proc. AAAS* 24 (1876): 203; Riley, "The Locust Plague," 221; Riley to LeConte, February 1, 1875 and June 18, 1875, LeConte Coll., APS.

52. At Riley's request, a bill incorporating their proposal was introduced in the *Congressional Record*, 44th Congress, 1st sess., Senate, vol. 4, pt. 2 (March 7, 1876), 1502.

53. "Professor Riley Addresses the Committee on Agriculture," *St. Louis Globe-Democrat*, February 24, 1876, box 2, scrapbook vol. 13, p. 26, SIA, CVR.

54. *Congressional Record*, 44th Cong., 1st sess., Senate, vol. 4, pt. 2 (March 7, 1876), 1502–4.

55. *Congressional Record*, 44th Cong., 1st sess., Senate, vol. 4, pt. 2 (March 7, 1876), 1502–4.

56. "Bibliography of the Locusts of America," in USEC, *First Annual Report* (1878), 274; Ralph W. Dexter, "The Organization and Work of the U. S. Entomological Commission (1877–82)," *Melsheimer Entomological Series* 26 (1979): 28.

57. *Congressional Record*, 44th Cong., 1st sess., Senate, vol. 4, pt. 2 (March 7, 1876), 1502–3.

58. *Congressional Record*, 44th Cong., 1st sess., Senate, vol. 4, pt. 2 (March 7, 1876), 1502–4, 1508.

59. *Congressional Record*, 44th Cong., 1st sess., Senate, vol. 4, pt. 2 (March 7, 1876), 1502–4.

60. *Congressional Record*, 44th Cong., 1st sess., Senate, vol. 4, pt. 2 (March 7, 1876), 1502–4, 1504–5.

61. *Nation* 22 (March 16, 1876): 169.

62. *Nation* 22 (March 30, 1876): 208.

63. *Congressional Record*, 44th Cong., 1st Sess., Senate, vol. 4, pt. 2 (March 7, 1876), 1504–1505, 1510–1511, 1541, 1558. See LeConte's account of the bill's failure in *CE* 8 (September 1876): 177–78.

64. Riley, "The Rocky Mountain Locust," *Eighth Annual Report* (1876), 133–37.

65. Thomas to Hayden, October 12, 1876, Hayden Survey, Letters Received.

66. Thomas to Hayden, October 12, 1876, Hayden Survey, Letters Received.

67. "Goodbye Grasshoppers; Interview with Prof. Riley, State Bug Master of Missouri," *Omaha Herald,* October 28, 1876, box 15, scrapbook vol. 43, p. 15, SIA, CVR.

68. "Goodbye Grasshoppers," *Omaha Herald,* October 28, 1876, box 15, scrapbook vol. 43, p. 15, SIA, CVR.

69. Riley, "The Rocky Mountain Locust," *Ninth Annual Report* (1877), 108.

70. Thomas to Hayden, October 28, 1876, Hayden Survey, Letters Received.

71. Box 15, scrapbook vol. 43, p. 37, SIA, CVR. See also Dexter, "Organization and Work of the U. S. Entomological Commission," 28; Thomas to Hayden, October 28, 1876, and January 6, 1877, Hayden Survey, Letters Received.

72. Riley to Hayden, January 3, January 13, and January 18, 1877, and Thomas to Hayden, January 19, 1877, all in Hayden Survey, Letters Received.

73. *U.S. Statutes at Large* 19 (1877), 357.

74. Thomas assigned his assistants, Emilie A. Smith and George H. French, to handle state entomologist affairs while he devoted his attention to locusts. *PF* (May 6, 1877), box 15, scrapbook vol. 43, p. 70, SIA, CVR.

75. Thomas to Hayden, October 28, 1876, and March 10, 1877, Hayden Survey, Letters Received.

76. Riley to [Hayden] [January 1877] in Hayden Survey, Letters Received.

77. Thomas to Hayden, January 19, 1877, in Hayden Survey, Letters Received.

78. Lester D. Stephens, "The Appointment of the Commissioner of Agriculture in 1877: A Case Study in Political Ambition and Patronage," *Southern Quarterly* 15, no. 4 (1977): 383–84.

79. "The Entomologists: General Satisfaction with the Appointments," [*General Press Dispatch*], March 21, [1877], box 15, scrapbook vol. 43, pp. 55–56, SIA, CVR; Dupree, *Science in the Federal Government*, 95.

80. John G. Sproat, *"The Best Men": Liberal Reformers in the Gilded Age* (London and New York: Oxford University Press, 1968), 100; Hans L. Trefousse, *Rutherford B. Hayes* (New York: Times Books; Henry Holt and Co., 2002), 71, 93–94.

81. USEC, *First Annual Report* (1878), xiv.

82. USEC, *First Annual Report* (1878), 1–2.

83. USEC, *First Annual Report* (1878), xii and 2.

84. USEC, *First Annual Report* (1878), xiv, 133, 247; and "Prof. C. V. Riley," *St. Louis Dispatch*, September 15, 1877; A. S. Packard, "The Grasshopper Question," n.s., May 23, 1881, box 5, scrapbook vol. 40, p. 87, SIA, CVR.

85. USEC, *First Annual Report* (1878), 6, 8, and 319. Bruner to Riley, July 17, 1877, box 15, scrapbook vol. 43, p. 81, SIA, CVR. Bruner eventually became a permanent assistant and went on to a distinguished career in entomology. Mallis, *American Entomologists*, 191–95. U.S. Entomological Commission, *Third Report of the United States Entomological Commission relating to the Rocky Mountain Locust, the Western Cricket, the Army Worm, Canker Worms, and the Hessian Fly; together with descriptions of Larvae of Injurious forest Insects, Studies on the Embryological Development of the Locust and of other Insects, and on the Systematic Position of the Orthoptera in Relation to Other Orders of Insects, with Maps and Illustrations* (Washington, DC: GPO, 1883), 50, 55–56 (hereafter cited as USEC, *Third Report*); "The Grasshopper," *Dallas* [Texas] *Eagle*, n. d., box 15, scrapbook vol. 43, p. 57, SIA, CVR.

86. In the first report, 2,500 responses to questionnaires were printed. USEC, *First Annual Report* (1878), 25.

87. USEC, *First Annual Report* (1878), 2–5.

88. USEC, *First Annual Report* (1878), 25.

89. USEC, *First Annual Report* (1878), xiii–xiv, 6, 9–11, 17, 24.

90. USEC, *First Annual Report* (1878), xiv, 6–9.

91. USEC, *First Annual Report* (1878), 24.

92. USEC, *First Annual Report* (1878), xiii–ix, 17, 125.

93. US Entomological Commission, *Second Report for the Year 1878 relating to the Rocky Mountain Locust and the best means of preventing its injuries and of guarding against its invasions* (Washington, DC: GPO, 1880), 81 (hereafter cited as USEC, *Second Report*). Riley predicted that a new invasion would not occur for many years, probably not again in the nineteenth century. USEC, *Second Report* (1880), 125.

94. USEC, *First Annual Report* (1878), xiv.

95. USEC, *Second Report* (1880), 56–67.

96. C. V. Riley, "The Philosophy of the Movement of the Rocky Mountain Locust," [August 1878], *Proc.* AAAS 27 (1879): 276–77.

97. USEC, *Second Report* (1880), xv, 6–7, 59, 156.

98. USEC, *Third Report* (1883), xi–xi, 9. See Bruner's report in chaps. 2 and 3.

99. USEC, *Third Report* (1883), 11.

100. USEC, *Second Report* (1880), 15, 276.

101. USEC, *Second Report* (1880), maps between pp. xvii and 1.

102. USEC, *Second Report* (1880), 302.

103. N. Waloff and G. B. Popov, "Sir Boris Uvarov (1889–1970): The Father of Acridology," *Annual Review of Entomology* 35 (1990): 1–5 (hereafter cited as *Ann. Rev. Ent.*).

104. In 1920, Uvarov moved to London, England, were he developed the Anti-Locust Research Centre and created the locust management unit within the United Nations Food and Agriculture Organization that today provides global leadership in locust control. Waloff and Popov, "Sir Boris Uvarov (1889–1970)," 1–5; and Jeffrey A. Lockwood, *Locust: the Devastating Rise and Mysterious Disappearance of the Insect that shaped the American Frontier* (New York: Basic Books, 2004), 149.

105. M. L. Anstey, S. M. Rogers, S. R Ott, M. Burrows, and S. J. Simpson, "Serotonin mediates behavioral gregarization underlying swarm formation in desert locusts," *Science* 323 (January 30, 2009): 627–30; and P. A. Stevenson, "The Key to Pandora's Box," *Science* 323 (January 30, 2009): 594–95.

106. Boris Uvarov, "A Revision of the Genus *Locusta,* L. (= *Pachytylus,* Fieb.), with a New Theory as to the Periodicity and Migrations of Locusts," *Bulletin of Entomological Research* 12 (1921): 135–63; Boris Uvarov, *Grasshoppers and Locusts: a Handbook of General Acridology* (2 vols.; Cambridge: Cambridge University Press, 1966), 1:379–89, 2:321–23, 368–70, 522–23; D. L. Gunn, "The Biological Background of Locust Control," *Ann. Rev. Ent.* 5 (1960): 279–300; C. P. Friedlander, *The Biology of Insects* (New York: Pica Press, 1977), 162; John Stoddard Kennedy, "Continuous Polymorphism in Locusts," in Kennedy, ed., *Insect Polymorphism* (London: Royal Entomological Society, 1961), 80–90; Lockwood, *Locust*, 143–50.

107. USEC, *First Annual Report* (1878), 78, 160.

108. USEC, *Second Report* (1880), 91–93.

109. Based on the density of locust swarms in Africa, Jeffrey Lockwood, a modern locust investigator, estimates the swarm contained 3.5 trillion insects, approximately four times the next largest recorded swarm in Africa in 1954. Lockwood, *Locust*, 18–21.

110. See Craig Miner, *West of Wichita*, chap. 5, "The Ravaging Hopper."

111. USEC, *First Annual Report* (1878), 212–13. Compare with his description of locust destruction quoted earlier.

112. In 1863, Morris appointed Uhler assistant librarian at the Peabody Institute and, in 1864, Agassiz placed him in charge of the insect collection and library at the MCZ. Mallis, *American Entomologists*, 205–8; unless otherwise noted, the history of the nomenclature is based on USEC, *First Annual Report* (1878), 43–52.

113. The initial spelling "*spretis*" rather than the later, "*spretus,*" was changed in later publications. See Lockwood, *Locust*, 28–29.

114. USEC, *First Annual Report* (1878), 143; Riley, "Rocky Mountain Locust," *Seventh Annual Report* (1875), 128 note; and Riley, *Locust Plague*, 13 note.

115. Quoted in Riley, "Rocky Mountain Locust," *Seventh Annual Report* (1875), 128.

116. F. H. Snow, "The Rocky Mountain Locust: *Caloptenus spretus* Uhler," *Transactions of the Kansas Academy of Science* 4 (1875): 26–27.

117. Riley, "Rocky Mountain Locust," *Seventh Annual Report* (1875), 124–28, 167–73, 180; also cited in USEC, *First Annual Report* (1878), 45–52. For some reason the otherwise perfectionist Riley wrote "atlanis" when he obviously meant "atlantis," a reference to the Atlantic seaboard where the third species occurred. See discussion of this in Lockwood, *Locust*, 152.

118. USEC, *First Annual Report* (1878), 37–40.

119. USEC, *First Annual Report* (1878), 37–40; see also Riley, "Rocky Mountain Locust," *Seventh Annual Report* (1875), 188.

120. Sorensen, *Brethren of the Net*, chap. 12.

121. *Common Names of Insects Database*, Entomological Society of America, Annapolis, Maryland, www.entsoc.org/common-names (accessed October 27, 2016); and Lockwood, *Locust*, 28–29.

122. Riley, "Rocky Mountain Locust," *Seventh Annual Report* (1875), 187–88; and USEC, *First Annual Report* (1878), 52. Riley, however, often called nonmigrating species "locusts."

123. Riley, "Rocky Mountain Locust," *Eighth Annual Report* (1876), 114–15.

124. Riley, "Rocky Mountain Locust," *Eighth Annual Report* (1876), 115.

125. Riley, "Rocky Mountain Locust," *Eighth Annual Report* (1876), 115.

126. Riley, "Rocky Mountain Locust," *Eighth Annual Report* (1876), 116. We have retained the Riley-Thomas rendition of scientific names (including capitalization).

127. USEC, *First Annual Report* (1878), 41–42.

128. USEC, *Second Report* (1880), 69.

129. USEC, *Second Report* (1880), 69.

130. USEC, *First Annual Report* (1878), 52.

131. USEC, *First Annual Report* (1878), 245. From the context and similarity to other statements by Riley, we assign this statement to him, although all three commissioners were coauthors.

132. *Conference of Governors* (1876), 23–25, 36–37.

133. *Conference of Governors* (1876), 37.

134. USEC, *First Annual Report* (1878), 28–29, 37. Chaps. 18 and 19 of the report cover "locusts" in the East and in other countries.

135. USEC, *First Annual Report* (1878), 249–50.

136. USEC, *First Annual Report* (1878), 182–86.

137. USEC, *First Annual Report* (1878), 182, 201; USEC, *Second Report* (1880), 135–36. Lockwood identifies an extended drought in the years 1873–75 as a decisive factor in the buildup of swarms in 1874–75. The weather data he cites, however, refers to the central United States (Kansas, Missouri, Ohio). Unusually dry conditions in Kansas and Missouri very likely helped foster the massive hatch in 1875 that resulted in the record swarms of the second generation of locusts moving to the northwest (e.g., Albert's swarm). The commissioners cited rainfall data from the permanent zone, where there seems to have been "normal" rainfall in those years. If these records reflect actual conditions in the permanent zone, the question remains, what led to the population buildup in 1874? See Lockwood, *Locust*, 22–23.

138. USEC, *First Annual Report* (1878), 203; and USEC, *Second Report* (1880), 104.

139. USEC, *First Annual Report* (1878), 107–8.

140. USEC, *Second Report* (1880), 72.

141. USEC, *Second Report* (1880), 73.

142. USEC, *Second Report* (1880), 89, 104.

143. According to Jeffrey Lockwood, entomologist and historian at the University of Wyoming, every specimen of the Rocky Mountain Locust preserved in collections

represents the gregarious phase; not a single specimen of the solitary phase is known to exist. He also notes that the difference between solitary and gregarious Old World locusts (e.g., the Desert Locust) is apparently much greater than in the Rocky Mountain Locust. This perhaps explains why Uvarov was prompted to develop the phase theory, whereas Riley et al. did not observe similar dramatic changes in coloration and outward appearance in the solitary and gregarious phases of the Rocky Mountain Locust. Jeffrey Lockwood, personal communication to Conner Sorensen, December 20, 2012.

144. USEC, *Second Report* (1880), 106.

145. USEC, *Second Report* (1880), 178–83; USEC, *First Annual Report* (1878), 268–70.

146. USEC, *First Annual Report* (1878), 181–82; USEC, *Second Report* (1880), 83–84.

147. USEC, *Second Report* (1880), 96–99.

148. Conner Sorensen, "Uses of Weather Data by American Entomologists 1830–1880," *AgH* 63 (Spring 1989): 171–72; Sorensen, *Brethren of the Net*, 142–47.

149. USEC, *Second Report* (1878), 78–79.

150. Riley, "Rocky Mountain Locust," *Eighth Annual Report* (1876), 101–02, 106; The problem was partly semantic. Bruner stated that the "return migration" was due to the locust's "longing to get back to its native climes and home scenes." USEC, *Third Report* (1883), 33. Even Thomas, who was skeptical about a return instinct, sometimes referred to a "homeland" to the northwest.

151. USEC, *First Annual Report* (1878), 161.

152. Packard, "Migrations of the Destructive Locust," (1877), 22.

153. See summary of Köppens monograph under "Acrydiidae" in *Zoological Record* 4 (1867): 459–61.

154. Ibid., 181–82.

155. USEC, *Second Report* (1880), 91, 107–8.

156. Uvarov later concluded that Packard and Thomas, who he says both favored a theory of exclusively wind directed swarms, were correct and that Riley's theory of return instinct was wrong. A close reading of the commission reports, however, indicates that Uvarov likely overstated Thomas's preference for wind influence. In *Brethren of the Net*, Conner Sorensen followed Uvarov's distinction between Riley on one side and Thomas and Packard on the other; however, now having read the reports more closely and having read Lockwood, *Locust*, he has changed his mind, as indicated here. See Uvarov, *Grasshoppers and Locusts*, 2:136–38.

157. Lockwood, *Locust*, 23.

158. USEC, *First Annual Report* (1878), 285.

159. USEC, *First Annual Report* (1878), 317–23.

160. In modern terminology, tachinid flies would be called parasitoids, but Riley and his contemporaries employed the term parasite. USEC, *First Annual Report* (1878), 317–23.

161. USEC, *First Annual Report* (1878), 289–92.

162. USEC, *First Annual Report* (1878), 248–49, 306.

163. USEC, *First Annual Report* (1878), 314–26.

164. USEC, *Second Report* (1880), 259–61.

165. USEC, *Second Report* (1880), 262–70.

166. Riley, "Rocky Mountain Locust," *Eighth Annual Report* (1876), 120.

167. USEC, *First Annual Report* (1878), 245–47. Riley's experiments and the results he reports are puzzling. He does not explain why the hoppers hatched in captivity failed to develop into winged adults capable of flying, as did millions of locusts that hatched outside captivity.

168. USEC, *First Annual Report* (1878), 245–47.

169. Aughey's report is printed in USEC, *First Annual Report* (1878), 338–50.

170. USEC, *First Annual Report* (1878), 343–46.

171. USEC, *First Annual Report* (1878), 350.

172. USEC, *First Annual Report* (1878), 334–35, 341.

173. USEC, *First Annual Report* (1878), 334–39. Thomas pointed out that in the locust's native breeding grounds, where shooting and poisoning had not yet decimated bird populations, bird predation had not prevented the formation of destructive locust swarms. USEC, *Second Report* (1880), 25.

174. USEC, *First Annual Report* (1878), 334.

175. USEC, *First Annual Report* (1878), 341, 351; and Mark V. Barrow Jr., *A Passion for Birds: American Ornithology after Audubon* (Princeton, NJ: Princeton University Press, 1998), 127–34.

176. USEC, *First Annual Report* (1878), 417.

177. USEC, *First Annual Report* (1878), 271 note, 339. On the English Sparrow, see Michael J. Brodhead, "Elliott Coues and the Sparrow War," *New England Quarterly* 44 (September 1971): 420–32.

178. USEC, *First Annual Report* (1878), 362–92.

179. USEC, *First Annual Report* (1878), 392–93; and ARCA (1883), Plate IX.

180. USEC, *First Annual Report* (1878), 409–14.

181. USEC, *First Annual Report* (1878), 404.

182. USEC, *Second Report* (1880), 271–72.

183. USEC, *Second Report* (1880), maps.

184. USEC, *Second Report* (1880), 272.

185. USEC, *Third Report* (1883), 20.

186. USEC, *Second Report* (1880), 24.

187. USEC, *First Annual Report* (1878), 126–27; and USEC, *Second Report* (1880), 303. Aughey, for example, also believed the settlement would increase rainfall on the plains. See USEC, *Third Report* (1883), 50.

188. USEC, *Second Report* (1880), 27, 303–6. Thomas's reassurance to Governor Pillsbury seems overoptimistic as the commission's maps indicated that only a minimal portion of the breeding grounds were located in Dakota Territory. See USEC, *Second Report* (1880), Map 1, Middle Eastern Section.

189. USEC, *Second Report* (1880), 21.

190. Donald Worster, *A River Running West: The Life of John Wesley Powell* (New York and London: Oxford University Press, 2002), 350–58 (hereafter cited as Worster, *Powell*).

191. USEC, *Second Report* (1880), 22, 311–13. In 1879, Powell hired Gannett as chief cartographer for the Geological Survey. Worster, *Powell*, 418–19.

192. USEC, *Second Report* (1880), 273–74; and USEC, *First Annual Report* (1878), 128, 419–20.

193. USEC, *Third Report* (1883), xi, 49–50; and USEC, *Second Report* (1880), 319.

194. Riley, *Locust Plague* (1877), 165.

195. USEC, *First Annual Report* (1878), 401.

196. C. V. Riley, "Destructive Locusts: A Popular consideration of a few of the more injurious Locusts (or 'Grasshoppers') in the United States, together with the best means of Destroying them," in US Department of Agriculture, Division of Entomology, *Bulletin 25* (Washington, DC: GPO, 1891), 59 [hereafter cited as Riley, "Destructive Locusts," (1891)]; Mallis, *American Entomologists*, 390. See also chap. 9, "Assisting Nature's Balance."

197. USEC, *Third Report* (1883), 65–85.

198. USEC, *Third Report* (1883), 65 note. There is some evidence that sunspot cycles affect weather patterns and that these in turn affect locust population dynamics, so Swinton's thesis was perhaps not totally nonsensical. Jeffrey Lockwood, personal communication to Conner Sorensen, December 20, 2012.

199. Riley, "Destructive Locusts," (1891), 10; Paul W. Riegert, *From Arsenic to DDT: A History of Entomology in Western Canada* (Toronto: University of Toronto Press, 1980), 35, 68.

200. USEC, *Third Report* (1883), map in preface between pp. 1 and 3, 273. In 1880, Bruner reported the Northwest was comparatively free of locusts. USEC, *Third Report* (1883), 17–18.

201. Worster, *Powell*, 400; and Mallis, *American Entomologists*, 52.

202. Riley, "Destructive Locusts," (1891), 10; and Riegert, *From Arsenic to DDT*, 35, 68.

203. Riegert, *From Arsenic to DDT*, 22, 214.

204. See Jeffrey A. Lockwood, "The Fate of the Rocky Mountain locust, *Melanoplus spretus* Walsh: Implications for Conservation Biology," *Terrestrial Arthropod Reviews* 3 (2010): 129–60.

205. Lockwood, *Locust*, 233, 246.

206. Lockwood, *Locust*, 237–39.

207. Lockwood, *Locust*, 178.

208. Lockwood, *Locust*, 237–39. Lockwood's thesis, that agricultural settlement in the permanent zone destroyed the locusts' breeding grounds, is consistent with Uvarov's thesis that the decline of migrating locusts invading Europe was due to

agricultural development in its permanent breeding grounds. Lockwood, *Locust*, 249; and Uvarov, *Grasshoppers and Locusts*, 2:528–29.

209. Lockwood, *Locust*, 242, 248.

210. Lockwood, *Locust*, 245.

211. Lockwood, *Locust*, 243; and Riley, "Rocky Mountain Locust," *Eighth Annual Report* (1876), 121.

212. Lockwood, *Locust*, 178.

213. Arnold Van Huis, Joost Van Itterbeeck, Harmke Klunder, Esther Mertens, Afton Halloran, Giulia Muir, and Paul Vantomme, "Edible insects: future prospects for food and feed security," UN Food and Agriculture Organization, FAO Forestry Paper 171, Rome, 2013, http://www.fao.org/docrep/018/i3253e/i3253e.pdf (accessed February 15, 2018).

214. *IL* 1 (December 1888): 194–95; John Sterling Kingsley, ed., *The Riverside Natural History*, 6 vols. (New York: Houghton Mifflin, 1886), 2:194–95; C. F. G. Cumming, "Locusts and Farmers of America," *Nineteenth Century* 17 (January 1885): 134–52; "An Insect Plague," *All the Year Round* (March 28, 1885): 30–32; Howard, *History of Applied Entomology*, 83–103; Riley, "Destructive Locusts" (1891), 7.

215. Jeffrey A. Lockwood, *Grasshopper Dreaming: Reflections on Killing and Loving* (Boston: Skinner House Books, 2002), 104–15, 96–97, 36–37, 22–23.

216. Dupree, *Science in the Federal Government*, 151–62.

217. Quoted in Dexter, "U. S. Entomological Commission," 31.

Chapter 8

Epigraph. "LeDuc Again," *PF* (May 22, 1880), box 5, scrapbook vol. 39, p. 48, SIA, CVR. For a positive assessment of LeDuc, see Ben F. Rogers, "William Gates LeDuc: Commissioner of Agriculture," *Minnesota History* (Autumn 1955): 287–95.

1. One disgusted observer wrote, "locusts might be utilized for hog feed; but I expect a hog would turn up his nose at 'Missouri legislation!'" C. L. Gould, "Missouri and Her State Entomologist, [1877]," box 2, scrapbook vol. 15, p. 60, SIA, CVR; "Prof. C. V. Riley," *Manitoba Free Press* (April 13, 1878) [from *St. Louis Globe Democrat*], box 2, scrapbook vol. 15, p. 102, SIA, CVR; C. V. Riley, "Missouri Entomological Reports," *CRW* (February 19, 1879), box 3, scrapbook vol. 32, p. 102, SIA, CVR; "Agricultural Entomology," *PF* 48 (March 31, 1877): 100; and "Agriculture and Entomology in Missouri," *PF* 48 (May 5, 1877): 140.

2. Zachary Karabell, *Chester Alan Arthur* (New York: Henry Holt and Co., 2004), 2–27.

3. The department's ambiguous status was reflected in the letterhead of the Division of Entomology that read "Department of Agriculture," not "Commissioner of Agriculture." Until 1889, when the commissioner of agriculture was upgraded to secretary of agriculture, the agency was referred to variously as the Department of Agriculture, the Agricultural Department, and in other ways. We employ the term

Agricultural Department for the years up to 1889 and Department of Agriculture for the following years. Earle D. Ross, "The United States Department of Agriculture During the Commissionership: A Study in Politics, Administration, and Technology, 1862–1889," *AgH* 20 (July 1946): 129–43 (hereafter cited as Ross, "Department of Agriculture During the Commissionership"); and Benjamin F. Rogers, "The United States Department of Agriculture (1862–1889): A Study in Bureaucracy" (PhD diss., University of Minnesota, 1950), 45 (hereafter cited as Rogers, "USDA, 1862–89").

4. *St. Louis Republican* [1877] box 1, scrapbook vol. 9, p. 71, SIA, CVR; Riley's choices in 1877 were Willard C. Flagg (who died in 1878), J. P. Reynolds, and H. D. Emery, his former employer at the *Prairie Farmer.* When these candidates fell out, Riley supported LeConte and even delivered a petition to President Hayes on LeConte's behalf. Stephens, "Commissioner of Agriculture," 377.

5. William G. LeDuc, *Recollections of a Civil War Quartermaster* (St. Paul, MN: North Central Publishing Co., 1963), 151, 158–61.

6. Rogers, "USDA, 1862–89," 77.

7. *Annual Report of the Commissioner of Agriculture* (Washington, DC: GPO, 1877), 16 (hereafter cited as *ARCA*).

8. *ARCA* (1877), 16.

9. Cotton figured significantly in Missouri's Mississippi River bottomlands, where it comprised twenty percent of the cultivated area. C. V. Riley, "The Army Worm," *Second Annual Report* (1870), 39–46; Eugene A. Smith, "The Cotton Belt," US Entomological Commission, *Fourth Report . . . on the Cotton Worm. [and the] Boll Worm* (Washington, DC: GPO, 1885), 59 [hereafter cited as USEC, *Fourth Report* (1885)].

10. Riley, Preface, *Seventh Annual Report* (1875), v.

11. Riley, "The Army Worm," *Second Annual Report* (1870), 39.

12. Riley, "The Cotton Worm," *Sixth Annual Report* (1874), 17.

13. USEC, *Fourth Report* (1885), 339.

14. H.R. Rep. No. 46-1318, at 1 (1879–80) (hereafter cited as House Report 1318).

15. USEC, *Fourth Report* (1885), 3.

16. Samuel Lockwood to LeDuc, February 25, 1878, box 2, scrapbook vol. 15, between pp. 49–50, SIA, CVR.

17. Riley was also reported to be the choice candidate for the proposed state entomologist of Iowa, but most observers agreed that he would accept the position at the Agricultural Department. "Prof. C. V. Riley," *Scientific American* 38 (March 16, 1878): 161 (hereafter cited as *SciAm*); "Prof. C. V. Riley," *Manitoba Free Press* (Canada), April 13, 1878, copy from the *St. Louis Globe Democrat*, box 2, scrapbook vol. 15, p. 102, SIA, CVR.

18. USEC, *Fourth Report* (1885), xx.

19. Anna B. Comstock, *The Comstocks of Cornell: John Henry Comstock and Anna Botsford Comstock: An Autobiography by Anna Botsford Comstock*, ed. Glenn W. Herrick and Ruby Green Smith (Ithaca, NY: Comstock Publishing Associates, 1953), 92 (hereafter cited as *Comstocks of Cornell*).

20. Mallis, *American Entomologists*, 209–10.

21. Biographical sketch of Schwarz in Matthew C. Perry, ed., *The Washington Biologists' Field Club: Its Members and Its History (1900–2006)* (Washington, DC: Washington Biologists' Field Club, 2007), 238–39, https://www.pwrc.usgs.gov/resshow/perry/bios/WBFC_booksm.pdf (accessed March 17, 2018). Howard later wrote that Hagen brought Schwarz with him from Germany, but this is incorrect. Howard, *History of Applied Entomology*, 6.

22. On Bela Hubbard, see Sally Kohlstedt, *The Formation of the American Scientific Community: The American Association for the Advancement of Science, 1848–1860* (Urbana, Chicago, and London: University of Illinois Press, 1976), 67 and AAAS membership list in appendix.

23. Mallis, *American Entomologists*, 252–56.

24. Mallis, *American Entomologists*, 94–97.

25. ARCA (1878), 210. Riley referred to "Prof. A. R. Grote," though it is unclear how or where the title "Prof." came from. Riley's use of the title "special agent" reflects his creative approach to administration. He referred to Grote and Comstock (the next appointee) as special agents, whereas Pergande and Schwarz were considered permanent employees.

26. C. V. Riley, "History of the Literature," in USEC, *Fourth Report* (1885), 325.

27. Riley, "History of the Literature," in USEC, *Fourth Report* (1885), 327. Subsequent revision has resulted in the current scientific name *Alabama argillacea* (Hübner). David L. Wagner, "Ode to *Alabama*: the Meteoric Fall of a Once Extraordinarily Abundant Moth," *AE* 55 (Fall 2009): 170.

28. Riley identified four separate revivals of the migration theory dating back to 1846. He and Afflick rejected the migration theory in favor of winter hibernation of the adult moth. Riley, "History of the Literature," in USEC, *Fourth Report* (1885), 323. In a parallel dispute with Scudder and others over the migration or hibernation of the Monarch Butterfly, Riley championed the theory (later discredited) that Monarchs hibernated in the midwestern or southern states. Brower, "Understanding . . . the Monarch Butterfly," 307–16.

29. USEC, *Fourth Report* (1885), 20–21, 325–28, and note 20 in appendix, pp. [104–5].

30. C. V. Riley, "Hibernation of Cotton Worm Moth," *SciAm* 40 (June 14, 1879): 375.

31. See Grote's report for 1878 in the history of Cotton Worm investigations in USEC, *Fourth Report* (1885), xxvii.

32. USEC, *Fourth Report* (1885), xxx–xxxi.

33. USEC, *Fourth Report* (1885), xx.

34. Mallis, *American Entomologists*, 307.

35. ARCA (1878), 210.

36. *Comstocks of Cornell*, 96; Howard, *History of Applied Entomology*, 57.

37. Mallis, *American Entomologists*, 254; Howard, *History of Applied Entomology*, 84–85; and *Comstocks of Cornell*, 98.

38. F. I. Herriott, "William Stebbins Barnard," *Annals of Iowa* (July 1936): 25.

39. Herriott, "William Stebbins Barnard," 40 (Barnard bibliography).

40. USEC, *Fourth Report* (1885), xxxvi–xxxvii. According to Barnard's biographer, Riley recruited Barnard as assistant in 1878 on the Cotton Worm investigations, but this date is incorrect. Herriott, "William Stebbins Barnard," 13.

41. Herriott, "William Stebbins Barnard," 14–15, 40–41.

42. Herriott, "William Stebbins Barnard," 29, 31–32, 35–36.

43. Riley's introduction to Barnard's *Report on Machinery* [1881?], quoted in Herriott, "William Stebbins Barnard," 19.

44. Mallis, *American Entomologists*, 408–09. Mallis is incorrect when he states that Sullivan was hired about 1889. Riley credits her with illustrations in the Cotton Worm report. USEC, *Fourth Report* (1885), xx.

45. Congressmen perhaps reacted to Riley's dual positions as federal entomologist and chief of the Entomological Commission. The remaining $15,000 was approved later.

46. "Prof. C. V. Riley," *CRW* (April 3, 1878), box 2, scrapbook vol. 15, p. 89 SIA, CVR.

47. *PF* n.d., box 2, scrapbook vol. 15, p. 96, SIA, CVR; wedding invitation, Thursday, June 20, 1878, NAL, MC 143 Series III, Box 3.

48. *PF* quoted above.

49. USEC, *Fourth Report* (1885), xxiv.

50. Riley's agents included Dr. E. H. Anderson, Kirkwood, Mississippi; William J. Jones, Virginia Point, Texas; J. E. Willett, Macon, Georgia; Eugene A. Smith, Tuscaloosa, Alabama; E. H. Anderson, Canton, Mississippi; R. W. Jones, Oxford, Mississippi; J. P. Stelle, Mobile, Alabama; L. C. Johnson, Holly Springs, Mississippi; J. F. Bailey, Marion, Alabama; J. C. Neal, Archer, Florida; and James Roane, Washington, DC. USEC, *Fourth Report* (1885), xx, xxiii.

51. *Comstocks of Cornell*, 96–97.

52. Harriet Beecher Stowe, "Our Florida Plantation," *Atlantic Monthly* 43 (May 1879): 648–49.

53. *Comstocks of Cornell*, 98; USEC, *Fourth Report* (1885), xxvi–xxvii.

54. These are now called extrafloral nectaries and are common on certain cultivars of cotton.

55. C. V. Riley, "The Cotton Worm," *Atlanta Constitution,* September 8, 1878, box 3, scrapbook vol. 24, p. 28, SIA, CVR; *ARCA* (1878), 214.

56. *ARCA* (1878), 214.

57. *ARCA* (1878), 18–22, 39.

58. Howard, *History of Applied Entomology*, 85–86; *Comstocks of Cornell*, 105; St. Louis *Globe Democrat,* April 13, 1879.

59. *Rural New Yorker,* April 17, 1880.

60. *Rural New Yorker,* April 17, 1880.

61. Howard, *History of Applied Entomology*, 86–87.

62. It is unclear what material from the collection Riley took home and what remained at the Division of Entomology. See discussion of the collection in chap. 13. Howard, *History of Applied Entomology*, 86–87;. and *Comstocks of Cornell*, 116.

63. Howard, *History of Applied Entomology*, 87.

64. Howard, *History of Applied Entomology*, 116–18; and *ARCA* (1880), 235.

65. USEC, *Fourth Report* (1885), xxviii. In a somewhat contorted account of the dispute, Riley stated, "Owing to difficulties which grew out of this action [the transfer of Cotton Worm investigations to the Entomological Commission] we resigned . . . [effective May 1, 1879.]" Here, Riley reverses the sequence of events, the dispute having occurred before, and as a direct cause of, his resignation.

66. "Condensed Information" from Digest of Appropriations 1879, undated, box 4, scrapbook vol. 30, pp. 28–29, SIA, CVR (hereafter cited as "Condensed Information," 1879). Although the document was unsigned, the contents indicate that LeDuc and Comstock were its authors.

67. "Condensed Information," 1879; and *Comstocks of Cornell*, 107.

68. USEC, *Fourth Report* (1885), xxviii.

69. USEC, *Fourth Report* (1885), xviii, 85–86; and C. V. Riley, "The Cotton Worm," US Entomological Commission, Department of the Interior, Bulletin No. 3 (Washington, DC: GPO, 1880), 1 [hereafter cited as USEC, "Cotton Worm" (1880)].

70. USEC, *Fourth Report* (1885), xxx, 21.

71. *Comstocks of Cornell*, 107; *ARCA* (1879), 13; USEC, "Cotton Worm" (1880).

72. C. V. Riley, *SciAm* 40 (May 17, 1879): 313; and Riley, "Hibernation of the Cotton Worm," *SciAm* 40 (June 14, 1879): 375.

73. C. V. Riley, "Hibernation of the Cotton Worm (1879)," 375; and USEC, *Fourth Report* (1885), 20–21. In the twentieth century, entomologists established that the native home of the Cotton Worm is restricted to a band between latitude 20° north and latitude 20° south of the equator, placing the northernmost permanent habitat of the Cotton Worm at the latitude of Vera Cruz, Mexico. The Cotton Worm does not overwinter north of this latitude. Santin Gravena, Winfield Sterling, and Dean Allen, *Abstracts, References, and Key Words of Publications Relating to the Cotton Leafworm, Alabama argillacea* (*Huebner*) (*Lepidoptera: Noctuidae*). Thomas Say Foundation Monographs, no. 10 (College Park, MD: Entomological Society of America, 1985), 1, 78; Ross E. Hutchins, *Insects* (Engelwood Cliffs, NJ: Prentice-Hall, 1972), 86; US Department of Agriculture, *Insects: The Yearbook of Agriculture 1952* (Washington, DC: GPO, 1952), Plate V.

74. *ARCA* (1879), 280–81.

75. USEC, "Cotton Worm" (1880), preface, p. 5; and USEC, *Fourth Report* (1885), xix.

76. The Riley and Comstock versions of publication of Comstock's report differ significantly. Riley claimed Comstock's preliminary report was distributed only to

selected congressmen; Anna Comstock later claimed all members received copies; Riley states that the final version came out in August; Anna Comstock says it was produced a few days after the preliminary version, i.e., in May 1880. Comstock, in his report for 1879, wrote that the report was submitted June 30, 1880 (allowing for delays in printing of the 1879 Department of Agriculture report) and in his report for 1880, he gives a publication date of May 18, 1880. *ARCA* (1879), 186; *ARCA* (1880), 275; *Comstocks of Cornell*, 124; USEC, *Fourth Report* (1885), 328.

77. USEC, "Cotton Worm," (1880), 3.

78. *ARCA* (1879), 269.

79. USEC, *Fourth Report* (1885), 328.

80. USEC, "Cotton Worm," (1880), 38.

81. *ARCA* (1879), 298.

82. *ARCA* (1879), 298.

83. C. V. Riley, "Natural Enemies"; USEC, *Fourth Report* (1885), 101–2.

84. C. V. Riley, "Parasites of the Cotton Worm, *CE* 11 (September 1879): 161–62; and C. V. Riley, "Parasites bred from the Cotton Worm," *CE* 11 (November 1879): 205.

85. Riley, "Parasites of the Cotton Worm (1879), 116–17.

86. Riley, "Parasites of the Cotton Worm (1879), 118. Howard, *History of Applied Entomology*, 87–88, note 1.

87. *Comstocks of Cornell*, 105. His salary was raised to $2,000 in 1880. Howard, *History of Applied Entomology*, 180.

88. Howard, *History of Applied Entomology*, 115.

89. "Proceedings of the American Association," *SciAm* 41 (September 20, 1879): 180; *Cincinnati Daily Commercial*, August 19, 1881, box 3, scrapbook vol. 25, p. 120, SIA, CVR.

90. "The Entomological Club at the AAAS," *CE* 11 (September 1879): 163–76; "The American Science Association," *SciAm* 41 (September 13, 1879), box 4, scrapbook vol. 29, p. 88, SIA, CVR; "The Entomological Club," *SciAm* 41 (September 13, 1879): 168.

91. "The American Science Association," *SciAm* 43 (September 25, 1880), 196, and (October 16, 1880), 241; and "The Entomological Club," *CE* 12 (September 1880): 160–64.

92. "Of the Cotton Worm," *Colorado Citizen* (Columbia, CO) July 17, 1879, box 3, scrapbook vol. 23, pp. 91–92, SIA, CVR; "The Cotton Worm," *Galveston News*, October 30, 1879 in box 3, scrapbook vol. 23, pp. 166–67, SIA, CVR; Wm. J. Jones, "The Cotton Caterpillar," *Galveston News*, August 19, 1879, in box 3, scrapbook vol. 24, pp. 25–26, SIA, CVR.

93. *PF* (December 20, 1879), box 4, scrapbook vol. 29, p. 22, SIA, CVR; *American Agriculturalist*, January 18, 1880, box 4, scrapbook vol. 29, p. 22, SIA, CVR; "Agricultural Interests," n.s., box 4, scrapbook vol. 29, p. 20, SIA, CVR. Riley's paper was later printed in the first issue of the American Agricultural Association's Journal. See also "American Agricultural Association," *Ohio Farmer* 59 (April 30, 1881): 292.

94. "The Cotton Caterpillar," *Southern Cultivator* 38 (June 1880): 217; "The Entomological Commission and the Department of Agriculture," *American Agriculturist* (n. d.) box 4, scrapbook vol. 30, p. 12, SIA, CVR. See also C. V. Riley, "The Cotton Worm," *Galveston News*, October 24 and October 30, 1879, box 3, scrapbook vol. 23, pp. 164–67, SIA, CVR; "The Cotton Destroyers," *New Orleans Democrat*, September 21, 1880, box 3, scrapbook vol. 23, pp. 71–73, SIA, CVR; "The Cotton Worm Investigation," *Daily Constitutional* (Atlanta, GA), July 20, 1880, 3:24–12; *Florida Dispatch,* October 6, 1880, box 4, scrapbook vol. 30, p. 27, SIA, CVR; "Work of the US Entomological Commission in 1880," *Kansas Farmer,* October 6, 1880, box 4, scrapbook vol. 30, pp. 30–31, SIA, CVR; "Riley's Researches," *St. Louis Globe-Democrat,* October 5, 1880, box 4, scrapbook vol. 30, p. 40, SIA, CVR; "Of the Cotton Worm," (Columbia) *Colorado Citizen,* July 17, 1879, box 3, scrapbook vol. 23, pp. 16–17, SIA, CVR.

95. *Bulletin of the Philosophical Society of Washington* 2 (1878): 201.

96. Philip J. Pauly, *Biologists and the Promise of American Life: From Meriwether Lewis to Kinsey* (Princeton, NJ: Princeton University Press, 2000), 438 (hereafter cited as Pauly, *Biologists*); *Bulletin of the Philosophical Society of Washington* 3 (April 7, June 6, June 21, 1879; June 11, 1881), 5 (December 16, 1882), 6 (October 13, 1883), 7 (February 2, March 1, 1884), 8 (November 21, 1885), 9 (November 6, 1886), 10 (February 26, 1887, March 31, 1888), and 11 (January 31, 1891).

97. George Crosette, *Founders of the Cosmos Club of Washington 1878: A Collection of Biographical Sketches and Likenesses of the Sixty Founders* (Washington, DC: Cosmos Club, 1966), 13.

98. Pauly, *Biologists*, 51–54; and Worster, *Powell*, 437–39.

99. Pauly, *Biologists*, 51, 254n19.

100. Paul Lawrence Farber, *Finding Order in Nature: The Naturalist Tradition from Linnaeus to E. O. Wilson* (Baltimore and London: Johns Hopkins University Press, 2000), 56–71.

101. Pauly, *Biologists*, 60 and note 38. See Riley's tribute to Darwin in chap. 5.

102. Ashley B. Gurney, "A Short History of the Entomological Society of Washington," *Proc. ESW* 78 (July 1976): 226–29.

103. C. V. Riley "The Cotton Worm," *SciAm* 40 (May 17, 1879), 313; and C. V. Riley "Hibernation of the Cotton Worm," *SciAm* 40 (June 14, 1879), 375.

104. C. V. Riley "The Cotton Worm," *SciAm* 40 (May 17, 1879), 313; and Riley "Hibernation of the Cotton Worm," *SciAm* 40 (June 14, 1879) "The Sons of Science," *Mirror of the Far West*, June 29, 1878, quoting the *New York Times*, n.d., box 2, scrapbook vol. 15, pp. 90–91, SIA, CVR. The writer was unaware that Riley was candidate for membership, not a member of the National Academy of Sciences.

105. Clipping, n.s., n.d., in box 10, scrapbook vol. 72, p. 128; and box 10, scrapbook vol. 73, p. 132, both in SIA, CVR. On Baird's sponsorship of Riley, see Riley to Hilgard, April 2, 1883, Hilgard Family Papers, Bancroft Library, University of California, Berkeley. The tendency of academicians to rate science in categories with "pure" at the

top and "applied" at the bottom persists. J. Kirkpatrick Flack, in his otherwise excellent account of post–Civil War intellectual life in Washington, comments that Riley, one of the founders of the Cosmos Club, was not a scientist at all but an "agriculturist" [!] Flack, *Desideratum in Washington: The Intellectual Community in the Capital City 1870–1900* (Cambridge, MA: Schenkman Publishing Co., Inc., 1975), 81.

106. Howard, *History of Applied Entomology*, 86.

107. In 1886, Howard became a member of the Cosmos Club and was very prominent there. *The Twenty-fifth Anniversary of the Founding of the Cosmos Club* (Washington, DC: Cosmos Club, 1904), 314.

108. House Report 1318, p. 1.

109. House Report 1318, pp. 1, 2–4.

110. House Report 1318, pp. 1, 2–4.

111. Riley to Packard, May 3, 1882, MCZ.

112. Riley to Bessey, March 31, 1880, University of Nebraska Archives (hereafter cited as UNA), RG 12. See Richard Overfield, *Science with Practice: Charles E. Bessey and the Maturing of American Botany* (Ames: Iowa State University Press, 1993), 20–21.

113. Karabell, *Arthur*, 51–52.

114. Quoted in Karabell, *Arthur*, 52.

115. *Telegraph* (Mentor, OH), February 19, [1881], box 5, scrapbook vol. 39, p. 98, SIA, CVR.

116. In February 1881, Iowa senator William Allison, chairman of the Appropriations Committee, assured Riley that Garfield would replace LeDuc. Riley to Packard, February 6, 1881, MCZ.

117. Riley to Packard, May 3, May 12, and June 15, 1881, MCZ.

118. Ross, "Department of Agriculture During the Commissionership," 135.

119. "The Washington Agricultural Convention," *PF* 43 (March 2, 1872): 65; and "Late Agricultural convention at Washington," n.s., n.d., box 1, scrapbook vol. 5, p. 39, SIA, CVR.

120. Riley to Packard, February 6, 1881, MCZ.

121. Riley to Packard, March 19, 1881, MCZ.

122. Riley to Packard, June 2, 1881, MCZ.

123. *Washington Post*, February 27, 1881, box 3, scrapbook vol. 24, p. 45, SIA, CVR; "Occasional," *SciAm* 36 (April 28, 1877): 260.

124. "The Entomological Commission," *Daily Inter Ocean*, May 11, 1881.

125. Riley to Packard, June 15, 1881, MCZ.

126. Riley to Packard, June 2, 1881, MCZ.

127. "The Department Entomologist," *CRW* 34 (June 23 and July 9, 1881); and *Cleveland Herald*, July 9, 1881.

128. *Comstocks of Cornell*, 136.

129. Karabell, *Arthur*, 58–59.

130. Karabell, *Arthur*, 63–64.

131. Karabell, *Arthur*, 1.

132. *Comstocks of Cornell*, 134–35.

133. "Cotton Convention at Atlanta, Georgia," *American Wine and Grape Growers [Journal?]* (January 1882), box 9, scrapbook vol. 67, pp. 69–71, SIA, CVR; "Cotton and its Future—An Opportunity for Invention," *SciAm Supplement* (February 11, 1882), box 5, scrapbook vol. 38, p. 1, SIA, CVR.

Chapter 9

1. Schlebecker, *Whereby We Thrive*, 160.

2. Schlebecker, *Whereby We Thrive*, 166.

3. Schlebecker, *Whereby We Thrive*, 157–59.

4. During the two years of Comstock's administration, the budget was reduced to $5,000. "Statement of Appropriations and Expenditures for the U. S. Department of Agriculture from 1878 to 1892," *Annual Report of the Secretary of Agriculture* (1894), 48–54 (hereafter cited as *ARSA*). Howard's figures differ slightly from the figures cited in this 1894 compilation. He reports, for example, that Comstock's 1881 budget increased to $7,000. Howard, *History of Applied Entomology*, 167.

5. *ARCA* (1884), 5.

6. Riley to Lintner, April 17, 1885, Lintner Corr., NYSM. See also Ross, "Department of Agriculture During the Commissionership," 136; and Floyd G. Summers, "Norman J. Colman, First Secretary of Agriculture," *Missouri Historical Review* 19 (April 1925): 405.

7. *AE* 3 (September 1880): 224.

8. C. V. Riley, "General Truths in Applied Entomology," address delivered before the Georgia State Agricultural Society, Savannah, Georgia, February 12, 1884, reprinted in *ARCA* (1884), 324 [hereafter cited as Riley, "General Truths," (1884)].

9. C. V. Riley, "Insects in Relation to Agriculture," *Stoddart's Encyclopaedia Americana* 1 (1883):135–42; and C. V. Riley, "Insects in Relation to Agriculture," *Encyclopaedia Britannica, American Supplement* 1 (9th ed., Philadelphia, PA: Hubbard Bros., 1889), 135–41.

10. Riley, "General Truths," (1884), 324.

11. David Rich Lewis, "American Indian Environmental Relations," in Douglas Cazaux Sackman, ed., *A Companion to American Environmental History* (Chichester, UK: Wiley-Blackwell, 2010), 193–95, 198–202.

12. In his *Fifth Report*, for example, Riley recalled travelling through the Black Hills in 1867, a time when the Plains Indian culture was still dominant, but he confined his observations to yuccas and the insects associated with them. Riley, "New Genus," *Fifth Annual Report* (1873), 157. So far as is known, the only time Riley referred (even indirectly) to Indians in his publications was when he described the larvae of a *Scenopinus* species infesting the blanket of a Navajo Indian. C. V. Riley, "[*Scenopinus*]," *Proc. ESW* 1 (March 30, 1886): 16.

13. Leach, *Butterfly People*, xvii–xviii.

14. Steven Stoll, *Larding the Lean Earth: Soil and Society in Nineteenth Century America* (New York: Hill and Wang, 2002), 19–23, 32–35, 53–63.

15. Philip J. Pauly, *Fruits and Plains: The Horticultural Transformation of America* (Cambridge, MA: Harvard University Press, 2007), 7, 263–66.

16. Quoted in Pauly, *Fruits and Plains*, 133.

17. Marshall P. Wilder, "The Importance of Entomology to the Fruit Grower," *AE* 3 (January 1880): 19.

18. Riley, "General Truths," (1884), 324. Riley estimated losses due to injurious insects at $300 to $400 million annually.

19. C. V. Riley, "New Insects Injurious to Agriculture," *Proc. AAAS* (1881): 272–73.

20. C. V. Riley, "A Foe to Cottonwoods: The Streaked Cottonwood Beetle," *AE* 3 (July 1880): 159–60. The Apple Maggot is discussed in chap. 5.

21. C. V. Riley, "Change of Habit: Two New Enemies of the Egg-plant," *AN* 16 (August 1882): 678.

22. C. V. Riley, "The Grape Phylloxera in California," *AE* 3 (January 1880): 3.

23. E. W. Hilgard, "Further on the Grape Phylloxera in California," *AE* 3 (April 1880): 94–96.

24. *AE* 3 (September 1880): 224. About this time, Isidor Bush, Riley's vintner friend in Bushberg, Missouri, reported that the phylloxera there were now feeding on previously immune varieties and species of grapes. *AE* 3 (September 1880): 226.

25. Riley, "General Truths," (1884), 324.

26. James Whorton, *Before Silent Spring: Pesticides and Public Health in Pre-DDT America* (Princeton, NJ: Princeton University Press, 1974), 22; USEC, *Fourth Report* (1885), 149.

27. USEC, *Fourth Report* (1885), 149.

28. USEC, *Fourth Report* (1885), 150.

29. Riley did not endorse specific commercial products, instead recommending that users mix the basic ingredients themselves. Commercial producers of Paris Green and London Purple nevertheless regularly claimed endorsements by Riley and Bessey.

30. Riley, "Insecticides for the Protection of Cotton," *SciAm* 43 (October 16, 1880): 241.

31. Whorton, *Before Silent Spring*, 72, 223–25, 250. See also Robert J. Spear, *The Great Gypsy Moth War: A History of the First Campaign in Massachusetts to Eradicate the Gypsy Moth, 1890–1901* (Amherst and Boston: University of Massachusetts Press, 2005), 126, 148; and Thomas R. Dunlap, *DDT: Scientists, Citizens, and Public Policy* (Princeton, NJ: Princeton University Press, 1982).

32. Mallis, *American Entomologists*, 139.

33. C. V. Riley, "The Colorado Potato Beetle," *Seventh Annual Report* (1875), 8–10.

34. C. V. Riley, "Insecticides for the Protection of Cotton," (1880), 241.

35. C. V. Riley, "The Use of Poisons to Destroy Insects," *Farmers Review*

(September 13, 1880), box 3, scrapbook vol. 23, p. 148, SIA, CVR.

36. Whorton, *Before Silent Spring*, 33–34.

37. Whorton, *Before Silent Spring*, 32.

38. W. J. Daughtery, of Selma, Alabama, patented such a device in 1878, but it proved to be too elaborate, heavy, and expensive. USEC, Introduction, *Fourth Report* (1885), xxxv–xxxvii; and Barnard, "Machinery," USEC, *Fourth Report* (1885), 258. Barnard's biographer Herriott dates Barnard's hiring in 1881, but that is in error. F. I. Herriott, "William Stebbins Barnard: Professor of Biology, Drake University 1886–1887," reprint from *Annals of Iowa* (July 1936): 17.

39. E. G. Lodeman, *The Spraying of Plants; a Succinct Account of the History, Principles and Practice of the Application of Liquids and Powders to Plants, for the Purpose of Destroying Insects and Fungi*, preface by B. T. Galloway (New York and London: Macmillan and Co., 1896), 222 (hereafter cited as Lodeman, *Spraying of Plants*).

40. The story of Barnard and the cyclone nozzle appears in various places: USEC, *Fourth Report* (1885), xx, xxxv–xxxvii, 181, 191, 259, 361–62, and notes [114–19]; C. V. Riley, "The Cyclone Nozzle," *Rural New Yorker* (August 22, 1885): 567; and Herriott, "Barnard," 19.

41. Riley's address is reprinted in *ARCA* (1881–82), 153–57.

42. *ARCA* (1881–82), 156.

43. "Professor Riley's Lectures," *Atlanta Constitution*, November 5, 1881, box 3, scrapbook vol. 24, p. 34, SIA, CVR.

44. *ARCA* (1881–82), 157.

45. USEC, *Fourth Report* (1885), notes on pp. [117], [118], and [119].

46. USDA, Bureau of Entomology, *Catalogue of the Exhibit on Economic Entomology at the World's Industrial and Cotton Centennial Exposition, New Orleans 1884–85* (Washington, DC: Judd and Detweiler Printers, 1884), 452–70, "Pumps," 475–80.

47. *ARCA* (1887), 17.

48. Lodeman, *Spraying of Plants*, 51, 113, 185–88, 203–4.

49. USEC, *Fourth Report* (1885), 155; Riley, "General Truths," (1884), 327; Herriott, "Barnard," 18–19.

50. C. V. Riley, "Emulsions of Petroleum and their Value as Insecticides," *SciAm* 49 (November 10, 1883): 294; C. V. Riley, "Report of the Entomologist," *ARCA* (1881), 106: and Loring, "Report of the Commissioner," *ARCA* (1881), 11–12; C. V. Riley, "Report of the Entomologist," *ARCA* (1883), 100; and "Report of the Entomologist," *ARCA* (1884), 285; C. V. Riley, "The Kerosene Emulsion: Its Origins, Nature, and Usefulness," *Proc. Soc. Prom. Ag. Sci.* (1892): 83–98; Lodeman, *Spraying of Plants*, 83–84, 153–54. Howard later credited Cook in Michigan with the discovery of kerosene emulsion, but it is not clear why. Howard, *History of Applied Entomology*, 65, 91.

51. USEC, *Fourth Report* (1885), 158; and C. V. Riley, "Insecticides for the Protection of Cotton," (1880), 241.

52. C. V. Riley, "The Cyclone Nozzle," *Rural New Yorker* (August 22, 1885): 567.

53. Decisions of the commissioner of patents and of US courts in the patent cases: *Riley v. Barnard*, Decided June 2, 1892, SIA, Henry Guernsey Hubbard Papers 1871–1899, Record Unit (hereafter cited as RU) 7107, Box 2 [hereafter cited as *Riley v. Barnard* (1892)].

54. *Riley v. Barnard* (1892), pp. 1921 and 1922; and Herriott, "Barnard," 323.

55. *Riley v. Barnard* (1892), p. 1919.

56. *Riley v. Barnard* (1892), p. 1919. It is not clear why the Patent Office took so long to process Barnard's applications. Nan Meyers, librarian at Wichita State University who specializes in patent issues, states that such applications normally require less than a year to process. Personal communication to Conner Sorensen, Wichita, Kansas, August 25, 2011.

57. *Riley v. Barnard* (1892), p. 1919.

58. "Violation of Agreement by Barnard," in *Riley v. Barnard* (1892), p. 1920.

59. As late as October 1883, Riley was praising the work of Barnard. See Riley's letter of transmittal, October 10, 1883, with regard to W. S. Barnard in "Experimental Tests of Machinery Designed for the Destruction of the Cotton Worm," in "Reports and Observations and Experiments in the Practical Work of the Division," USDA, Division of Entomology, Bulletin No. 3 (Washington, DC: GPO, 1883).

60. Herriott, "Barnard," 323.

61. Again, it is unclear why the Patent Office took so long to rule on Riley's protest and in favor of Barnard's patent applications.

62. *Riley v. Barnard* (1892), pp. 1920–21.

63. *Riley v. Barnard* (1892), p. 1920.

64. Catherine L. Fisk, "Removing the 'Fuel of Interest' from the 'Fire of Genius': Law and the Employee-Inventor, 1830–1930," *University of Chicago Law Review* 65 (Fall 1998): 1129–32, 1151, 1164–65.

65. Fisk, "Removing the 'Fuel of Interest,'" 1168–69.

66. "That which [the employee] has been employed and paid to accomplish becomes . . . the property of his employer," cited in Fisk, "Removing the 'Fuel of Interest,'" 1174–75, and also in *Riley v. Barnard* (1892), p. 1920. For background on the American patent system and a comparison to the patent system in Great Britain, see B. Zorina Khan, *The Democratization of Invention: Patents and Copyrights in American Economic Development, 1790–1920* (Cambridge: Cambridge University Press, 2005). The *Solomons* decision and related cases are treated on pp. 103–4.

67. The commissioner of patents quoting *Solomons v. United States*, in *Riley v. Barnard* (1892), p. 1920. The commissioner pointed out that Barnard's signed statement declining any right to patent was dated May 22, 1884, and therefore not applicable to his applications for patent in 1882.

68. *Riley v. Barnard* (1892), p. 1921.

69. *Riley v. Barnard* (1892), p. 1922.

70. "Emulsion and Method of Making Same," Patent No. 580,150 (April 6, 1897)

and "Fluid Distributor," Patent No. 580,151 (April 6, 1897). It is not clear why it took so long to issue the patents.

71. Barnard's son asserted that Barnard's application was delayed primarily because of counter claims by Riley, who insisted that he (Riley) or the Division of Entomology be granted 50 percent of any profit from the patents. This statement runs counter to everything known about Riley, in particular Riley's opinion that government agents should not be granted patents for discoveries made while in government service. Herriott, "Barnard," 338.

72. Lodeman, *Spraying of Plants*, 85.

73. C. V. Riley, "Control Measures," USEC, *Fourth Report* (1885), 138–39.

74. Quotation from a letter addressed to him from Dr. Gustav Radde, director of the Imperial Museum of Natural History, Tiflis [Tblisi], Transcaucasia, n.d., in Riley, "Control Measures," USEC, *Fourth Report* (1885), 164.

75. Whorton, *Before Silent Spring*, 15–16.

76. Whorton, *Before Silent Spring*, 168.

77. Riley, "Insecticides for the Protection of Cotton," (1880), 241; Riley, "Control Measures," USEC, *Fourth Report* (1885), 166–67, 177.

78. Riley, "Insecticides for the Protection of Cotton," (1880), 241.

79. Riley, "Insecticides for the Protection of Cotton," (1880), 241; and Riley, "General Truths," (1884), 327.

80. Riley, "General Truths," (1884), 327.

81. Riley, "General Truths," (1884), 327.

82. Riley, "General Truths," (1884), 327.

83. USDA, Bureau of Entomology, *Catalogue of the Exhibit* (1884–1885).

84. Whorton, *Before Silent Spring*, 16; Lodeman, *Spraying of Plants*, 79.

85. C. V. Riley, "Ox-Eye Daisy as an Insecticide," *AE* 3 (August 1880): 196.

86. Among other insecticides subject to experimentation in 1880–81 were oil of creosote, oil of tar, gas-tar water, carbolic acid, and cottonseed oil. Riley, "Control Measures," USEC, *Fourth Report* (1885), 162–63, 183; Riley, "Insecticides for the Protection of Cotton," (1880), 241.

87. Margaret Rossiter, "The Organization of the Agricultural Sciences," in *The Organization of Knowledge in Modern America, 1860–1920*, ed. Alexandra Oleson and John Voss (Baltimore and London: Johns Hopkins University Press, 1976), 231.

88. Lodeman, *Spraying of Plants*, 100–102.

89. Rossiter, "Organization of the Agricultural Sciences," 231–32; and Pauly, *Biologists*, 81.

90. Lodeman, *Spraying of Plants*, 96–103.

91. Lodeman, *Spraying of Plants*, 108–9, 113, 122, 136.

92. Pauly, *Biologists*, 10–11, 56–61, 70.

93. Rossiter, "Organization of the Agricultural Sciences," 232.

94. USEC, *Fourth Report* (1885), 34–36, 120–88.

95. USEC, *Fourth Report* (1885), 123.

96. C. V. Riley, "On some Interactions of Organisms," *AN* 15 (April 1881): 323. Riley regularly quoted Forbes and reported his answers to Forbes's queries in the *American Entomologist*. See for example, S. A. Forbes, "The Food Habits of Thrushes," *AE* 3 (January 1880); S. A. Forbes, "The Food of the Bluebird," *AE* 3 (September 1880): 203; S. A. Forbes, "The Food of the Blackbird," *AE* 3 (October 1880): 232–34; and C. V. Riley, "Insects from Stomach of Lark, Robins, and Sunfish," *AE* 3 (November 1880): 278.

97. Riley's further involvement with parasites (parasitoids) as control agents is discussed in chap. 12.

98. See the report of his paper at the AAAS, "The Men of Science," *Daily Minnesota Tribune*, August 16, 1883, box 7, scrapbook vol. 50, p. 29, SIA, CVR; Robert A. Croker, *Stephen Forbes and the Rise of American Ecology* (Washington and London: Smithsonian Institution Press, 2001), 106–8.

99. C. V. Riley, "Entomology: The Use of Contagious Germs as Insecticides," *AN* 17 (November 1883): 1169–70; USEC, *Fourth Report* (1885), 188–90. See also C. V. Riley, "Bacterial Disease of the Imported Cabbage Worm," *SciAm* 49 (December 1, 1883): 337; and David Moore, "Fungal Control of Pests," in *Encyclopedia of Pest Management*, ed. David Pimentel (New York: Marcel Dekker, 2002), 320–24.

100. C. V. Riley, "Beneficial Insects: Silkworms," *Fourth Annual Report* (1872), 90 (hereafter cited as Riley, "Silkworms"). The quotation is from Alexander Pope, *An Essay on Man* (1733–34).

101. USEC, *Fourth Report* (1885), 120–23.

102. USEC, *Fourth Report* (1885), 126.

103. USEC, *Fourth Report* (1885), 127.

104. USEC, *Fourth Report* (1885), 136.

105. In the early twenty-first century, growers, targeting other pest species, continue to spread massive doses of pesticides on cotton fields. In 2003, growers applied 7 million pounds of insecticides, averaging over ten pounds per acre, on California cotton fields. Wagner, "Ode to Alabama," 170–71; Gravena et al., *Abstracts*, 53.

106. Wagner, "Ode to Alabama," 170.

107. In Texas and elsewhere, growers are subject to fines if cotton is left in the field after September 1. Wagner, "Ode to Alabama," 170–71; and Gravena et al., *Abstracts*, 1.

108. Howard, *History of Applied Entomology*, 65.

Chapter 10

1. Pauly, *Biologists*, 47; Worster, *Powell*, 386; *Harper's Weekly*, May 20, 1882, 315; and De B. Keim, *Washington and Its Environs: Descriptive and Historical Hand-Book* (Washington, DC: Dr. Keim, 1888), 244.

2. Pauly, *Biologists*, 47–51.

3. Worster, *Powell*, 391.

4. In April 1878, newspapers reported that Riley was offered an appointment as

Iowa state entomologist but speculated that he would accept the position as entomologist at the Agricultural Department. "Prof. C. V. Riley," *Manitoba Free Press* (Canada), April 13, 1878, copy from the *St. Louis Globe Democrat*, box 2, scrapbook vol. 15, p. 102, SIA, CVR.

5. "The Worm that Works," *Post*, December 26, 1878, box 2, scrapbook vol. 15, p. 100, SIA, CVR. This reference to Riley as a "new citizen" is the only known documentation of Riley's change of citizenship.

6. Worster, *Powell*, 384–85.

7. Worster, *Powell*, 386; and Mrs. John A. Logan, *Thirty years in Washington; or Life and scenes in our national capital: Portraying the wonderful operations in all the great departments, and describing every important function of our national government . . . With sketches of the presidents and their wives . . . from Washington's to Roosevelt's Administration* (Hartford, CT: A. D. Worthington & Co., 1901), 521.

8. Constance McLaughlin Green, *Washington: Village and Capital, 1800–1878* (Princeton, NJ: University of Princeton Press, 1962), 3–4; Logan, *Thirty Years*, 523.

9. Logan, *Thirty Years*, 523.

10. Unless noted, birth and death dates are based on grave markers in the family plot in Glenwood Cemetery, Lot 108, Site 3, 2219 Lincoln Road NE, Washington, DC.

11. Undated family portraits from the early 1890s. NAL, MC 143, Box 14, Folder 346.

12. See chap. 1.

13. Mira to Charlie, April 20, 1877, Folder 7C, RFR, NAL.

14. Mira to Charlie, April 26, 1874, and Mira to Emilie Riley, April 27, 1886, in Folder 7C, RFR, NAL. Mira later wrote to Emilie that Nina had always been the child of her affections. Mira to Emilie, April 27, 1886, in Folder 7C, RFR, NAL. There is some indication that Nina lived with Mira and Tim after 1870, as Riley reported visiting Nina, Mira, and Tim when he was in London in 1871 and 1875.

15. Handwritten poem, entitled "Nina," by C. V. Riley, April 15, 1876, box 2, scrapbook vol. 12, between pp. 36–37, SIA, CVR.

16. Quoted in a letter from Mira to Charlie, April 20, 1877, Folder 7C, RFR, NAL.

17. Mary Cannon's death is reported in "Hopeful Kansas," n.s., Topeka, May 21, 1877; and *New York Evangelist* 48 (May 31, 1877), both in box 8, SIA, CVR; and in *Commonwealth*, May 12, 1877, box 15, scrapbook vol. 43, p. 70, SIA, CVR. The circumstances of her death are detailed in Emilie Wenban-Smith Brash to Edward and Janet Smith, n.d., Family Tree and Genealogy, Folder 7D, RFR, NAL.

18. Nina Lafargue, replying to a correspondent regarding "Wheat Worms," from the US Entomological Commission's St. Louis headquarters, June 15, 1877, in envelope labeled "scraps to notice or reply to," box 13, scrapbook vol. 95, p. 179, SIA, CVR; Riley to Mary Murtfeldt, November 3, 1886, and Riley to J. C. Pearsall, July 30, 1887, both in box 15, Outgoing Correspondence, SIA, CVR.

19. Riley to Howard, September 17, 1887, and Riley to Viala, January 28, [1888], both in box 15, Outgoing Correspondence, SIA, CVR.

20. Nina to Mrs. C. V. Riley, May 3, 1897, postmarked Albert House, Addington Square, Margate, Folder 7C, RFR, NAL.

21. Mira to Emilie, January 27, 1883, Folder 7C, RFR, NAL; and Richard Lay, Bellefontaine Cemetery, St. Louis, to Conner Sorensen, January 30, 2012, Sorensen files, Eschbach, Germany.

22. Emilie Wenban-Smith Brash (Helen's daughter) to Janet and Edward Smith, April 25, 1987, Riley Children, Folder 7A, RFR, NAL.

23. *St. Louis Globe*, November 12 189[4], box 13, scrapbook vol. 95, p. 179 (in envelope), SIA, CVR.

24. See map with residences of naturalists in Pauly, *Biologists*, 54–55. An 1882 directory listing the addresses of Agricultural Department personnel indicates that most of Riley's departmental colleagues also lived in northwest Washington. *Boyd's Directory to Washington, DC* (1882).

25. Keim, *Washington and Its Environs*, 162.

26. Col. Robert Ingersoll, lecture "On Abraham Lincoln," at the National Theater, *Washington Post*, January 8, 1894, box 13, scrapbook vol. 94, p. 97, SIA, CVR.

27. Keim, *Washington and Its Environs*, 162.

28. Keim, *Washington and Its Environs*, 158.

29. Keim, *Washington and Its Environs*, 159–60.

30. Keim, *Washington and Its Environs*, 159.

31. Keim, *Washington and Its Environs*, 161.

32. "The Department of Agriculture's Seeds," *Rural New Yorker*, June 24, 1882, box 5, scrapbook vol. 34, p. 104, SIA, CVR; "The Government Seed Bureau," *Florida Dispatch*, September 29, 1884, box 8, scrapbook vol. 59, p. 10, SIA, CVR; and "Letter from Washington," *New England Farmer*, May 30, 1885, in box 8, scrapbook vol. 64, n.p., SIA, CVR; Keim, *Washington and Its Environs*, 156–62. For earlier criticism of the seed program see, "The Government Seed Swindle, " *Tribune* (Kansas), April 29, 1874, box 2, scrapbook vol. 2, p. 93, SIA, CVR; "The Agricultural College," *St. Louis Democrat*, February 17, 1874, SIA, CVR, 13:90; and "Government Seeds," [1874?], SIA, CVR, 13:95.

33. "LeDuc vs. Riley," *Washington Post*, July 29, 1880, SIA, CVR, 5:39–99 and 100.

34. See "Froebel's Birthday," n.s., n.d., SIA, CVR, 9:69–180. The kindergarten was located at 1127 Thirteenth Street NW.

35. Riley obituary, *CRW*, September 19, 1895, NAL, MC 143, Series IV, Box 4, 142–43.

36. Logan, *Thirty Years*, 524–26.

37. "The Six O'clock Club," *Washington Star*, November 9, 1888, box 12, scrapbook vol. 89, p. 7, SIA, CVR.

38. "Is LeDuc to be Removed?" *Rural New Yorker*, December 1, 1880, box 5, scrapbook vol. 39, pp. 57–58, SIA, CVR.

39. Worster, *Powell*, 394–95.

40. "Settled by the First Comptroller," n.s., n.d., box 9, scrapbook vol. 69, p. 182, SIA, CVR.

41. [Norman Colman], "Prof. Riley and the Riley Nozzle," *Philadelphia* [?] July 28 [1891], box 11, scrapbook vol. 84, p. 182, SIA, CVR.

42. In 1880, a supporter of the Agricultural Department noted that, although farm production amounted to over five times that of manufacturing production, the commissioner of agriculture was not a full cabinet officer, and his salary of $3,000 was one-third of full cabinet members, one-half of the United States geologist, and three-fifths of a US congressman. "Is LeDuc to be Removed?" *Rural New Yorker*, December 1, 1880, box 5, scrapbook vol. 39, pp. 57–58, SIA, CVR. In 1887, Commissioner Colman complained that the department had lost three chemists to other agencies that were able to offer higher pay. *ARCA* (1887), 10.

43. Worster, *Powell*, 385.

44. "Entomology Pays," *Washington Post*, March 24, 1882; and "District Government Affairs," [building permits for Professor C. V. Riley to erect six dwellings at Thirteenth and R Streets], both in box 9, scrapbook vol. 69, pp. 157, 158, SIA, CVR.

45. Anne E. Peterson, "Hornblower & Marshall, Architects," ([Washington, DC]: National Trust for Historic Preservation, The Preservation Press, 1978), 37.

46. Riley probably met Joseph C. Hornblower at the Cosmos Club, where Hornblower was elected to membership in 1883, the year he and Marshall formed their partnership. Peterson, "Hornblower & Marshall, Architects," 5, 9, 37. Peterson's "Catalogue of Hornblower & Marshall Buildings" lists Riley residences at 1303, 1305, and 1307 Thirteenth Street NW.

47. "New Buildings," *Republican* (Washington, DC), April 9, 1884, in box 13, scrapbook vol. 95, p. 179, SIA, CVR. Riley to Dr. Corey, July 19, 1889, in box 16, Outgoing Correspondence, Paris, SIA, CVR.

48. Riley to Coquillett, August 13, 1887, in box 15, Outgoing Correspondence, SIA, CVR.

49. Advertisement in *Colman's Rural World* 34 (December 1, 1881): 384; Riley resigned as editor of the entomological department at the *American Naturalist* in 1883, following a dispute with E. D. Cope, Packard's coeditor, involving Cope's refusal to publish a notice of a publication by LeConte. Riley to Lintner, January 4, 1883, Lintner Corr., NYSM. In 1887, the *Century Dictionary* paid Riley $36, based on a rate of $1 per hour of composition, for an article he had worked on from time to time over the preceding two years. Riley to Century Co. August 3, 1887, box 15, Outgoing Correspondence, SIA, CVR. See also "The Rural's Special Contributors for 1886," *Rural New Yorker*, January 9, 1886, in box 9, scrapbook vol. 66, p. 135, SIA, CVR. Another note indicates that Riley received $8 per column for writing in agricultural journals. "Prof. C. V. Riley, $8, Jan. Feb. March, 1879," n.s., in box 3, scrapbook vol. 23, p. 112, SIA, CVR; Isidor Bush to Riley regarding catalogue revision, [ca. 1894], NAL, MC 143, Series I, Box 1.

50. Riley to Bessey, March 8, 1873, UNA, RG 12. In 1876, Riley inquired (in confidence), how much Bessey had received for a recent lecture series in California. Riley to Bessey, February 3, 1876, UNA, RG 12.

51. J. D. Putnam, urged Riley's appointment as state entomologist but advised against combining the university and state positions. Putnam, letter to *Davenport Daily Gazette*, [February 20], 1880, box 5, scrapbook vol. 39, p. 93, SIA, CVR; Riley to Bessey, March 31, 1880, UNA, RG 12.

52. "Opinions of Dr. Riley," *Missouri Statesman*, March 1, 1895, box 13, scrapbook vol. 91, p. 43, SIA, CVR; and R. H. Jesse (President of the university), "The College of Agriculture," *CRW* 47 (November 22, 1894), 8; Riley's lecture notes, Folder 9A, RFR, NAL.

53. Riley to Bessey, October 22, 1873, UNA, RG 12.

54. Riley to Lintner, May 16, 1888, Lintner Corr., NYSM.

55. In 1891, for example, Mary Murtfeldt purchased illustrations from Riley. Mary E. Murtfeldt, *Outlines of Entomology* (Jefferson City, MO: Tribune Printing Co., 1891), preface, ii.

56. L. O. Howard, *The Insect Book* (New York: Doubleday, Page, and Co., 1905), preface, viii.

57. "Memorandum for Howard," [July 1888], box 15, Outgoing Correspondence, SIA, CVR.

58. Riley to Pearsall, August 14, 1887, box 15, Outgoing Correspondence, SIA, CVR; Riley to McNally [August 1887], box 15, Outgoing Correspondence, SIA, CVR. Riley had originally planed to sail on August 9. He apparently caught a later ship because of unfinished business.

59. Riley to Howard (from New York), August 11, 1887, box 15, Outgoing Correspondence, SIA, CVR.

60. Riley to Pearsall, August 14, 1887; and Riley to Coquillett, August 13, 1887, box 15, Outgoing Correspondence, SIA, CVR.

61. Riley to Comstock, July 31, 1887; and Riley to Putnam, August 2, 1887, box 15, Outgoing Correspondence, SIA, CVR.

62. C. V. Riley, "On the Phengodini and their Luminous Larviform Females"; and C. V. Riley, "The Buffalo-Gnat Problem in the Lower Mississippi Valley," both in *Proc. AAAS* 36 (1888): 362; "Scientists taking a Sail," *New York Sunday World*, August 14, 1887, box 2, scrapbook vol. 15, p. 90–91, SIA, CVR.

63. Ground Plan, George Trusedell's Addition to Washington Heights and Surroundings, NAL, MC 143, Series III, Box 3, Map Case A, Drawer 19. See also Kathryn Schneider Smith, ed., *Washington at Home: An Illustrated History of Neighborhoods in the Nation's Capital* (Northridge, CA: Windsor Publications, 1988).

64. Riley to Saunders, June 22, 1889; Riley to Howard, July 19, 1889; and Riley to J. B. Miner, July 19, 1889, all in box 16, Outgoing Correspondence, Paris, SIA, CVR.

65. Riley to Howard, June 3, 1889, in box 16, Outgoing Correspondence, Paris, SIA, CVR.

66. "Sketch of House for Prof. C. V. Riley, Hornblower and Marshall, Archt.," March 30, 1889, and Lot Plan and Ground Plan, George Trusedell's addition to Washington Heights and surroundings, all in NAL, MC 143, Series III, Box 3, Map Case A, Drawer 19; on the architectural style, see Martin Filler, "Our Grand and Randy Great Architects," *NYRB* 58 (May 26–June 8, 2011): 21; Peterson describes the Romanesque Revival or Colonial Revival style employed by Hornblower and Marshall as having "a lack of extraneous ornament that verges on abstraction," a description that does not apply to the Riley residence. Peterson, "Hornblower & Marshall," 15. One source indicates that the firm of Lane and Associates was paid $20,350 for house plans and/or construction. Riley to J. B. Miner, July 19, 1889, box 16, Outgoing Correspondence, Paris, SIA, CVR.

67. In 1875, Riley wrote to Hilgard that his eyes were giving him problems. Riley to Hilgard, October 18, 1875, Hilgard Family Papers, Bancroft Library, University of California, Berkeley.

68. *ARCA* (1884), 13. In Edinburgh, Riley exhibited eight cases of insects with special reference to forest trees, for which he was awarded a gold medal (the only gold medal awarded to an American). *Catalogue of the Exhibit* (1884), 3–4; and "Prof. Riley's Gold Medal," *Republican* (St. Louis, MO), December 8, 1884, both in box 9, scrapbook vol. 66, p. 10, SIA, CVR.

69. Riley to Lintner, February 12, 1886, NYSM.

70. Riley to Colman, May 7, [1886], box 15, Outgoing Correspondence, SIA, CVR.

71. Riley to Howard, November 13, 1887, in box 15, Outgoing Correspondence, SIA, CVR.

72. Mary to Papa, [January 1894], Folder 7C, RFR, NAL.

73. Riley to Osten Sacken, January 30, 1888, box 15, Outgoing Correspondence, SIA, CVR. See also Riley to Saunders, August 12, 1889, in box 15, Outgoing Correspondence, SIA, CVR.

74. A reporter on the excursion to the Rocky Mountains commented that Riley could converse on practically any subject but theology. "N. A. Agricultural Editorial Excursion, 1873," *Rural New Yorker*, August 9, 1873, box 1, scrapbook vol. 11, pp. 12–13, SIA, CVR.

75. Notes of Gottlieb Conzelman, 1873–1880 (typewritten), 33, in Conzelman Family, Folder 7A, RFR, NAL; transcript of interview with Emerson and James Conzelman (nephews of Emilie Conzelman Riley), January 15, 1985, Nokomis, Florida, in Conzelman Family, Folder 7A, RFR, NAL.

76. Sermon by J. C. Learned, December 10, 1871, box 1, scrapbook vol. 5, p. 187, SIA, CVR.

77. Notice of Charles V. Riley funeral, NAL, MC 143, Series IV, Box 4, Folder 145.

78. John S. Haller Jr., *Outcasts from Evolution: Scientific Attitudes of Racial Inferiority, 1859–1900* (Urbana: University of Illinois Press, 1975), vii–viii. For the earlier nineteenth century, see William Stanton, *The Leopard's Spots: Scientific Attitudes toward*

Race in America, 1815–59 (1960; reprint, Chicago and London: University of Chicago Press, 1966).

79. C. V. Riley, "The Great Leopard Moth," *Fourth Annual Report* (1872), 142, 144; and Riley, "Grape Phylloxera," *Sixth Annual Report* (1874), 60.

80. "N. A. Agricultural Editorial Excursion," (1873) box 1, scrapbook vol. 11, pp. 13–14, SIA, CVR.

81. Cotton Worm Sprayer, *ARCA* (1881–1882), plate IX following p. 159.

82. Muehl, "Isidor Bush and the Bushberg Vineyards of Jefferson County," 42–58. See pp. 49–53 for Bush's association with Riley and the importance of his catalogues in the Phylloxera campaign.

83. He wrote this in 1894, in response to a questionnaire circulated by J. S. Billings, an army physician. Riley to Dr. J. S. Billings, US Army, December 26, 1894, box 27, SIA, CVR. Copy in Misc. Corr., RFR, NAL.

84. "Lecture Notes, 1868–69," NAL, MC 143, Series II, Box 21, Folder 49.

85. William W. Turnbull, *The Good Templars: A History of the Rise and Progress of the Independent Order of Good Templars* ([Milwaukee, WI]: B. F. Parker, 1901), 68.

86. See C. W. Murtfeldt, "Prohibition Our only Safety," *CRW* 34 (March 24, 1881): 91.

87. Riley concluded the questionnaire by noting that he enjoyed an occasional drink in a social setting, but he believed the negative aspects of habitual drinking outweighed any temporary benefits. Riley to Billings, December 26, 1894, box 27, SIA, CVR.

88. Transcript of interview with Emerson and James Conzelman, January 30, 1985, Conzelman Folder 7A, RFR, NAL.

89. "Scientists Taking a Sail," *N. Y. Sunday World*, August 14, 1887, box 9, scrapbook vol. 70, p. 115, SIA, CVR.

90. See floor plan, Riley Residence, NAL, MC 143, Series III, Map Case A, Drawer 19.

91. Charles V. Riley, "Mr. Maxim's Flying Machine," *SciAm* (October 6, 1894), box 13, scrapbook vol. 91, p. 38, SIA, CVR; "Aeronautics," SIA, CVR, 12:86.

92. "Aeronautics," SIA, CVR, 12:86. In October 1890, during a stopover in Chicago between a bee culture conference in Lansing, Michigan, and a conference on flying machines near Chicago, Riley shared his views on flight with a reporter. "Believes in Air Ships," *Chicago Journal*, October 10, 1890, box 12, scrapbook vol. 86, p. 57, SIA, CVR.

93. Typed copy of handwritten will, received from George Ordish, Bohemia, Nancledra, Penzance, Cornwall, England, December 1987, Folder 7F, RFR, NAL (hereafter cited as Will and Testament).

94. Riley, "Outlook for Applied Entomology," *IL* 3 (January 1890): 209.

95. Will and Testament.

96. The familial relationship between Charles V. Riley and his "ward," Charles

Fewtrell Wilde, is not clear. In a letter Riley wrote to Albert Koebele, who was bound for Australia on the Cottony Cushion Scale project in 1891, Riley recommended that Koebele seek assistance from his "half uncle," Dr. Robert T. Wylde of Adelaide, who, Riley said, was something of a naturalist and getting along in years. Although Charles Fewtrell Wilde was not mentioned in the letter, he may have been the son of Robert T. Wylde, which would make Charles Riley and Charles Fewtrell Wilde cousins. Riley to Koebele, July 21, 1891, Koebele Correspondence (1881–1893), California Academy of Sciences, San Francisco, CA (hereafter cited as CAS).

Chapter 11

Philip Walker, assistant director of the Division of Entomology Silk Division, describing his vision of the potential for silk production in the United States, "A Great Big Silk Farm," *Star* (Washington, DC), November 30, 1889, box 11, scrapbook vol. 79, p. 145, SIA, CVR.

1. For histories of silk culture in America, see L. O. Howard, "The United States Department of Agriculture and Silk Culture," *Yearbook of the United States Department of Agriculture* (Washington, DC: GPO, 1904), 137–48; and Nelson Klose, "Sericulture in the United States," *AgH* 37 (October 1963): 225–34.

2. C. V. Riley, "Beneficial Insects: Silkworms," *Fourth Annual Report* (1872), 73 (hereafter cited as "Silkworms").

3. Ailanthus is a tree, known as "Tree of heaven" in English.

4. Riley, "Silkworms," *Fourth Annual Report* (1872), 112–20. See also his account in *PF* (April 28, 1866): 289. Riley records his rearing of silkworms in *Memorandum Entomological*, box 15, scrapbook vol. 4, p. 126, SIA, CVR.

5. Riley, "Silkworms," 83.

6. Riley, "Silkworms," 98.

7. The most recent failure had been the *multicaulis* debacle of the 1830s, when entrepreneurs invested vast sums in the cultivation of *Morus multicaulis* (many-stemmed) mulberry in the expectation that abundant food for the projected silkworm population would bring huge profits. Riley, "Silkworms," 72.

8. Riley, "Silkworms," 98–99, 181.

9. Riley, "Silkworms," 81.

10. Michael Ruse, "Charles Darwin and Artificial Selection," *JHI* 36 (April–June 1975): 339–50.

11. Riley, "Silkworms," 75, 85.

12. In 1889, Riley recommended the original silkworm as the best moth for interbreeding and experimentation. He noted that breeding of this moth had produced "many well-marked races," but he did not say whether he or others had performed the experiments. C. V. Riley, "Pedigree Moth-Breeding," *Entomologist's Monthly Magazine* (London) 23 (May 1889), 277–78.

13. In 1862, Packard, then twenty-three years old, witnessed Trouvelot's breeding

experiments. Trouvelot's involvement with the Gypsy Moth is discussed in Spear, *Gypsy Moth War*, 7–27.

14. L[éopold] Trouvelot, "The American Silkworm," [part 1] *AN* 1 (March 1867): 32. See also part 2, (April 1867): 85–94 and part 3 (May 1867): 145–49.

15. É. Léopold Trouvelot, "On a Method of Stimulating Union Between Insects of Different Species," [read February 27, 1867] *Proceedings of the Boston Society of Natural History* 11 (1866–68): 136–37 (hereafter cited as *Proc. BSNH*). There is apparently no independent record of the success or failure of this procedure.

16. Spear, *Gypsy Moth War*, 125, 129.

17. Riley, "Silkworms," 103–7. See footnote p. 103 for Riley's discussion of the scientific name.

18. Riley, "Silkworms,"107.

19. Riley, "Silkworms,"130–36. *Antheraea yamamai* is the current scientific name for the species. See footnote p. 130 for the history and culture of this species in Europe.

20. Riley, "Silkworms," 134.

21. Riley, "Silkworms," 137.

22. Riley, "Silkworms," 101.

23. Riley, "Silkworms," 100–103.

24. Riley's figures for the Oneida Community seem high. Matthew Cooper lists the number of women and girls employed at Oneida's Wallingford silk works at thirty-five. See Matthew Cooper, "Relations of Modes of Production in Nineteenth-Century America: The Shakers and Oneida," in *American Culture: Essays on the Familiar and Unfamiliar*, ed. Leonard Plotnicov (Pittsburgh, PA: University of Pittsburgh Press, 1990), 54–57. Silk production, however, did constitute one of Oneida's four main enterprises in the 1860s and 1870s.

25. Riley, "Silkworms," 77–78.

26. Riley referred to silkworm disease only by the term pébrine; however, silkworms were also being affected by flacherie, a form of viral dysentery. Crocker, *Forbes*, 106–7.

27. Ibid., 77, 81; and Klose, "Sericulture," 227.

28. Riley, "Silkworms," 82–83; and Klose, "Sericulture," 228.

29. See for example Riley's talk on silk culture at Kansas Agricultural College, "Silk Culture in Kansas," n.s., n.d, box 2, scrapbook vol. 14, pp. 1–2, SIA, CVR.

30. C. V. Riley, "Silk Culture: a new source of Wealth to the United States," *Proc. AAAS* 27 (1879): 277–83; and *St. Louis Globe-Democrat*, August 28, 1878, box 3, scrapbook vol. 19, p. 203, SIA, CVR.

31. *ARCA* (1878), 215–16.

32. Klose, "Sericulture," 229.

33. "The Agricultural Department," *SciAm* 38 (June 1, 1878): 340.

34. *ARCA* (1872), 216.

35. *ARCA* (1872), 217–18.

36. *ARCA* (1872), 217–18.

37. *ARCA* (1880), 24–25.

38. C. V. Riley, "A New Source of Wealth to the United States," *Kansas Farmer*, November 12, 1879, box 5, scrapbook vol. 39, pp. 91–92, SIA, CVR. See testimonials praising Riley's silkworm manual in box 3, scrapbook vol. 73, pp. 62–63, SIA, CVR.

39. Klose describes the 1880s as the fourth silk craze in America. Klose, "Sericulture," 228.

40. *American Farmer*, September 1, 1882, box 5, scrapbook vol. 33, pp. 132–33, SIA, CVR. At the close of the Philadelphia exhibit, the Women's Silk Culture Association presented Mrs. Garfield with the dress.

41. "The Women's Silk Culture Association," *Philadelphia Tribune*, January 31, [1882], box 5, scrapbook vol. 34, p. 15, SIA, CVR; "The Philadelphia Silk Fair," *Rural New Yorker*, April 22, 1882, in box 5, scrapbook vol. 33, p. 110, SIA, CVR; "Osage Orange Silk," *Gardener's Monthly* (June 1882) in box 5, scrapbook vol. 33, p. 120, SIA, CVR.

42. "Exhibition of Silk Culture," *New York Dry Goods Bulletin*, June 17, 1882, in box 5, scrapbook vol. 33, p. 115, SIA, CVR; "The American Silk Exchange," *Rural Record*, January 10, 1882, in box 5, scrapbook vol. 33, pp. 116, 120, SIA, CVR.

43. "Exhibition of Silk culture." *New York Dry Goods Bulletin*, June 17, 1882, in box 5, scrapbook vol. 33, p. 115, SIA, CVR.

44. "Silk Culture," *Farmer's Review*, September 7, 1882, box 5, scrapbook vol. 33, p. 117, SIA, CVR.

45. "Southern Silk Culture," *PF* (August 6, 1882), box 5, scrapbook vol. 33, p. 116, SIA, CVR.

46. *ARCA* (1883), 104–5.

47. "Work of the California Silk Culture Association," *PRP* (August 12, 1882), box 5, scrapbook vol. 34, p. 44, SIA, CVR; Klose, "Sericulture," 250; ARCA (1883), 101–2.

48. "Exhibition of Silk Culture," *New York Dry Goods Bulletin* (June 17, 1882), box 5, scrapbook vol. 33, p. 115, SIA, CVR; and "American Silk," *New England Farmer* (July 1882), in box 5, scrapbook vol. 33, p. 118, SIA, CVR.

49. "The American Silk Exchange," *Rural Record*, June 10, 1882 in box 5, scrapbook vol. 33, p. 120, SIA, CVR.

50. "Silk Culture in the Mississippi Valley," *National Scientific Journal*, August 1, 1882, in box 6, scrapbook vol. 48, p. 41, SIA, CVR; and *National Farmer*, July 13, 1882, in box 6, scrapbook vol. 48, pp. 32–36, SIA, CVR.

51. *Catalogue of the Exhibit* (1884).

52. *ARCA* (1883), 102–4.

53. *ARCA* (1884), 13–14.

54. He had arranged with Loring to spend the months from May to September in Europe. *ARCA* (1884), 13, 287.

55. *ARCA* (1884), 286, 361.

56. Compiled from *ARCA* (1884–1891) and Klose, "Sericulture," 229.

57. In the peak year 1888, the Silk Division reported 6,975 correspondents and distributed 8,000 silkworm manuals. *ARCA* (1889), 50, 111.

58. "The Agricultural Congress," *Globe*, May 15, 1874, box 1, scrapbook vol. 1, p. 106, SIA, CVR; and C. V. Riley, "Economic Entomology," *AN* n.d., box 1, scrapbook vol. 8, p. 104, SIA, CVR.

59. On the emergence of economics as a discipline in this period, see Dorothy Ross, "The Development of the Social Sciences," in *The Organization of Knowledge in Modern America, 1860–1920*, ed. Alexandra Oleson and John Voss (Baltimore and London: Johns Hopkins University Press, 1979), 107–38.

60. "Tired Scientists," *Daily Minnesota Tribune*, August 22, 1883, box 6, scrapbook vol. 41, p. 1, SIA, CVR. The Economics and Statistical Section was organized in 1882. "The American Association for the Advancement of Science," *SciAm*, September 16, 1882, box 5, scrapbook vol. 35, p. 108, SIA, CVR.

61. J. R. Dodge, "The American Farmer, His Condition and Prospects," (1891), reprinted in *SciAm*, September 12, 1891, box 11, scrapbook vol. 85, p. 1, SIA, CVR.

62. Riley to Dodge, October 8, [1887], box 15, Outgoing Correspondence, 1886–1895, SIA, CVR.

63. Riley, "Silkworms," 73, 81, 83, 92.

64. Mrs. C. Thompson to Riley, February 5, 1883, box 16, Incoming Correspondence, SIA, CVR. It is unclear where Mrs. Thompson lived. Kamm is perhaps a misspelling of a town in Illinois.

65. *ARCA* (1885), 215–16.

66. *ARCA* (1885), 549–50.

67. B. F. Peixotto, "Silk Culture in the United States," *PF* 55 (July 28, 1883), box 6, scrapbook vol. 48, pp. 59–60, SIA, CVR.

68. Philip Walker, "A Great Big Silk Farm," *Star* (Washington, DC), November 30, 1889, box 11, scrapbook vol. 79, p. 145, SIA, CVR.

69. [Anonymous], "Silk Culture for Women," *Farmer's Review*, June 29, 1882, box 5, scrapbook vol. 33, p. 122, SIA, CVR.

70. *ARCA* (1887), 19.

71. *ARCA* (1885), 216.

72. *ARCA* (1884), 286; and Edward Wellman Serrell entry in *The Twentieth Century Biographical Dictionary of Notable Americans*, vol. 11 (Boston: The Biographical Society, 1904). The secretary of state was not named in the sources, but from the chronology it must have been Frelinghuysen. See Karabell, *Chester Arthur*, 70.

73. *ARCA* (1884), 360–61.

74. *ARCA* (1885), 216–17.

75. *ARCA* (1885), 218–19; and US Patent Office, Edward W. Serrell Jr., "Device for Reeling Silk From the Cocoon," Patent No. 317,222 (May 5, 1885). Application filed February 27, 1884. Patent with sketch available on the US Patent and Trademark Office website, https://patents.google.com/ (accessed March 20, 2018); *ARCA* (1886), 462,

546. Officially, support for reeling in Philadelphia and New Orleans was discontinued, but reeling seems to have continued, at least in Philadelphia, under the special $5,000 appropriation noted above.

76. *ARCA* (1887), 20–21.

77. *ARCA* (1887), 119–20.

78. *ARCA* (1887), 49.

79. *ARCA* (1887), 87, 121.

80. *ARCA* (1888), 22, 111.

81. *ARCA* (1888), 111, 116–19.

82. *ARCA* (1888), 55–56.

83. Philip Walker, "A Great Big Silk Farm," *Star* (Washington, DC), November 30, 1889, box 11, scrapbook vol. 79, p. 145, SIA, CVR.

84. Philip Walker, "A Great Big Silk Farm," *Star* (Washington, DC), November 30, 1889, box 11, scrapbook vol. 79, p. 145, SIA, CVR; see also "Threads of Golden Hae," n.s. March 17, 1889, in box 11, scrapbook vol. 83, p. 49, SIA, CVR.

85. In February 1889, Congress upgraded the office of commissioner of agriculture to full cabinet status. Colman thus became the first secretary of agriculture in the last month of his tenure, and Rusk was the second secretary.

86. *ARCA* (1890), 37–38.

87. *ARCA* (1889), 333.

88. "The Art of Silk Making," *Topeka Daily Capital*, July 18, 1891, box 11, scrapbook vol. 85, p. 64, SIA, CVR.

Chapter 12

1. Pauly, *Fruits and Plains*, 198–204, 222. For Saunders, see Pauly, *Fruits and Plains*, 311n28; and "William Saunders," *Yearbook of the U. S. Department of Agriculture* (Washington, DC: GPO, 1900), 628. The first shipment of citrus fruit from southern California via the transcontinental railroad, in 1877, was destined for St. Louis. Riley, then in the midst of the locust campaign, may have sampled an orange from that shipment. See Robert F. Luck, "Notes on the Evolution of Citrus Pest Management in California," in *Proceedings of the Fifth California Conference on Biological Control*, ed. Mark S. Hoddle and Marshall W. Johnson (Riverside, CA: n.p., 2006), 2.

2. *ARCA* (1879), 202–7; Pauly, *Fruits and Plains*, 135–36, 140; and *Comstocks of Cornell*, 118–20.

3. *ARCA* (1880), 15, 277.

4. *ARCA* (1880), 277; and D. W. Coquillett, "Scale Insects," *Rural Californian* (June 1883), box 6, scrapbook vol. 47, pp. 102–4, SIA, CVR.

5. Among these were the notorious Oyster Shell Bark Louse of Apple and mealybugs that infested greenhouse plants. *ARCA* (1880), 276.

6. Comstock supplemented his field investigations by breeding scale insects on potted plants at the Agriculture Department.

7. *ARCA* (1879), 208; and *ARCA* (1880), 235, 279–81.

8. *ARCA* (1880), 286–89.

9. *ARCA* (1880), 289–90.

10. Although Riley visited Tennessee, Georgia, the Carolinas, and Virginia in the late summer and fall of 1878, he apparently did not visit Florida.

11. Comstock cited Riley's recommendation of Barnard's milk-kerosene solution in his 1880 report, but apparently he had not yet learned of Hubbard's kerosene-soap emulsion. *ARCA* (1880), 288 (quoting Riley in *SciAm*, October 16, 1880).

12. Riley to Koebele, October 29 and June 1, 1881, Koebele Corr., CAS; and Mallis, *American Entomologists*, 352. Riley frequently disregarded his own distinction between locusts and grasshoppers, referring here to grasshoppers as locusts.

13. D. W. Coquillett, "Scale Insects," *Rural Californian* (June 1883), box 6, scrapbook vol. 47, pp. 102–4, SIA, CVR.

14. "Prof. Riley in Florida," *Florida Dispatch*, [month?] 27, 1882, box 8, scrapbook vol. 39, p. 129, SIA, CVR; "Victory over Scale Bug [*sic*]," *Wine and Fruit Grower* (May 1884), in box 8, scrapbook vol. 57, pp. 29–31, SIA, CVR; and [C. V. Riley], "Riley's researches," *Florida Daily Times*, March 29, 1882, in box 8, scrapbook vol. 42, pp. 28–29, SIA, CVR.

15. C. V. Riley, "Successful Management of the Insects Most Destructive to the Orange," *SciAm* 46 (May 27, 1882): 335–36.

16. Howard, *History of Applied Entomology*, 95.

17. "Insect Pests: The Cottony Cushion Scale," *Rural Californian* (December 1883): 245; and [C. V. Riley], "The Cottony Cushion-Scale (*Icerya purchasi* Maskell)," *ARCA* (1886), 468.

18. Riley to Koebele, January 28, 1886, Koebele Corr., CAS.

19. *ARCA* (1878), 4.

20. Riley, "The Cottony Cushion-Scale," *ARCA* (1886), 468.

21. Riley, "The Cottony Cushion-Scale," *ARCA* (1886), 471.

22. Riley, "The Cottony Cushion-Scale," *ARCA* (1886), 466, 482–83.

23. Riley, "The Cottony Cushion-Scale," *ARCA* (1886), 476–81.

24. C. V. Riley, "Fluted Scale and Enemies," *PF* (August 9, 1890), box 9, scrapbook vol. 71, p. 31, SIA, CVR.

25. Riley, "The Cottony Cushion-Scale," *ARCA* (1886), 485–88.

26. "Semi-Annual State Convention of Fruit Growers, April 12, 1887," *PRP* (April 23, 1887), box 8, scrapbook vol. 61, p. 47, SIA, CVR.

27. "Semi-Annual State Convention of Fruit Growers, April 12, 1887," *PRP* (April 23, 1887), box 8, scrapbook vol. 61, p. 47, SIA, CVR.

28. C. V. Riley, "Parasitic and Predacious Insects in Applied Entomology," *SciAm* 36, Supplement 937 (December 16, 1893): 14978–81.

29. Comstock advised a similar strategy of placing infested chrysalides (pupae) of the Imported Cabbage Worm in boxes with wire covers, allowing the parasites to

escape. Riley, LeBaron, and Comstock admitted they could not prove that the transplantations worked because it was impossible to say whether the parasites already existed in these locations. Riley, "Parasitic and Predacious Insects in Applied Entomology," 14978–81.

30. Riley, "Parasitic and Predacious Insects in Applied Entomology," 14978–81; *ARCA* (1884), 323; and *ARCA* (1889), 236–37.

31. Richard L. Doutt, "Vice, Virtue, and Vedalia," *Bulletin of the Entomological Society of America* 4 (December 1958): 120.

32. Riley to Koebele, June 28, 1887, Koebele Corr., CAS.

33. Charles V. Riley, "On the Original Habitat of *Icerya purchasi*," *PRP* 35 (May 12, 1888), box 8, scrapbook vol. 61, pp. 15–16, SIA, CVR; and C. V. Riley, "The Fluted Scale," *ARCA* (1888), 80.

34. Riley, "The Fluted Scale," *ARCA* (1888), 81.

35. Riley, "The Fluted Scale," *ARCA* (1888), 81–82.

36. Later, when the controversy over credit for saving the California orange groves heated up, some Californians asserted that Klee's parasites had survived and helped control the Cottony Cushion Scale. Riley agreed that this was a possibility, but by then it was impossible to ascertain when and where each parasite had been released. C. V. Riley, "The Imported Parasite of the Fluted Scale," *PRP* (July 21, 1888), box 9, scrapbook vol. 71, p. 13, SIA, CVR.

37. Riley, "The Imported Parasite of the Fluted Scale," *PRP* (July 21, 1888), box 9, scrapbook vol. 71, p. 13, SIA, CVR; and *ARCA* (1888), 89. The genus *Lestophonus* has since been changed to *Cryptochetum*.

38. C. V. Riley, "Parasitic and Predacious Insects," 14980.

39. Riley to Koebele, June 6, 1888, Koebele Corr., CAS.

40. Riley to Koebele, June 6, 1888, Koebele Corr., CAS; and C. V. Riley, "The Entomological Mission to Australia," *PRP*, February 8, 1890, SIA, 10:78–192.

41. C. V. Riley, "Importation of *Icerya* Enemies from Australia," *PRP*, December 21, 1889, box 10, scrapbook vol. 61, pp. 75–76, SIA, CVR; and Mallis, *American Entomologists*, 94.

42. C. V. Riley, "The Imported Parasite of the Fluted Scale," *PRP*, July 21, 1888, box 9, scrapbook vol. 71, p. 13, SIA, CVR; and *ARCA* (1888), 19.

43. Riley to Koebele, January 7, 1889, CAS.

44. *ARCA* (1889), 91.

45. "More Foes of Icerya," *PRP* (December 29, 1888), box 10, scrapbook vol. 73, pp. 95–96, SIA, CVR.

46. *ARCA* (1888), 90. Riley's report is dated December 1, 1888, but in it he refers to events up to January 1889.

47. Nineteenth-century entomologists used the spelling *Cryptochaetum*. The spelling *Cryptochetum* is now preferred. L. E. Caltagirone and R. L. Doutt, "The History of the Vedalia Beetle Importation to California and Its Impact on the Development of

Biological Control," *Ann. Rev. Ent.* 34 (1989): 4n1, 7 (hereafter cited as Caltagirone and Doutt, "Vedalia Beetle Importation"); and Doutt, "Vice, Virtue, and Vedalia," 121.

48. It is unclear, however, whether the northern California population originated with Crawford's shipment to Klee or Koebele's later shipment to Coquillett. Doutt, "Vice, Virtue, and Vedalia," 121; and Caltagirone and Doutt, "Vedalia Beetle Importation," 6–8.

49. Doutt, "Vice, Virtue, and Vedalia," 121

50. Coquillett, "Imported Australian Ladybird," 73–74.

51. Riley to Koebele, January 3, 1889, Koebele Corr., CAS.

52. Luck, "Notes on the Evolution of Citrus Pest Management in California," 3.

53. Coquillett, "Imported Australian Ladybird," 73–74.

54. Correspondence from J. R. Dobbins, "The Spread of the Australian Lady Bird," [July 2, 1889] *IL* 3 (October 1889): 112.

55. *ARCA* (1889), 335–36; and Coquillett, "Imported Australian Ladybird," (1889), 74.

56. Riley, "Parasitic and Predacious Insects" (1893), 14980.

57. *ARCA* (1889), 334.

58. Koebele collected an additional six thousand Vedalia beetles in various stages in New Zealand on his return trip, but the disposition of these is uncertain, and by the time he returned to California, the original Vedalia beetles had already produced a multitude of beetles in southern California. Caltagirone and Doutt, "Vedalia Beetle Importation," 13.

59. C. V. Riley, "The Fluted Scale and Enemies," *PRP* (August 9, 1890), box 9, scrapbook vol. 71, p. 31, SIA, CVR; ARSA (1892), 38.

60. As noted above, Riley's experience with transplanting natural enemies involved primarily parasites. In modern usage, these are called parasitoids; however, Riley and his contemporaries used the term parasites to include all parasites and parasitoids.

61. Cited in Caltagirone and Doutt, "Vedalia Beetle Importation," 13.

62. R. G. Van Driesche, Mark Hoddle, and Ted Center, *Control of Pests and Weeds by Natural Enemies: An Introduction to Biological Control* (Malden, MA: Blackwell Publishing, 2008), 127–33, 186–87.

63. Daniel Coquillett, "Report on Gas Treatment," *ARCA* (1887), 124.

64. Mallis, *American Entomologists*, 390.

65. "Semi-Annual State Convention of Fruit Growers, April 2, 1887," *PRP*, April 23, 1887, box 8, scrapbook vol. 61, p. 47, SIA, CVR; and "Statement of Appropriations and Expenditures for the U. S. Department of Agriculture from 1878 to 1892," *ARSA* (1894), 52–53; the dispute between Riley and C. Hart Merriam over budgets is discussed in Keir B. Sterling, *Last of the Naturalists: The Career of C. Hart Merriam* (New York: Arno Press, 1977), 60–70; and Barrow, *A Passion for Birds*, 60.

66. Coquillett, "Report on Gas Treatment," *ARCA* (1887), 124; and *ARCA* (1887), 129–33.

67. [F. W. Morse], "The Uses of Gases Against Scale Insects," University of California Agricultural Experiment Station, Bulletin 71 [June 1887]; and C. W. Woodworth, "Orchard Fumigation," University of California Agricultural Experiment Station, Bulletin 122 (January 1899), 4–6, both available at http://catalog.hathitrust.org/Record/002137876 (accessed March 20, 2018). According to Woodworth, some growers recognized the almond odor of HCN used by Coquillett and passed this information on to Morse.

68. Coquillett's account of his work on HCN fumigation makes no reference to his patent application. Growers in southern California were generally aware of his discovery and, likely, of his patent application. Riley, in 1891, referred to Coquillett's patent application with disapproval. *ARCA* (1887), 125; and "Prof. Riley on the History of the Gas Treatment," *PRP* (September 19, 1891), box 10, scrapbook vol. 78, pp. 28–30, SIA, CVR.

69. Woodworth, "Orchard Fumigation," (January 1899), 4; and A. Scott Chapman to Morse, June 8, 1887, printed in newspaper clipping bound with [F. W. Morse], "The Uses of Gases Against Scale Insects," [June 1887] in Woodworth, "Orchard Fumigation," 4, http://catalog.hathitrust.org/Record/002137876 (accessed March 20, 2018).

70. Coquillett, "Report on Gas Treatment," *ARCA* (1887), 123–24, quote on pg. 123. There is no indication that Riley objected to Hilgard's sending his agent Morse to conduct scale insect investigations in southern California, although he likely considered this his turf.

71. *ARCA* (1892), 168.

72. Coquillett, "Report on Gas Treatment," *ARCA* (1887), 126–29; and *ARSA* (1891), 266.

73. It was eventually learned that during the day, sunlight and heat decompose HCN into other gasses that are harmful to trees. *ARSA* (1891), 265.

74. The other patent applicants included A. H. Alward, W. B. Wall, and M. S. Jones. According to Woodworth, Bishop's wife first suggested fumigating at night. Woodworth, "Orchard Fumigation," (January 1899), 5, http://catalog.hathitrust.org/Record/002137876 (accessed March 20, 2018).

75. "Prof. Riley on the History of the Gas Treatment," *PRP*, September 19, 1891, box 10, scrapbook vol. 78, pp. 28–30, SIA, CVR.

76. They charged that Coquillett and Riley had attempted to keep the process a secret. Coquillett had attempted to keep his patent application secret, but Riley certainly did not attempt to keep the process secret. He was committed to making useful knowledge available to the public. "A Rejoinder: Open Letter to Messrs. Riley and Coquillett," *Santa Ana Weekly* [?], February 27, 1890, box 11, scrapbook vol. 79, p. 143, SIA, CVR.

77. *ARCA* (1891), 266.

78. D. W. Coquillett, "The Patent on the Hydrocyanic Acid Gas Process Declared Invalid," *IL* 7 (December 1894): 257–58.

79. Woodworth, "Orchard Fumigation," (January 1899), 5–6, http://catalog.hathitrust.org/Record/002137876 (accessed March 20, 2018).

80. Mallis, *American Entomologists*, 352; and Nick Price, "Fumigants," in Pimentel, *Encyclopedia of Pest Management*, 318.

81. In reaction to these articles, Riley urged Koebele to counter with a public statement giving credit to him (Riley), referring to himself as "the engineer at headquarters." Riley to Koebele, December 5, 1889, Koebele Corr., CAS.

82. C. V. Riley, "Importation of Icerya Enemies from Australia," responding to articles in the San Luis Obispo *Tribune* (September 27, 1889) and other papers. *PRP* (December 21, 1889), box 8, scrapbook vol. 61, pp. 75–76, SIA, CVR.

83. Riley to Koebele, March 6 and March 27, Koebele Corr., CAS; and C. V. Riley, "The Entomological Mission to Australia," *PRP* (February 8, 1890), box 10, scrapbook vol. 78, pp. 1–2, SIA, CVR.

84. "Entomological," *PRP* (September 24, 1891), box 10, scrapbook vol. 87, p. 175, SIA, CVR. After Riley's death, Howard continued to support Riley's position, estimating that the Californians' overemphasis on biological control set back the progress of effective pest control in that state by at least a decade. Howard, *History of Applied Entomology*, 155. See also Caltagirone and Doutt, "History of Vedalia Beetle Importation," 8, 10.

85. Riley to Koebele, May 29, 1891, Koebele Corr., CAS; and *ARCA* (1891), 233–34, 251.

86. Riley to Koebele, June 18, 1891, Koebele Corr., CAS; and *ARCA* (1891), 234.

87. Riley to Koebele, June 18, 1891, Koebele Corr., CAS.

88. Lelong to Koebele, December 9, 1891, Koebele Corr., CAS.

89. Lelong to Koebele, February 4, 1892, Koebele Corr., CAS.

90. Lelong to Koebele, March 31 and April 1, 1892, Koebele Corr., CAS.

91. Riley to Koebele, May 23, 1892, Koebele Corr., CAS.

92. Riley to Koebele, May 23, 1892, and Riley to Koebele, August 4 and August 31, 1892, both in Koebele Corr., CAS.

93. *ARSA* (1892), 153.

94. *ARSA* (1892), 38.

95. Doutt, "Vice, Virtue, and Vedalia," 122.

96. Handwritten note on *San Francisco Chronicle* article (May 16, 1890), box 10, scrapbook vol. 82, p. 76, SIA, CVR.

97. *L. A. Evening Express*, May 17, 1890, box 10, scrapbook vol. 82, p. 14, SIA, CVR.

98. "A Raid on the University," *Oakland Tribune*, February 31, 1895, box 10, scrapbook vol. 94, p. 124, SIA, CVR.

99. E. O. Essig, *A History of Entomology* (New York and London: Hafner Publishing Co., 1965), 685–87.

100. Morton to Ellwood Cooper, September 22, 1893, NA, RG 16, Entry 15, Morton Letterpress Book, vol. 3 (September 20–November 14, 1893), pp. 19–23, microfilm no. 440; and Mallis, *American Entomologists*, 391.

101. Riley to Koebele, October 2, 1893, Koebele Corr., CAS. Riley expressed surprise at Koebele's intention to resign. It apparently did not occur to him that a salary nearly double what he was receiving constituted a good reason. Riley to Koebele, October 20, 1893, Koebele Corr., CAS.

102. *ARSA* (1891), 235; *ARSA* (1892), 39; and Riley, "Parasitic and Predaceous Insects," 14979.

103. Riley to Koebele, July 3, 1891, Koebele Corr., CAS.

104. Riley to Koebele, July 8, 1891, Koebele Corr., CAS. He also advised Koebele which parasitic species he should consider taking from America to Australia and New Zealand to combat introduced species there (Apple Root Louse, Oyster Shell Bark Louse, Codling Moth, Cabbage Plutella, Red Scale, and Black Scale).

105. *ARSA* (1890), 250–51.

106. Doutt, "Vice, Virtue, and Vedalia," 119; and William L. Bruckart III, "History of Biological Controls," in Pimentel, ed., *Encyclopedia of Pest Management*, 374.

107. C. V. Riley, "Australia," *PRP* 38 (December 21, 1889), 70, box 8, scrapbook vol. 61, pp. 175–76, SIA, CVR.

108. "Statement of Appropriations and Expenditures for the U. S. Department of Agriculture from 1878 to 1892," *ARSA* (1894), 52–53.

109. Caltagirone and Doutt, "History of the Vedalia Beetle Importation," 6. In 1893, in response to the secretary of agriculture's request for the total cost of importing "the lady bug" to California, a clearly exasperated Riley replied that the expenses of Koebele and Webster in 1888 were paid by the Department of State and that Koebele's expenses for the second expedition were paid by the State of California. Morton to Riley, December 23, 1893, NA, RG 16, E 15, Morton Letterpress Book 4 (November 14–December 27, 1893), p. 431, microfilm roll no. 2; and Riley to Morton, December 26, 1893, NA, RG 7, E 3, Box 35, Morton-Moyer File Folder.

Chapter 13

Epigraph. SIA, United States National Museum (hereafter cited as USNM), Accession 10.

1. H. B. Weiss, *The Pioneer Century of American Entomology* (New Brunswick, NJ: self-published, 1936), 155–56, 277. The development of insect collections in America is discussed in Sorensen, *Brethren of the Net*, chap. 3, "Of Cabinets and Collections."

2. LeConte to Baird, May 10, 1859, SIA, Office of the Assistant Secretary, Incoming Correspondence (1850–1877), RU 52, 17:309.

3. William Goetzman, *Exploration and Empire: the Explorer and the Scientist in the Winning of the American West* (New York: Alfred A. Knopf, 1966), 474–75, 495, 497, 502, 527–28; Mike Foster, "Ferdinand Vandiveer Hayden as Naturalist," *American Zoologist* 26 (1986): 343.

4. Powell to Carl Schurz, August 1, 1878, NA, RG 48, Hayden Survey, Letters Received; Worster, *Powell*, 258.

5. Dupree, *Science in the Federal Government*, 84–86.

6. E. F. Rivinus and E. M. Youssef, *Spencer Baird of the Smithsonian* (Washington and London: Smithsonian Institution Press, 1992), 44.

7. Dupree, *Science in the Federal Government*, 85.

8. Smithsonian Institution, *Annual Report* (1858), 30–31.

9. Rivinus and Youssef, *Baird*, 78; and Sorensen, *Brethren of the Net*, 56–57.

10. Dupree, *Science in the Federal Government*, 85.

11. L. O. Howard, "On the History of the Division of Insects of the United States National Museum [unpublished ms]," SIA, RU 55, Box 18, p. 1.

12. Howard, "On the History of the Division of Insects"; and Mallis, *American Entomologists*, 343–48.

13. Sorensen, *Brethren of the Net*, 55–56. The MCZ also obtained the collections of Osten Sacken, Jacob Böll, Jean Théodore Lacordiaire, and Herman Loew.

14. Sorensen, *Brethren of the Net*, 54–56.

15. Riley to Baird, July 2, 1878, SIA, Office of the Secretary, Incoming Correspondence (hereafter cited as OSIC) (1863–79), RU 26, 179: 59.

16. Riley to Baird, May 21, 1879, SIA, OSIC (1879–1882), RU 28, Box 5, 6:2600.

17. Baird to Riley, May 24, 1879, SIA, Office of the Secretary, Outgoing Correspondence (hereafter cited as OSOC) (1865–1891), RU 33, letter 81.

18. The collection has since been renamed the National Entomological Collection.

19. Crossett, *Founders of the Cosmos Club*, 22–24.

20. Rivinus and Youssef, *Baird*, 47–55.

21. On Kennicott's arctic exploration see Morgan B. Sherwood, *Exploration of Alaska, 1865–1900* (New Haven and London: Yale University Press, 1965), 17–24. On Riley's relationship to Kennicott (and also to Kennicott's father, who died in 1863), see Riley, Diary, June 5, July 14, and September 12, 1863, and March 12, 1864, RFR, NAL; and *Rockford Seminary Magazine* 1, no. 2 (February 1873), box 1, scrapbook vol. 9, p. 157, SIA, CVR.

22. Baird to Riley, January 11, 1881, RU 112, Vol. L6, Reel 6, and Riley to Baird, January 13, 1881, OSIC (1879–1882), RU 28, Box 39, 4:18593, SIA.

23. Although the Entomological Commission was officially terminated in June 1881, Riley and Packard continued to issue commission publications up to the early 1890s. Understandably, many people confused Riley's various positions as chief of the Division of Entomology, chief of the Entomological Commission, and curator of insects at the National Museum.

24. Smithsonian Institution, *Annual Report* (1881), 106.

25. USNM, *Proceedings* (1881), Appendix, Circular no. 4.

26. USNM, *Annual Report* (1886), Part 2, 186.

27. Rivinus and Youssef, *Baird*, 120.

28. USNM, *Proceedings* (1881), Appendix, Circular no. 4.

29. USNM, *Proceedings* (1881), Appendix, Circular no. 4.

30. Diary, October 3, 9, 24, 1861, and May 13, 1863, RFR, NAL. In 1863, Emery purchased a more powerful microscope, evidently for Riley's use. Diary, October 23, 1863, RFR, NAL.

31. C. V. Riley, "Queries Answered: Apple Tree Caterpillars," *PF* 34 (September 8, 1866): 152.

32. "The Illinois State Fair," *PF* 34 (October 6, 1866): 218.

33. "The Illinois State Fair," *PF* 36 (October 12, 1867): 226. Riley's description of the Walsh cabinet following Walsh's death in 1869 confirms the count of eight thousand to ten thousand species. C. V. Riley, "The Walsh Entomological Collection," *AE* 2 (December–January, 1869–1870): 93–94.

34. Anonymous, "Prof. Charles V. Riley and His Entomological Museum," *Rockford Seminary Magazine* 1 (February 1873), box 1, scrapbook vol. 9, p. 157, SIA, CVR.

35. "Entomology: The State Bug Shop: Prof. C. V. Riley's Insect Observatory," n.s. (July 5, 1873), box 1, scrapbook vol. 7, pp. 161–62, SIA, CVR.

36. "The St. Louis Fair," *PF* 47 (October 21, 1876): 537; and "The St. Louis Fair," *Western Rural*, October 14, 1876, box 1, scrapbook vol. 9, pp. 161–62, SIA, CVR.

37. Riley, *Seventh Annual Report* (1875), preface, v; and Mary E. Murtfeldt, "A Protest," n.s., box 2, scrapbook vol. 15, p. 98, SIA, CVR. Murtfeld protested against the state's treatment of Riley, and she mentioned the collection in passing. See "Prof. C. V. Riley," *CRW* (April 3, 1878), box 2, scrapbook vol. 15, p. 89, SIA, CVR.

38. Riley emphasized his personal ownership of the collection in an interview in 1873: "The bugs belong to Prof. Riley. *They are his private property* [emphasis in original]," "Entomology: The State Bug Shop."

39. Riley to Baird, July 11, 1885, SIA, Assistant Secretary, United States National Museum (hereafter cited as ASUSNM) (1875–1902), RU 201, Box 4.

40. "Glover's Museum," *PF* 36 (August 24, 1867): 120.

41. Riley to Baird, October 23, 1885, SIA, RU 305, Box 63, Accession 16738.

42. Baird to Riley, November 1, 1885, SIA, RU 305, Box 63, Accession 16735.

43. Riley to Baird, October 23, 1885, SIA, RU 305, Box 63, Accession 16738.

44. C. V. Riley, "Division of Insects," USNM, *Annual Report*, part 2 (1886), 196.

45. "The Entomological Club at the AAAS," *CE* 11 (September 1879): 163–76.

46. Comstock had not made a new appointment as no additional staff, except for Comstock's wife, was hired in 1879–1880. *ARCA* (1879); *ARCA* (1880); *Comstocks of Cornell*, 111.

47. J. H. Comstock, "Report of the Entomologist," *ARCA* (1879), 186.

48. The next to last column, No. of Species, represents the maximum possible number of species, because many duplicates would occur in such a set of large collections.

49. Baird to LeConte, December 26, 1881, LeConte Coll., ANSP.

50. Worster, *Powell*, 400, 405.

51. Rivinus and Youssef, *Baird*, 181.

52. Riley to Baird, July 16, 1885, SIA, ASUSNM (1875–1902), RU 201, Box 4.

53. Mallis, *American Entomologists*, 315.

54. C. L. Marlatt, "The Entomological Club," *CE* 35 (March 1903): 53–54.

55. Comstock recognized the problem posed by staff-owned private collections, and he forbade members of the Division of Entomology to add specimens from work-related activities to their private collections. *Comstocks of Cornell*, 117.

56. Smith to Riley, undated [1887?], SIA, Temp. Loc. No. 2–4, Box 1. On another occasion, Riley delayed one of Smith's publications on Lepidoptera, giving as grounds his desire "to check and somewhat curb the species grinding mania which had materially interfered with your usefulness as an Assistant Curator." Riley to Smith, October 29, 1887, Division of Insects, Incoming Correspondence (1878–1906), SIA, RU 138, Box 4.

57. Smith to Riley, May 17, 1887, and Smith to Riley, May 22, 1887, both in Division of Insects, Incoming Correspondence (1878–1906), SIA, RU 138, Box 1.

58. Riley to Smith, July 30, 1882, in Division of Insects, Incoming Correspondence (1878–1906), SIA, RU 138, Box 4.

59. Riley to Smith, October 25, 1887, in Division of Insects, Incoming Correspondence (1878–1906), SIA, RU 138, Box 4.

60. Smith to Langley, March 18, 1889, SIA, ASUSNM (1875–1902), RU 201, Box 4.

61. Mallis, *American Entomologists*, 315–20.

62. USNM, *Annual Report* (1886), 166.

63. USNM, *Annual Report* (1894), 58.

64. Mallis, *American Entomologists*, 260; and Riley to Casey, January 27, 1892, Thomas Lincoln Casey Papers (1870–1897), SIA, RU 7134, Box 2, Folder R.

65. USNM, *Annual Reports* (1883, 1884, 1885, 1886, 1887, 1890, and 1891).

66. Peale to J. Rhees, October 9, 1884, OSOC (1865–1891), RU 33, vol. 90, p. 239; and Riley to Baird, November 15, 1884, OSIC (1882–90), SIA, RU 30, Box 10:281.

67. USNM, *Annual Report* (1886), 188.

68. USNM, *Annual Report* (1883), 243.

69. USNM, *Annual Report* (1887), 119.

70. "Collected Notebooks of Entomologists, circa 1881–1931," RU 7135, SIA.

71. In one letter to Koebele, Riley identified eleven specimens of parasites on the basis of specimens he examined in the National Insect Collection. Riley to Koebele, September 9, 1890, Koebele Corr., CAS.

72. B. D. Walsh and C. V. Riley, "Ticks and Texas Fever," *AE* 1 (October 1868): 28.

73. Special Collections, NAL. Riley's microscope, his desk, and other office equipment are also on display at the NAL.

74. [Editorial], *Science* 4 (December 19, 1884): 540.

75. C. H. Fernald, "On the Care of Entomological Museums," *Science* 5 (January 9, 1885): 25.

76. C. V. Riley, "The Collection of Insects in the National Museum," *Science* 5 (March 6, 1885): 188.

77. Riley, "The Collection of Insects," 188.

78. *Comstocks of Cornell*, 115.

79. Riley, "Insecticides for the Protection of Cotton," 241. See also Riley's deprecation of Hagen, the "eminent entomologist," and his yeast-fermented insecticide in USEC, *Fourth Report* (1885), 188–90.

80. Mallis, *American Entomologists*, 248, 384.

81. Riley to Goode, January 18, 1894, SIA, ASUSNM (1875–1902), RU 201, Box 4.

82. USNM, *Annual Report* (1894), 58.

83. Riley to Goode, August 2, 1893, SIA, ASUSNM (1875–1902), RU 201, Box 4.

84. Riley to Herman Strecker, December 21, 1894, SIA, RU 139, Box 3; and Riley to Scudder, May 5, 1894, Scudder Corr., MSB.

85. Riley to Goode, November 13, 1894, SIA, ASUSNM (1875–1902), RU 201, Box 4.

86. Riley to Hilgard, February 19, 1885, Hilgard Family Papers, Bancroft Library, University of California, Berkeley.

87. J. M. Aldrich, "The Division of Insects in the United States National Museum," Smithsonian Institution, *Annual Report for 1919* (Washington, DC: GPO, 1921), 368.

88. Anonymous, USNM Entomological Collections, https://entomology.si.edu/Collections.html (accessed March 22, 2018).

Chapter 14

1. C. V. Riley, "Some Interrelations of Plants and Insects," *Proc. BSW* 7 (May 1892): 81–104; and C. V. Riley, "Further Notes on Yucca Insects and Yucca Pollination," *Proc. BSW* 8 (June 1893): 41–54. Riley's discoveries regarding Yucca Moths, mimicry in butterflies, and other "non-economic" species would seem to qualify him as one of William Leach's "Butterfly People" rather than a killer of insects. Leach, *Butterfly People*, 253.

2. *Boston Transcript* (January 12, 1892), NAL, MC 143, Series IV, Box 4, Folder 120 "Clippings"; Riley's lecture notes in NAL, MC 143, Series II, Box 26, "Lowell Lectures"; and Flack, *Desideratum in Washington*, 158–59.

3. See clipping, n.s., n.d., regarding *IL* in box 10, scrapbook vol. 76, p. [10], SIA, CVR.

4. On Riley's preparations for the exhibition, see "The Paris Exposition Next Year," *CRW*, October 25, 1888; and "American Agriculture at the Paris Exposition," *CRW*, November 1, 1888, both in box 10, scrapbook vol. 74, SIA, CVR.

5. *Rural New Yorker*, July 12, [1889]; and clipping from *Otar* [?], September 30, 1889, both in box 10, scrapbook vol. 74, SIA, CVR.

6. Riley to Howard, March 21, 1895, Records of the Bureau of Entomology and Plant Quarantine, NA, RG 7, Entry 3, Box 39. Riley's exhibit was a part of the overall Division of Entomology exhibit prepared by the staff following his resignation in April 1894. Riley told reporters the Columbian Exhibition was the finest he had attended since the Crystal Palace Exposition in London in 1851. He regretted, however, that the physical structures housing the Chicago exposition were to be removed rather

than forming a permanent forum, as the Crystal Palace structures had been. He also objected to the high admission charge that, he feared, discouraged common farmers and laborers from attending. "Beyond Description—Is What Prof. Riley Says of World's Fair," *Evening Star*, July 9, 1893, box 13, scrapbook vol. 91, p. 6, SIA, CVR.

7. See chap. 8.

8. Riley to Westwood, December 1886, cited in Audrey Z. Smith, *A History of the Hope Entomological Collections in the University Museum Oxford: With Lists of Archives and Collections* (Oxford: Clarendon Press, 1986), 46; and R. McLachlan, "[Riley] Obituary," *Ent. Mon. Mag.* 31 (1895): 270.

9. See Riley's plan to write a text for schools and colleges in C. V. Riley, "Introduction to Entomology: Its Relations to Agriculture and its Advancement," *Fifth Annual Report* (1873), 5. Howard stated that as early as 1880 Riley contemplated issuing a revised edition of the Missouri Reports but later decided to write a textbook of entomology instead. "Charles V. Riley, Ph.D." *Proc. ESW* 3 (October 6, 1896): 297.

10. See, for example, the London Purple Co., Ltd., advertisement that quoted supposed endorsements of its product by Riley, Charles E. Bessey, and A. J. Cook in *CRW* 34 (June 2, 1881): 176.

11. C. V. Riley, "The Cotton Worm," written in Oxford, Mississippi, October 8, 1879, to editor, *News* (Oxford, MS), October 24, 1879, box 5, scrapbook vol. 39, pp. 100–101, SIA, CVR.

12. *St. Louis Globe*, July 8, 1874, box 1, scrapbook vol. 9, p. 71, SIA, CVR; and "A Satisfactory Grasshopper Machine," *SciAm* 37 (September 15, 1877): 169. The Riley locust catcher is pictured in *ARCA* (1883), plate IX.

13. Report of the meeting of the National Grange, Washington, DC, *PRP* (December 3, 1881), box 6, scrapbook vol. 47, p. 86, SIA, CVR.

14. Riley to Lamar, January 18, 1887, box 15, Outgoing Correspondence, SIA, CVR.

15. Riley to John B. Cotton, October 22, [1890], 4;27, SIA, CVR; Riley to Morton, May 29, 1894 [referring to Riley's earlier assurance to McKnight that he would be paid], NA, RG 7, Entry 3, Morton Letterpress Book; Morton to Riley, May 29, 1894, [declining payment of $1,500 to McKnight], NA, RG 7, Entry 3, Morton Letterpress Book.

16. Riley to Morton, May 31, 1894, NA, RG 7, Entry 3, Box 5, no. 50. As noted earlier, the judge ruled in favor of Barnard in 1892. When Morton refused to pay McKnight from departmental funds, Riley and McKnight continued, unsuccessfully, to press for payment. In 1897, two years after Riley's death, McKnight was still attempting to get reimbursement for his services. The outcome is not known. Riley to Morton, May 15 [or 16], 1894, no. 304; May 29, no. 497; and May 31, 1894, no. 50, all in NA, RG 7, Entry 3, Box 5; and McKnight to Congressman John Dalzell, November 20, 1897, NA, RG 16, Entry 8, Box 8.

17. "Improved Method of Using Hydrocyanic Acid Gas," *California Fruit Grower*, February 15, 1890, box 10, scrapbook vol. 78, pp. 11–13, SIA, CVR; and "Prof. Riley

on the Gas Treatment Patent," *PRP* (July 25, 1891), in box 10, scrapbook vol. 78, p. 32, SIA, CVR.

18. C. V. Riley, "Economic Entomology," *New York Tribune*, June 28, 1890, box 9, scrapbook vol. 71, pp. 25–26, SIA, CVR.

19. Philadelphia [name missing], July 28 [1891], box 11, scrapbook vol. 84, p. 132, SIA, CVR. See also Riley's rebuttal, "Dr. C. V. Riley," *SciAm* 65 (August 22, 1891), NAL, MC 143, Series IV, Box 4.

20. Clipping (source unclear) regarding charges in the *Sun* (July 20, 1891), NAL, MC 143, Series IV, Box 4, Folder 122.

21. "Meeting of the American Association," *SciAm* 65 (September 12, 1891).

22. See chap. 9.

23. *Rural New Yorker* (June 13, 1885), box 9, scrapbook vol. 66, p. 128, SIA, CVR; and "Kerosene and Soap Mixture," *Pardy's Fruit Recorder* (July 1, 1885) in box 8, scrapbook vol. 63, p. 12, SIA, CVR.

24. "Important Modifications of the Cyclone Nozzle," n.s. (June 1889), box 10, scrapbook vol. 73, p. 175, SIA, CVR.

25. *ARSA* (1894), 159.

26. Riley to Morton, May 31, 1894, NA, RG 7, Entry 3, Box 35, Morton-Moyer File Folder (microfilm).

27. See chap. 9.

28. Roy V. Scott, *The Reluctant Farmer: The Rise of Agricultural Extension to 1914* (Urbana, Chicago, and London: University of Illinois Press, 1970), 32.

29. Dupree, *Science in the Federal Government*, 170.

30. C. V. Riley, "Agricultural Advancement in the U. S.," *Journal of the American Agricultural Association* 1 (April 1881): 47–53.

31. Riley, "Agricultural Advancement in the U. S.," 47–49.

32. Riley, "Agricultural Advancement in the U. S.," 50–51, 53.

33. For background on the Hatch Act, see Alan I. Marcus, *Agricultural Science and the Quest for Legitimacy: Farmers, Agricultural Colleges, and Experiment Stations, 1870–1890* (Ames: Iowa State University Press, 1985); Scott, *Reluctant Farmer*, 32; Dupree, *Science in the Federal Government*, 170; and Rossiter, "Organization of the Agricultural Sciences," 215.

34. Dupree, *Science in the Federal Government*, 171–72.

35. Gould P. Colman, "Pioneering in Agricultural Education: Cornell University, 1867–1890," *AgH* 36 (October 1962): 205; and *Comstocks of Cornell*, 160.

36. Dupree, *Science in the Federal Government*, 172.

37. Edwards to Lintner, April 26, 1878, Lintner Corr., NYSM; and Sorensen, *Brethren of the Net*, chap. 3, "Cabinets and Collections."

38. *Comstocks of Cornell*, 92.

39. Riley to Howard, June 21 and August 5, 1889, box 16, Outgoing Correspondence, Paris, SIA, CVR. On the Entomological Club see C. L. Marlatt, "The Entomological

Club of the American Association for the Advancement of Science," *CE* 35 (March 1903): 53–54; and Sorensen, *Brethren of the Net*, 90, 133, 248–52.

40. Riley to Howard, June 21, 1889, box 16, Outgoing Correspondence, Paris, SIA, CVR.

41. "Special Notes," *IL* 1 (January 1889): 201–2.

42. "Special Notes," *IL* 1 (January 1889): 201–2.

43. Riley to Howard, June 3, 1889, box 16, Outgoing Correspondence, Paris, SIA, CVR.

44. "The Association of Official Economic Entomologists," *IL* 2 (September 1889): 87–88.

45. Forbes to Riley, [September or October 1890], Stephen A. Forbes Correspondence, University of Illinois, Urbana, Archives (hereafter cited as Forbes Corr.).

46. "Association of Economic Entomologists: First Annual Meeting [November 1889, Department of Insects, USNM]," *IL* 2 (November 1889): 181.

47. Forbes to Riley, [May?] 26, 1885, Box 2, No. 378, Forbes Corr.; and Croker, *Forbes*, 115.

48. *IL* 3 (January 1891): 204.

49. *IL* 4 (August 1892): 410.

50. A decade later, in 1903, Schwarz, William Ashmead, and other Washington-based entomologists mounted an unsuccessful effort to revive the Entomological Club. Marlatt, "The Entomological Club of the American Association for the Advancement of Science," including a review of the club's history by E. A. Schwarz entitled "A Sketch of the History of the Entomological Club of the American Association," *CE* 35 (March 1903): 54–58.

51. Riley, presidential address, "The Outlook for Applied Entomology," *IL* 3 (January 1891): 209.

52. Riley to Howard, May 18, 1889, box 16, Outgoing Correspondence, Paris, SIA, CVR; and "The Assistant Secretary of Agriculture," *Rural New Yorker*, April 2, 1889, box 10, scrapbook vol. 75, p. 57, SIA, CVR.

53. For example, Riley toasted Willits at his sixtieth birthday party, "Those were unique toasts with comments on 'Bugs and Humbugs,'" *Washington Post*, April 28, 1890, box 11, scrapbook vol. 80, p. [70], SIA, CVR. See also *Comstocks of Cornell*, 118, 157, 122–23.

54. Henry F. Graf, *Grover Cleveland* (New York: Time Books, 2002), 114.

55. Howard, *History of Applied Entomology*, 93.

56. *Chicago Times*, Washington, DC, bureau, February 18, 1893, 14:SB H, SIA, CVR.

57. *American Farmer*, July 15, 1893, in 14:SB H, SIA, CVR.

58. "For the Farmers," *Washington Evening Star*, March 22, 1893, 14:SB H-11, SIA, CVR.

59. "Asst. Secretary of Agriculture," *American Farmer*, July 15, 1893, 14:SB-H, SIA, CVR.

60. *American Farmer*, January 1, 1894, 14:SB-H, SIA, CVR.

61. "Does Not Want Appointment," *Herald* (Washington, DC), August 17, 1893, box 13, scrapbook vol. 95, p. 196, SIA, CVR.

62. Graf, *Grover Cleveland*, 114–17.

63. Morton to Riley, March 25, 1893, NA, RG 16, Entry 15, Morton Letterpress Book (Microfilm 440, roll 1).

64. Riley to Morton, April 1, 1893, NA, RG 16, Entry 8, Box 4, Moroney-Morton File Folder.

65. Morton to Riley, August 11, 1893, NA, RG 16, Entry 15, Morton Letterpress Book (Microfilm 440, roll 1).

66. Morton to Riley, December 12, 1893, NA, RG 16, Entry 15, Morton Letterpress Book (Microfilm 440, roll 2).

67. Riley to Lintner, December 1, 1893, Lintner Corr., NYSM. *The Bibliography of American Entomology* (1889) lists 2,033 publications by Riley up to that time, including 478 authored jointly with Walsh.

68. Dupree, *Science in the Federal Government*, 172–74.

69. Riley to Packard, February 6, 1881, MCZ.

70. *Comstocks of Cornell*, 162; and Pamela Henson, "The Comstock Research School in Evolutionary Entomology," *Osiris* 8 (1993), 159–60.

71. Mallis, *American Entomologists*, 136.

72. "The Descent of the Lepidoptera," presented at the AAAS in 1893, and "Evolution and Taxonomy," both in the *Wilder Century Book*; *Comstocks of Cornell*, 186–87; and Henson, "The Comstock Research School in Evolutionary Entomology," 160–67.

73. Comstock's salary increased from $2,500 in 1888 to $3,000 in 1890. In the latter year, Jordan offered Comstock $4,000 to become the chair of the new entomology department at Stanford University. *Comstocks of Cornell*, 156, 169, 172.

74. Riley to Howard, August 19, 1887, written onboard the *City of Rome*, August 19, 1887, box 16, Outgoing Correspondence, Paris, SIA, CVR.

75. *Comstocks of Cornell*, 159.

76. *Comstocks of Cornell*, 117, 170.

Chapter 15

1. Riley to Morton, April 27, 1894, with attached memoranda dated February 12, 1894. The letter but not the memoranda is in Riley to Morton, April 27, 1894, box 27, SIA, CVR. The letter with the memoranda is in NA, RG 16, E 8, Dabney Letterpress Book, Box 4.

2. Dabney to Riley, April 23, 1894, NA, RG 16, E 8, Dabney Letterpress Book, vol. 18, pp. 436–37.

3. Memorandum (February 12) in Riley to Morton, April 27, 1894, box 27, SIA, CVR; and Morton to Riley, May 3, 1894, NA, RG 7, E 3, Box 35, Morton Letterpress Book, vol. 7, pp. 323–25 (microfilm). Riley explained that no mention was made of

Hubbard's travel because these expenses were paid by the Montserrat Company. He also explained that delays due to unreliable transportation had prevented Hubbard from completing his report within the time set in the memorandum.

4. Riley to Morton, April 26, 1894, NAL, MC 143, Series I, Box 1; and Riley to Morton, May 3 and May 5, 1894 (reply to Morton April 27), NA, RG 7, E 3, Box 35 (Morton-Moyer file), microfilm 440, Roll 4 (vol. 7–8), Morton Letterpress Book, pp. 323–35 (microfilm).

5. Riley to Morton, February 7, 1895, NA, RG 16, E 8, Box 7; and Morton to Riley, February 13, 1895, NA, RG 16, Morton Letterpress Book.

6. Riley to Packard, May 9, 1894; and Riley to Scudder, May 9, 1984, MCZ.

7. Morton to Howard, May 25, 1894, declining invitation for June 1 (not sent?), NA, RG 7, E 3, Box 35, Morton Letterpress Book, vol. 7, pp. 468–70 (microfilm); Edward B. Southwick to Riley, responding to his invitation, in NAL, MC 143, Series I, Box 1; and *Washington Star*, June 4, 1894, in NAL, MC 143, Series IV, Box 4, "News Clippings."

8. Helen to Papa, August 5, 1894, Folder 7C, RFR, NAL.

9. David Miller, "Apropos C. V. Riley," *Pan-Pacific Entomologist* 22 (1946): 28–29.

10. *ARSA* (1894), 5.

11. See endorsements of Howard in Dabney correspondence dated May 19 to May 23, 1894, NA, RG 16, E 8, Dabney Letterpress Book, vol. 19 (April 24–May 25, 1894).

12. Riley to Morton, February 7, 1895, NA, RG 16, E 8, Moroney-Morton Folder, Box 7; In an interview with a San Francisco reporter, Riley said he was considering purchasing a small fruit farm near Pasadena as a retirement home. Nothing more was heard of this project. "Dr. Riley Could Tell Much," *San Francisco Examiner*, January 21, 1895, 13:95, SIA, CVR.

13. R. H. Jesse [president of the university], "The College of Agriculture," *CRW* 47 (November 22, 1894), 8. Riley's lectures were scheduled from January 7 to January 19, 1895.

14. "[Riley] Obituary," *Entomologist's Record and Journal* 7 (1895): 72; and R. McLachlan, "[Riley] Obituary," *Entomologist's Monthly Magazine* 31 (1895): 269–70.

15. Howard, "Charles Valentine Riley," 111; and Riley Obituary, *CRW*, September 19, 1895.

16. Compiled from Certificate of Death and the following obituaries: *Washington Post*, September 15, 1895; *Washington Globe*, September 14, 1895; *Washington Evening Times*, September 14, 1895; *Washington Star*, September 14 and September 17, 1895; and *CRW*, September 19, 1895, all in NAL, MC 143, Series IV, Box 4, 142–53.

17. "Scientists Attend His Funeral," *Washington Post*, September 17, 1895.

18. Their infant son, Harold Gottlieb (1884–1885) was the first family member interred in the family plot.

19. Riley Residences Folder 9B, RFR, NAL; and grave markers in Glenwood Cemetery.

20. Janet Smith's notes of the Conzelman family in Conzelman Family Folder 7A

and Riley Residences Folder 9B, both in RFR, NAL; and grave markers in the family plot in Glenwood Cemetery.

21. Helen's daughter, Emilie Wenban-Smith Brash, died in 2015. She had no descendants.

22. Howard, *History of Applied Entomology*, 56–57.

23. Howard, *History of Applied Entomology*, 90–91.

24. Mallis, *American Entomologists*, 69–79.

25. L. O. Howard, *Fighting the Insects: The Story of an Entomologist* (New York: Macmillan Company, 1933), 29.

26. Henshaw, *Bibliography of American Economic Entomology*.

27. Nina to Emilie, February 4, 1897, Folder 7C, RFR, NAL.

28. Cathryn's notes on p. 95 of Howard, *History of Applied Entomology* (1930), Library, Field Museum of Natural History, Chicago.

29. Edward H. Smith and Janet R. Smith, "Charles Valentine Riley: The Making of the Man and His Achievements," *AE* 42 (Winter 1996): 228–38; W. Conner Sorensen and Edward H. Smith, "Charles Valentine Riley: Art Training at Bonn, 1858–1860," *AE* 43 (Summer 1997): 92–104; Carol A. Sheppard and Edward H. Smith, "Entomological Heritage," *AE* 43 (Fall 1997): 142–46; Carol A. Sheppard and Richard A. Oliver, "Yucca Moths and Yucca Plants: Discovery of 'the Most Wonderful Case of Fertilization,'" *AE* 50 (Spring 2004): 32–46; W. Conner Sorensen, Edward H. Smith, Janet Smith, and Yves Carton, "Charles V. Riley, France, and *Phylloxera*," *AE* 54 (Fall 2008): 134–49; Gene Kritsky, "Entomological Reactions to Darwin's Theory in the Nineteenth Century," *Ann. Rev. Ent.* 53 (2008): 345–60; Gene Kritsky, "Darwin, Walsh and Riley: The Entomological Link," *AE* (Summer 1995): 89–95; Edward Smith and Gene Kritsky, "Charles Valentine Riley: His Formative Years," *AE* 57 (Summer 2011): 74–80; and Sorensen, *Brethren of the Net*.

30. Alfred G. Wheeler Jr., E. Richard Hoebecke, and Edward H. Smith, "Charles Valentine Riley: Taxonomic Contributions of an Eminent Agricultural Entomologist," *AE* 56 (Spring 2010): 14–30.

31. Wheeler et al., "Taxonomic Contributions," 15–16; and C. V. Riley, "The Number of Insects in the World," *Nature* (February 23, 1893), box 10, scrapbook vol. 78, pp. 130–31, SIA, CVR; the figure of 10 million species appeared a year earlier in C. V. Riley, "Directions for Collecting and Preserving Insects," USNM, Bulletin No. 39 (1892), 7. Modern estimates are summarized and cited in: Nigel E. Stork, James McBroom, Claire Gely, and Andrew J. Hamilton, "New approaches narrow global species estimates for beetles, insects, and terrestrial arthropods," *Proceedings of the National Academy of Sciences* 112 no. 24 (2015): 7519–7523.

32. Wheeler et al., "Taxonomic Contributions," 18–19.

33. Wheeler et al., "Taxonomic Contributions," 19–20.

34. Wheeler et al., "Taxonomic Contributions," 21.

35. C. V. Riley, "The Bark Lice of the Apple-Tree," *First Annual Report* (1868), 13.

36. C. V. Riley, "Hackberry butterflies," *Sixth Annual Report* (1874), 14.

37. Riley to Forbes, April 10, 1885, Box 3, Incoming Corr., Forbes Corr.

38. Riley to Lintner, January 28, 1888, Lintner Corr., NYSM.

39. C. V. Riley, "Canker Worms," *Seventh Annual Report* (1875), 80–90.

40. C. V. Riley, "Canker Worms," *Eighth Annual Report* (1876), 18–19.

41. C. V. Riley, "The Colorado Potato Beetle," *Eighth Annual Report* (1876), 2.

42. C. V. Riley, "The Colorado Potato Beetle," *Eighth Annual Report* (1876), 10.

43. Croker, *Forbes*, 48, 62–66.

44. Croker, *Forbes*, 68–73, 79–83, 85–86, 111.

45. Croker, *Forbes*, 71, 84–85, 109–14.

46. Croker, *Forbes*, 111.

47. Croker, *Forbes*, 115.

48. James Fletcher, "Charles Valentine Riley," *CE* 27 (October 1895): 274.

Bibliography

The most important sources for Riley's early years in England, France, Germany, Kankakee, and Chicago are his Journal (1850–57), his Diary (1861–65), and family correspondence (1860–99), designated here as the Riley Family Records. These records were located up to the time the manuscript was completed (April 2018) at the residence of Janet R. Smith, in Asheville, NC. These records were then transferred to the Special Collections of the National Agricultural Library, Beltsville, Maryland (see RFR below). Riley's granddaughter, Emilie Wenban-Smith Brash (deceased 2015), of Headley Down, Hampshire, England, made these documents available. Riley used the term "Journal" during the years 1850–57 (in England and France) and "Diary" for the years 1861–65 (in America), and the citations correspond to Riley's usage.

The most important source for Riley's career during his Missouri and Washington, DC, years are his scrapbooks, comprised of more than 113 volumes, located in the Smithsonian Institution Archives (RU 7076, see manuscripts below). The scrapbooks consist of notices cut from newspapers and periodicals relating to Riley or authored by him (together with some correspondence) that are pasted onto the pages of Commissioner of Agriculture Reports with only perfunctory organization according to date or topic. Shortly before submitting the manuscript for this biography to the University of Alabama Press, Conner Sorensen, together with Tad Bennicoff at the Smithsonian Institution Archives, confirmed that notations in Samuel Henshaw, *Bibliography of . . . American Economic Entomology* (1890), which Riley initiated, refer to clippings in the scrapbooks. This cross-referencing is an indication of Riley's passion for reducing masses of information into order, whether in the realm of insect form and function or bibliographic information.

Digitized and Internet Databases

Common Names of Insects Database. Entomological Society of America, Annapolis, Maryland. http://www.entsoc.org/common-names/.

Agricultural and Environmental Science Database. ProQuest.

America's Historical Newspapers: Early American Newspapers (1690–1922). Readex.

Nineteenth-Century US Newspapers. Gale.

FAOSTAT 2016, latest statistics on unreeled cocoons from 2013. Accessed October 31, 2016. http://faostat.fao.org/.

Riley Insect Wall Charts. Special Collections website. Kansas State University, Manhattan, Kansas. http://www.lib.k-state.edu/depts/spec/archives.html/.

Bibliographies, Reference Works, and Publications in Series

American Entomologist: An Illustrated Magazine of Popular and Practical Entomology. Volumes 1–2 (1868–1870), Volume 3 (1880).

Bulletin de la Société centrale d'agriculture de l' Hérault (1870–1885).

Bulletin de la Société entomolique de France (1870–1885).

Bulletin of the Philosophical Society of Washington. Volumes 1–13 (1871–1894).

Canadian Entomologist. Volumes 1–27 (1868–1895).

Commissioner [beginning in 1889, Secretary] of Agriculture, *Annual Report.* Washington, DC: Government Printing Office, 1862–1895.

Darwin, Charles. *The Correspondence of Charles Darwin.* Cambridge: Cambridge University Press. The correspondence is being published in chronological order. Currently volumes 1–25 (through 1877) are completed.

Fitch, Asa. *First [–Fourteenth] Report on the Noxious, Beneficial, and other Insects of the State of New York.* Albany: State Agricultural Society, 1855–1872. Title varies. For publication history of Fitch's *Reports,* see Appendix A, "Entomological Publications by Dr. Asa Fitch," in J. K. Barnes, *Asa Fitch and the Emergence of American Entomology, with an Entomological Bibliography and a Catalog of Taxonomic Names and Type Specimens.* New York State Museum Bulletin 46. Albany: New York State Museum, 1988.

Gilbert, Pamela. *A Compendium of the Biographical Literature on Deceased Entomologists.* London: British Museum (Natural History), 1977.

Henshaw, Samuel, ed. *Bibliography of the More Important Contributions to American Economic Entomology: Parts 1, 2, and 3, the more important writings of Benjamin Dann Walsh and Charles Valentine Riley.* Washington, DC: Government Printing Office, 1890.

Illinois State Agricultural Society. *Transactions.* Volumes 4–8 (1859–1870). Springfield: State Printer, 1861–1871.

Illinois State Horticultural Society. *Transactions* (1861–1870). Published in Illinois State Agricultural Society, *Transactions,* Volumes 4–8 (1859–1870). Springfield: State Printer, 1861–1871.

Insect Life: Devoted to the Economy and Life Habits of Insects, Especially in their Relations to Agriculture. Volumes 1–7. Washington, DC: US Department of Agriculture, 1888–1895.

LeBaron, William. *First [–Fourth] Annual Report on the Noxious Insects of the State of Illinois.* Springfield: State Printer, 1871–1874.

Lintner, Joseph A. *First [–Second] Annual Report on the Injurious and other Insects of the State of New York.* Albany: Weed, Parsons, 1882–1885.

Missouri State Board of Agriculture. *First [–Sixth] Annual Report.* Jefferson City: State Printer, 1865–1870.

Missouri State Horticultural Society. *Proceedings.* Jefferson City: State Printer, 1866–1871.

Packard, A. S., ed. *Record of American Entomology for the Year 1868 [–1873]*. Salem, MA: Naturalist's Book Agency, Essex Institute Press, 1869–1874.

Philosophical Society of Washington. *Bulletin*. Volumes 1–15 (1871–1913).

Practical Entomologist. Volumes 1–2. Philadelphia, PA: Entomological Society of Philadelphia, 1865–1867.

Prairie Farmer. Volumes 1–29 (1841–1870).

Proceedings of the American Association for the Advancement of Science. Volumes 19–42 (1870–1894).

Proceedings of the Entomological Society of Philadelphia. Volumes 1–6 (1863–1867).

Psyche: Organ of the Cambridge Entomological Club. Cambridge, Massachusetts. Volumes 1–3 (1874–1882).

Riley, Charles V. *First [–Ninth] Annual Report on the Noxious, Beneficial, and Other Insects of the State of Missouri*. Jefferson City: State Printer, 1869–1877.

Riley, C. V. "General Index and Supplement to the Nine Reports of the Insects of Missouri." Bulletin No. 6, US Entomological Commission, Department of the Interior. Washington, DC: Government Printing Office, 1881.

Riley, Charles V., ed. *Report on Agriculture, volume 5 of Reports of the United States Commissioners to the Universal Exposition of 1889 at Paris*. Washington, DC: Government Printing Office, 1891.

Smithsonian Institution. *Annual Report* (1858–1895). Washington, DC: Smithsonian Institution Press, 1858–1895.

US Entomological Commission. *First Annual Report for the Year 1877 relating to the Rocky Mountain Locust and the best means of preventing its injuries and of guarding against its invasions*. Washington, DC: Government Printing Office, 1878.

US Entomological Commission. *Second Report for the Year 1878 relating to the Rocky Mountain Locust and the best means of preventing its injuries and of guarding against its invasions*. Washington, DC: Government Printing Office, 1880.

US Entomological Commission. *Third Report of the United States Entomological Commission relating to the Rocky Mountain Locust, the Western Cricket, the Army Worm, Canker Worms, and the Hessian Fly; together with descriptions of Larvae of Injurious forest Insects, Studies on the Embryological Development of the Locust and of other Insects, and on the systematic Position of the Orthoptera in Relation to Other Orders of Insects, with maps and Illustrations*. Washington, DC: Government Printing Office, 1883.

US Entomological Commission. *Fourth Report . . . on the Cotton Worm. [and the] Boll Worm*. Edited by Charles V. Riley, PhD. Washington, DC: Government Printing Office, 1885.

US National Museum, *Annual Report* (1880–1895). Washington, DC: United States National Museum, 1880–1895.

US National Museum, *Proceedings* (1880–1895). Washington, DC: n.p., 1880–1895.

Walsh, Benjamin D. *First Annual Report on the Noxious Insects of the State of Illinois*. Appendix to the *Transactions of the Illinois State Horticultural Society for 1867*, volume 1 (1868). Springfield: State Printer, 1868 [Separate printing by the Prairie Farmer Company, Chicago, 1868].

Zoological Record. Volumes 1–12 (1864–1880). London: J. V. Voorst, 1864–1880.

Manuscripts

Albany, New York

New York State Museum.

J. A. Lintner Correspondence.

Asa Fitch Correspondence.

Beltsville, Maryland

National Agricultural Library, Special Collections.

Charles Valentine Riley Collection, Manuscript Collection No. 143.

Riley Family Records.

C. V. Riley, Bonn Art Book, 1858–1860.

C. V. Riley Diary, 1861–1865.

C. V. Riley Journal, 1850–1857.

Riley family correspondence and related records, ca. 1800–1950.

Edward H. Smith and Janet R. Smith correspondence with genealogists, archives, and other individuals and institutions.

Berkeley, California

University of California, Berkeley, Bancroft Library.

Hilgard Family Papers, C-B 972 (Eugene W. Hilgard letters).

Boston, Massachusetts

Museum of Science, Boston (Formerly Boston Society of Natural History).

Samuel Hubbard Scudder Correspondence.

Cambridge, Massachusetts

Museum of Comparative Zoology Archives, Ernst Mayr Library, Harvard University Museum of Natural History.

Correspondence of Entomologists.

Chicago, Illinois

Field Museum of Natural History.

Charles Valentine Riley Collection.

Eschbach, Germany

W. Conner Sorensen files.

Patricia Wickham Mills, "The Bug Invasion." Unpublished manuscript, September 14, 2010.

Correspondence with representatives of Bellefontaine Cemetery, St. Louis, regarding graves of George Riley and Gottlieb Conzelman.

Ithaca, New York

Cornell University.

Cornell University Library, Division of Rare and Manuscript Collections.

John Henry and Anna Botsford Comstock Papers, 1833–1955.

Department of Entomology, Correspondence (historical).

Church of the Latter Day Saints, Burleigh Drive, Ithaca.

Genealogical records, microfilm.

Kankakee, Illinois

Kankakee County Recorder of Deeds and Tax Assessor.

Lincoln, Nebraska

University of Nebraska Archives.

Charles E. Bessey Collection. Record Group 12.

London, England

Public Records (birth, death, marriage, baptism), Census, Church Directories, prison records, and other records searched by E. P. Stanton, genealogist.

Manhattan, Kansas

Kansas State University Library.

Morse Department of Special Collections. Charles V. Riley Insect Wall Charts.

History Index.

Philadelphia, Pennsylvania

Academy of Natural Sciences of Philadelphia.

Entomological Society of America Collection.

John L. LeConte Collection.

American Philosophical Society.

LeConte Collection (John L. LeConte and his father John Eatton LeConte).

St. Louis, Missouri

Bellefontaine Cemetery, St. Louis. Grave markers and related documentation for George Riley and Gottlieb Conzelman.

San Francisco, California

California Academy of Sciences

Albert Koebele Correspondence, 1881–1893.

Urbana, Illinois

University of Illinois, Urbana, Archives

Stephen A. Forbes Correspondence.

Washington, DC

Smithsonian Institution Archives

Assistant Secretary, U. S. National Museum (1875–1902). Record Unit 201.

Assistant Secretary in Charge of the U.S. National Museum (Richard Rathbun) (1897–1918). Record Unit 55.

Charles Valentine Riley Papers, 1866–1895 and undated. Record Unit 7076.

Collected Notebooks of Entomologists (circa 1881–1931). Record Unit 7135.

Division of Insects, Incoming Correspondence (1878–1906). Record Unit 138.

Henry Guernsey Hubbard Papers (1871–1899). Record Unit 7107 [Commissioner of Patents, *Riley v. Barnard*. This item cross-listed under government documents].

Office of the Assistant Secretary, Incoming Correspondence (1850–1877). Record Unit 52.

Office of the Secretary, Incoming Correspondence (1863–1879). Record Unit 26.

Office of the Secretary, Incoming Correspondence (1879–1882). Record Unit 28

Office of the Secretary, Incoming Correspondence (1882–1890). Record Unit 30.

Office of the Secretary, Outgoing Correspondence (1865–1891). Record Unit 33.

Thomas Lincoln Casey Papers (1870–1897). Record Unit 7134.

United States National Museum, Division of Insects, Outgoing Correspondence and Records (1882–1918). Record Unit 139.

United States National Museum, Division of Insects, Accession Records. Record Unit 305.

National Archives

RG 7, Records of the Bureau of Entomology and Plant Quarantine.

Entry 3: J. Sterling Morton Letterpress Book (microfilm) and Moroney-Morton File (microfilm).

Entry 24: Notes, Division of Entomology 1863–1905, Insect Rearing Notebooks, History and Description of Insects Raised by C. V. Riley (1863–1866).

RG 16, Records of the Secretary of Agriculture.

Entry 8: Assistant Secretary of Agriculture Charles W. Dabney Jr. Letterpress Book and Moroney-Morton Folder.

Entry 15: J. Sterling Morton Letterpress Book (microfilm).

RG 48, Records of the Secretary of the Interior, Hayden Survey, Letters Received Relating to the Entomological Commission.

Unpublished Works

Brunette, Donna A. "Charles Valentine Riley and the Roots of Modern Insect Control: The Tension between Biological Control and Chemical Insecticides." Master's thesis, University of Missouri-Columbia, 1988.

Cremer, Sabine Gertrud. "Nicholas Christian Hohe (1798–1868): Universitätszeichenlehrer in Bonn." PhD diss., Rheinischen Friedrich-Wilhelms-Universität zu Bonn, 2001.

Geong, Hae-Gyung. "Exerting Control: Biology and Bureaucracy in the Development of American Entomology, 1870–1930." PhD diss., University of Wisconsin-Madison, 1999. UMI no. 9937208.

Greenfield, Theodore John. "Variation, Heredity, and Scientific Explanation in the Evolutionary Theories of Four American Neo-Lamarckians, 1867–1897." PhD diss., University of Wisconsin, Madison, 1986. UMI no. 8614370.

Rogers, Benjamin F. "The United States Department of Agriculture (1862–1889): A Study in Bureaucracy." PhD diss., University of Minnesota, 1950.

Other Unpublished Sources

Riley Family Graves, Glenwood Cemetery, Lot 108, Site 3, 2219 Lincoln Road NE, Washington, DC.

Government Documents

United States Government

Congressional Record, 44th Congress, 1st Session, Senate, Volume 4, Part 2 (March 7, 1876), 1502–58. [Debate on Ingalls locust bill].

US Commissioner of Patents. Decisions of the Commissioner of Patents and of United States Courts in the Patent Cases: *Riley v. Barnard*, Decided June 2, 1892. Copy in Smithsonian Institution Archives, Henry Guernsey Hubbard Papers 1871–1899, Box 2. RU 7107.

US Commissioner of Patents, [Barnard application for patent]. "Emulsion and Method of Making Same," [Kerosene-Milk emulsion] Patent No. 580,150 (April 6, 1897).

US Commissioner of Patents, [Barnard application for patent], "Fluid Distributer." [Cyclone Nozzle] Patent No. 580, 151 (April 6, 1897).

US Commissioner of Patents, Edward W. Serrell Jr. "Device for Reeling Silk From the Cocoon." Patent No. 317,222 (May 5, 1885). Application filed February 27, 1884.

US Congress. Forty-Sixth Congress, Second Session (1879–80), House Report 1318, serial 1937. [Appropriation for Entomological Commission to investigate Cotton Worm].

US Department of Agriculture. *Insects: The Yearbook of Agriculture 1952*. Washington, DC: Government Printing Office, 1952.

US Department of Agriculture. Bureau of Entomology. *Catalogue of the Exhibit on Economic Entomology at the World's Industrial and Cotton Centennial Exposition, New Orleans 1884–85*. Washington, DC: Judd and Detweiler Printers, 1884.

US Department of Agriculture. Division of Entomology. "Reports and Observations and Experiments in the Practical Work of the Division." *Bulletin No. 3*. Washington, DC: Government Printing Office, 1883.

US Department of the Interior. National Park Service. National Register of Historic Places Registration Form, Crunden-Martin Manufacturing Company, alias Conzelman-Crunden Realty Co., Bowman Stamping Co., Swayzee Glass Co., St.

Louis, Missouri, Section 8, including biographical sketch of Theophilus Conzelman from the Missouri Historical Society Collections 5:3 (1928). Accessed March 13, 2018. https://dnr.mo.gov/shpo/nps-nr/05000013.pdf.

US Statutes at Large 19 (1877), 357. [Establishment of the US Entomological Commission].

State Government Documents

Illinois, State of. *Report of the Adjutant General of the State of Illinois, 1886, containing Reports for the Years 1861–1866.* 8 vols. Springfield, Illinois: H. W. Rokker, 1886.

Journal of the Missouri State Senate at the Adjourned Session of the 27th General Assembly, Commencing on Wednesday, January 7, 1874. Jefferson City: Regan and Carter, Public Printers, 1875.

Missouri, State of. *Laws of the State of Missouri passed at the Adjourned Session of the 24th General Assembly.* Jefferson City: Elwood Kirby, Public Printer, 1868.

Missouri State Senate. *Journal of the Missouri State Senate at the Adjourned Session of the 27th General Assembly, Commencing on Wednesday, January 7, 1874.* Jefferson City: Regan and Carter, Public Printers, 1875.

Books and Articles Published before 1900

Riley's articles in the Missouri Reports and the *American Entomologist* are cited in the footnotes, but not separately in the bibliography. These and other works are referenced in C. V. Riley, "General Index and Supplement to the Nine Reports of the Insects of Missouri," and Samuel Henshaw, ed., *Bibliography of the More Important Contributions to American Economic Entomology*, both listed in "Bibliographies, Reference Works, and Publications in Series." Published articles in the Riley Scrapbooks are likewise not listed separately in the bibliography (see Manuscripts).

1858 Illinois Gazetteer and Business Directory. Chicago: G. W. Hawes, 1858.

"A Satisfactory Grasshopper Machine." *Scientific American* 37 (September 15, 1877): 169.

"The Agricultural Department." *Scientific American* 38 (June 1, 1878): 340.

"Alton Horticultural Society Meeting at Kirkwood, Missouri." *Prairie Farmer* 42 (December 6, 1871): 381.

"The American Science Association." *Scientific American* 43 (September 25, 1880): 196 and (October 16, 1880): 241.

"An Insect Plague." *All the Year Round* (March 28, 1885): 30–32.

Anonymous. "A State Entomologist." *Colman's Rural World* 19 (December 1, 1867): 359.

Anonymous [W. C. Hewitson]. *Entomologist's Monthly Magazine* 15 (1878): 44–45.

Anonymous. "Moritz Schuster (1823–1894)." *Entomological News* 5 (1894): 96.

"The Association of Official Economic Entomologists." *Insect Life* 2 (September 1889): 87–88.

"Association of Economic Entomologists: First Annual Meeting [November 1889, Department of Insects, USNM]." *Insect Life* 2 (November 1889): 181.

Balbiani, E. G. "Sur la prétendue migration des *Phylloxera* ailés sur les chênes à kermes." *Compte rendus de l'Académie des Sciences, Paris* 79 (1874): 640–45.

Barns, C. R., ed. *The Commonwealth of Missouri, A Centennial Record.* St. Louis: Bryan, Brand & Co., 1877.

Bates, Henry Walter. [Obituary of W. C. Hewitson]. *Proceedings of the Entomological Society of London* 1878, lxiii–lxiv.

Bazille, G., J. É. Planchon, and F. Sahut. "Sur une maladie de la vigne actuellement régnante en Provence." *Compte rendus de l'Académie des Sciences, Paris* 67 (August 3, 1868): 333–36.

Beckwith, H. W. *History of Iroquois County, Together with Historic Notes on the Northwest.* Chicago, Illinois: H. H. Hill and Co., 1880.

Boyd's Directory to Washington, D.C. 1882.

Boyer de Fonscolombe, L. J. H. "Description des Kermès qu'on trouve aux environs d'Aix." *Annales de la Société entomologique de France* 3 (1834): 201–24.

Bush, I[sidor]., and [George E.] Meissner. *Les Vignes Américaines: Catalogue Illustré et descriptif avec de brèves indications sur leur culture.* Montpellier: C. Coulet; Paris: V.-A. Delahaye, 1876.

Bush, Isidor, & Son. *Illustrated Descriptive Catalogue of Grape Vines, Potatoes, Cultivated and For Sale at the Bushberg Vineyards & Orchards, Jefferson Co., Missouri, with Brief Directions for Planting and Cultivating.* St. Louis: R. P. Studley Co., 1869.

Bush, Isidor, & Son & Meissner. *Illustrated Descriptive Catalogue of American Grape Vines, with Brief Directions for their Culture.* St. Louis: R. P. Studley & Co., 1875.

Campbell's New Atlas of Missouri with Descriptions Historical, Scientific, and Statistical. St. Louis: R. A. Campbell, 1873.

Cazalis, F. "Maladie de la vigne." Minutes of Scientific Congress of France, Montpellier. *Bulletin de la Société central d'agriculture de l'Hérault* 56 (1869): 196–214.

"Chicago." *Encyclopedia Britannica.* 9th ed. 1878.

Claudey, G. "Victor Vermorel, connu et méconnu." *Bulletin de l'Académie de Villefranche sur Saône* (1991): 65–76.

"The College of Agriculture." *Colman's Rural World* (November 22, 1894).

Colman, Norman J. "Prof. C. V. Riley, U. S. Entomologist." *Colman's Rural World* (May 12, 1892): 3–8.

Conference of Governors. *The Rocky Mountain Locust, or Grasshopper, Being the Report of Proceedings of a Conference of the Governors of Several Western States and Territories . . . held at Omaha, Nebraska, on the Twenty-fifth and Twenty-sixth Days of*

October, 1876, to consider the Locust Problem; also a Summary of the Best Means now known for Counteracting the Evil. St. Louis: R. R. Studley Company, 1876.

Crolas, Dr., and V. Vermorel. *Manuel pratique des sulfurages. Guide du vigneron pour l'emploi du sulfure de carbone contre le Phylloxera.* Villefranche, Montpellier, and Lyon: Bibliothèque du progrès agricole et viticole, 1886.

Cumming, C. F. G. "Locusts and Farmers of America." *Nineteenth Century* 17 (January 1885): 134–52.

Davis, Walter Bickford, and Daniel S. Durrie. *An Illustrated History of Missouri, Comprising Its Early Record, and Civil, Political, and Military History from the First Exploration to the Present Time, Including . . . Biographical Sketches of Prominent Citizens.* St. Louis: A. J. Hall & Co., 1876.

"The Department Entomologist." *Colman's Rural World* 34 (June 23 and July 9, 1881).

Dunning, J. W. [Obituary of W. C. Hewitson]. *The Entomologist* 11 (1878): 166–68.

Eddy, T. M. *The Patriotism of Illinois: A Record of the Civil and Military History of the State in the War for the Union, with a History of the Campaigns in which Illinois Soldiers have been Conspicuous.* 2 vols. Chicago: Clarke and Co., 1865.

Editorial, *Science* 4 (December 19, 1884): 540.

Embleton, D. "Memoir of the Life of W. C. Hewitson." *Transactions of the Natural History Society of Northumberland, Durham and Newcastle upon Tyne* 7 (1880): 223–35.

Engelmann, George. "Notes on the genus *Yucca*, no. 2." *Transactions of the Academy of Science of St. Louis* 3 (1873): 210–14.

"The Entomological Club at the AAAS." *Canadian Entomologist* 11 (September 1879): 163–76.

"The Entomological Club." *Canadian Entomologist* 12 (September 1880): 160–64.

"The Entomological Club." *Scientific American* 41 (September 13, 1879): 168.

"The Entomological Commission: LeDuc and Professor Riley." *Daily Inter Ocean* (May 11, 1881).

Fernald, C. H. "On the Care of Entomological Museums." *Science* 5 (January 9, 1885): 25.

Fitch, Asa. "Address on our most Pernicious Insects." *Transactions of the New York State Agricultural Society* 19 (1860): 588–98.

Fitz-James, Comtesse de. *La viticulture franco-américaine (1869–1889).* Montpellier: Coulet; Paris: Masson, 1889.

Flagg, W. C. "History of the Illinois State Horticultural Society." *Transactions of the Illinois State Horticultural Society* (1868). Printed in *Transactions of the Illinois State Agricultural Society* (1867–68): 486–90.

Fletcher, James. "Charles Valentine Riley." *Canadian Entomologist* 27 (1895): 274.

Foëx, Gustave. *Manuel pratique de viticulture pour la reconstitution des vignobles méridionaux-vignes américaines, submersion, plantation dans les sables.* Montpellier: Coulet; Paris: Delahaye et Lecrosnier, 1887.

Golding, F. W. "Pen Sketch of Cyrus Thomas." *Transactions of the Illinois State Agricultural Society* 22 (1888): 106.

Hall, J. W. "In Memoriam—Vactor T. Chambers." *Journal of the Cincinnati Society of Natural History* 6 (1883): 239–44.

Hendrickson, W. B. "An Illinois Scientist Defends Darwinism: A Case Study in the Diffusion of Scientific Theory." *Transactions of the Illinois Academy of Science* 65 (1972): 25–29.

Henshaw, Samuel, ed. "The Entomological Writings of Dr. Alpheus Spring Packard." US Department of Agriculture, Division of Entomology, Bulletin No. 16. Washington, DC: Government Printing Office, 1887.

Hewitson, William C., and John O. Westwood. *The Genera of Diurnal Lepidoptera: Their Generic Characters, a Notice of their Habits and Transformations, and a Catalogue of the Species of Each Genus.* 2 vols. London: Longman, Brown, Green, and Longmans, 1846–1852.

Hewitson, William C., and W. Wilson Saunders. *Illustrations of New Species of Exotic Butterflies, Selected Chiefly from the Collections of W. Wilson Saunders and William C. Hewitson.* 5 vols. London: Van Voorst, 1851, 1852–1877, 1878.

"Horticultural." *Prairie Farmer* 47 (December 9, 1876): 394.

Howard, L. O. "A Brief Account of the Rise and Present Condition of Official Economic Entomology." *Insect Life* 7 no. 2 (1894): 55–108.

———. "Charles V. Riley, Ph.D." *Proceedings of the Entomological Society of Washington* 3 (October 6, 1896): 293–98.

———. "Charles Valentine Riley." *Proceedings of the 17th Meeting for the Promotion of Agricultural Science* (1896): 108–12.

"The Illinois State Fair." *Prairie Farmer* 34 (October 1866): 218.

"Illinois State Horticultural Society." *Prairie Farmer* 36 (July 11, 1868): 11.

"Independent order of Good Templars." *Brooklyn Daily Standard Union* (February 17, 1871).

"Insect Pests: The Cottony Cushion Scale." *Rural Californian* (December 1883): 245.

"Iowa State Horticultural Society." *Prairie Farmer* 46 (December 25, 1875): 410.

"Kansas Horticultural Society." *Prairie Farmer* 47 (January 22, 1876): 26.

Keim, De B. *Washington and Its Environs: Descriptive and Historical Hand-Book.* Washington, DC: Dr. Keim, 1888.

Kingsley, John Sterling, ed. *The Riverside Natural History.* 6 vols. New York: Houghton Mifflin, 1886.

Kirby, William, and William Spence. *An Introduction to Entomology; or, Elements of the Natural History of Insects.* 4 vols. London: Longman, Hurst, Rees, Orme, and Brown, 1816–26.

Kirby, W. F. *Catalogue of the Collection of Diurnal Lepidoptera Formed by the Late William Chapman Hewitson.* London: West Newman & Co., 1879.

Laliman, Léo. *Etudes sur les divers travaux phylloxériques et les vignes américaines, notamment sur les études faites en Espagne par Don Miret y Tarral, Vocal de la commission à Madrid*. Paris: Librairie Agricole, 1879.

LeBaron, William. "Will it Pay to Employ a State Entomologist?" *Prairie Farmer* 41 (October 1, 1870): 306.

LeConte, John L. "Hints for the Promotion of Economic Entomology." *Proceedings of the American Association for the Advancement of Science* 22 (1874): 10–22. Also published with extended title in *American Naturalist* 7 (November 1873): 710–22.

LeConte, John L. "Methods of Subduing Insects Injurious to Agriculture." *Canadian Entomologist* 7 (September 1875): 167–72.

———. "On the Method of Subduing Insects Injurious to Agriculture," *Proceedings of the American Association for the Advancement of Science* 24 (September 1876): 202–7.

Lichtenstein, Jules. *Histoire du Phylloxera*. Montpellier: Coulet; Paris: Baillére, 1879.

———. *Riley et l'Entomologie aux Etats Unis*. Montpellier: Imprimerie Centrale du Midi, Hamelin frères, 1883.

———. "Sur quelques nouveaux points de l'histoire naturelle du Phylloxera vastratix." *Compte rendus de l'Académie des Sciences, Paris* 79 (1874): 598–99.

Lichtenstein, Jules, and J. É. Planchon. "De l'identité spécifique du phylloxera des feuilles et du phylloxera des racines de la vigne." *Journal d'Agriculture Pratique* 34 (August 11, 1870): 181–82.

Lodeman, E. G. *The Spraying of Plants; a Succinct Account of the History, Principles and Practice of the Application of Liquids and Powders to Plants, for the Purpose of Destroying Insects and Fungi*. Preface by B. T. Galloway. New York and London: Macmillan and Co., 1896.

McLachlan, R. "[Riley] Obituary." *Entomologist's Monthly Magazine* 31 (1895): 269–70.

Meehan, Thomas. "On the Fertilization of Yucca." *North American Entomologist* 1 (November 1879): 34.

"Meeting of the American Association." *Scientific American* 65 (September 12, 1891).

"Missouri State Horticultural Society." *Prairie Farmer* 39 (December 12, 1868): 187.

"M. J. B." *Gardeners' Chronicle and Agricultural Gazette* 43 (October 24, 1868): 1113–14 and 44 (October 31, 1868): 1138.

Morse, Edward S. "What American Zoologists Have Done for Evolution." *Proceedings of the American Association for the Advancement of Science* 25 (1877): 137–76.

Morse, F. W. "The Uses of Gases Against Scale Insects." *University of California Agricultural Experiment Station, Bulletin 71* (June 1887). Accessed March 20, 2018. http://catalog.hathitrust.org/Record/002137876.

Murtfeldt, Mary E. *Outlines of Entomology*. Jefferson City, MO: Tribune Printing Co., 1891.

[Obituary of Riley]. *Colman's Rural World* (September 19, 1895).

"Obituary [Riley]." *Entomologist's Record and Journal* 7 (1895): 72.

"Occasional." *Scientific American* 36 (April 28, 1877): 260.

Packard, A. S., Jr. "The Cave Fauna of North America, with Remarks on the Anatomy of the Brain, and Origin of the Blind Species." *Memoirs National Academy of Sciences* 4 (1888). An abstract under the same title appeared in *AN* 21 (1887): 82–83.

———. "A Half Century of Evolution, with Special Reference to the Effects of Geological Change on Animal Life." *American Naturalist* 32 (September 1898): 623–74.

———. *Lamarck, the Founder of Evolution: His Life and Work*. New York: Longmans, Green, 1901.

———. "Migrations of the Destructive Locust of the West." *American Naturalist* 11 (January 1877): 22–29.

———. "A Monograph of the Geometrid Moths or Phalaenidae of the United States." In *Report of the United States Geological and Geographical Survey of the Territories*. Edited by F. V. Hayden. Washington, DC: Government Printing Office, 1876.

———. "On Certain Entomological Speculations: A Review." *Proceedings of the Entomological Society of Philadelphia* 6 (November 1866): 209–18.

———. "Rapid as well as Slower Evolution." *Independent* 29 (August 23, 1877): 6–7.

Planchon, J. É. "Le *Phylloxera* en Europe et en Amérique." *Revue des deux Mondes* (February 1–15, 1874): 51–52.

———. *Les vignes américaines, leur culture, leur résistance au Phylloxera et leur avenir en Europe*. Montpellier: Coulet; Paris: Delahaye, 1875.

———. "Nouvelles observations sur le Puceron de la vigne (*Phylloxera vastratix (nuper Rhizaphis)*, Planch.)." *Compte rendus de l'Académie des Sciences, Paris* 67 (September 14, 1868): 588–94.

Planchon, J. É., and Jules Lichtenstein. "De l'identité spécifique du phylloxera des feuilles et du phylloxera des racines." *Compte rendus de l' Académie des Sciences, Paris* 69 (1870): 298–300.

———. "Des modes d'invasion des vignobles par le Phylloxera." *Messager du Midi* 25 (August 25, 1869).

———. "Le *Phylloxera*, faits acquis et revue bibliographique (1868–1871)." *Congrés Scientifique de France*. Montpellier: J. M. Ainé, 1872.

———. "Notes entomologiques sur le *Phylloxera*." *Insectologie Agricole* 12 (1869): 315–24.

"The Pomological Exhibition." *Prairie Farmer* 46 (September 18, 1875): 298.

"Proceedings of the American Association." *Scientific American* 41 (September 20, 1879): 180.

"Prof. C. V. Riley." *Scientific American* 38 (March 16, 1878): 161.

Rennie, James. *Insect Architecture, The Library of Entertaining Knowledge*. London: Knight, 1830.

Rennie, James, and J. G. Wood. *Insect Architecture*, revised ed. London: G. Bell & Daldy, 1869.

Report of the Adjutant General of the State of Illinois, 1886, Containing Reports for the Years 1861–1866. 8 vols. Springfield: H. W. Rokker, 1886.

Riley, C. V. "Agricultural Advancement in the U. S." *Journal of the American Agricultural Association* 1 (April 1881): 47–54.

———. "Bacterial Disease of the Imported Cabbage Worm." *Scientific American* 49 (December 1, 1883): 337.

———. "The Biological Society of Washington." *Proceedings of the Philosophical Society of Washington* 12 (February 18, 1893): 555–58.

———. "The Buffalo-Gnat Problem in the Lower Mississippi Valley." *Proceedings of the American Association for the Advancement of Science* 36 (1888): 362.

———. "Change of Habit: Two New Enemies of the Egg-plant." *American Naturalist* 16 (August 1882): 678.

———. "The Collection of Insects in the National Museum." *Science* 5 (March 6, 1885): 188.

———. Communication de M. Riley, séance du 30 juin 1884. *Société centrale d'agriculture de l' Hérault*. Montpellier: Grollier, 1884.

———. Communication de M. Riley. *Recherches de M. G. Foex sur la Mildew, etc., extrait du procés verbal du 30 juin 1884*. Montpellier: Grollier, 1884.

———. "The Cotton Worm." *Scientific American* 40 (May 17, 1879): 313.

———. "The Cotton Worm." Bulletin No. 2, US Entomological Commission, Department of the Interior. Washington, DC: Government Printing Office, 1880.

———. "The Cottony Cushion-Scale (*Icerya purchasi* Maskell)." *Annual Report of the Commissioner of Agriculture* (1886): 466–92.

———. "The Cyclone Nozzle." *Rural New Yorker* (August 22, 1885): 567.

———. "Darwin's Work in Entomology." *Proceedings of the Biological Society of Washington* (1882): 70–80.

———. "Destructive Locusts: A Popular consideration of a few of the more injurious Locusts (or 'Grasshoppers') in the United States, together with the best means of destroying them." Bulletin 25, US Department of Agriculture, Division of Entomology. Washington, DC: Government Printing Office, 1891.

———. "Directions for Collecting and Preserving Insects." Bulletin No. 39, Smithsonian Institution, United States National Museum. Washington, DC: Government Printing Office, 1892.

———. "Division of Insects." US National Museum, *Annual Report*, part 2 (1886), 196.

———. "Emulsions of Petroleum and their Value as Insecticides." *Scientific American* 49 (November 10, 1883): 294.

———. "Entomology." *Prairie Farmer Annual and Agricultural and Horticultural Advertiser.* Chicago: Prairie Farmer Co., 1868.

———. "Further Notes on the Pollination of Yucca and on Pronuba and Prodoxus." *Proceedings of the American Association for the Advancement of Science* 29 (1881): 617–39.

———. "Further Notes on Yucca Insects and Yucca Pollination." *Proceedings of the Biological Society of Washington, D. C.* 8 (June 1893): 41–54.

———. "Further Remarks on *Pronuba yuccasella* and on the Pollination of Yucca." *Transactions of the Academy of Science of St. Louis* 3 (1877): 568–73.

———. "General Truths in Applied Entomology." Address delivered before the Georgia State Agricultural Society, Savannah, Georgia, February 12, 1884. Reprinted in *Annual Report of the Commissioner of Agriculture*, 323–29. Washington, DC: Government Printing Office, 1884.

———. "Grape Deterioration: On the Cause of Deterioration in some of our Native Grape-vines, and the Probable Reason Why European Vines Have so Generally Failed in the Eastern Half of the United States." *Moore's Rural New Yorker* 24 (October 21, 1871): 251; (October 28, 1871): 268; and (November 4, 1871): 283.

———. "Grape-leaf Gall-louse (*Phylloxera vitifoliae*, Fitch)." *American Entomologist and Botanist* 2 (December 1870): 354–56.

———. "Grasshoppers and Locusts." *Prairie Farmer* 34 (November 24, 1866): 333.

———. "Hibernation of Cotton Worm Moth." *Scientific American* 40 (January 14, 1879): 375.

———. "In Memoriam, [B. D. Walsh]." *American Entomologist* 2 (December–January 1869–1870): 65–68.

———. "Insecticides for the Protection of Cotton." *Scientific American* 43 (October 16, 1880): 241.

———. "Insects nuisibles à la vigne, le puceron de la vigne." *Bulletin de la Société centrale d'Agriculture de l' Hérault* 58 (February 20, 1871): 172–80.

———. "Insects nuisibles à la vigne, le puceron de la vigne." *Le Messager Agricole du Midi* 2 (April 10, 1871): 84–89.

———. "Insects in Relation to Agriculture." In vol. 1 of *Stoddart's Encyclopaedia Americana: A Dictionary of Arts, Sciences, and General Literature*, 135–42. New York: J. M. Stoddart, 1883.

———. "Insects in Relation to Agriculture." In vol. 1 of *Encyclopaedia Britannica, American Supplement*, 135–41. 9th ed. Philadelphia: Hubbard Bros., 1889.

———. "The Kerosene Emulsion: Its Origins, Nature, and Usefulness." *Proceedings of the Society for the Promotion of Agricultural Science* (1892): 83–98.

———. "Les espèces américaines du genre *Phylloxera*." *Compte rendus de l'Académie des Sciences, Paris* 79 (1874): 1384–88.

———. "Locusts." *Prairie Farmer* 34 (November 3, 1866): 290.

———. "The Locust Plague; How to Avert it." *Proceedings of the American Association for the Advancement of Science* 24 (1876): 215–21.

———. *The Locust Plague in the United States: Being More Particularly a Treatise on the Rocky Mountain Locust or so-called Grasshopper, as it occurs East of the Rocky Mountains, with Practical Recommendations for its Destruction*. Chicago, Illinois: Rand McNally, 1877.

———. "New Insects Injurious to Agriculture." *Proceedings of the American Association for the Advancement of Science* 29 (1881): 272–73.

———. "On the Causes of Variation in Organic Forms." *Proceedings of the American Association for the Advancement of Science* 37 (1889): 225–73.

———. "On a New Genus in the Lepidopterous Family Tineidae: with Remarks on the Fertilization of Yucca." *Transactions of the Academy of Science of St. Louis* 3 (June 1873): 55–64.

———. "On the Oviposition of the Yucca Moth," *American Naturalist* 7 (October 1873): 619–23.

———. "On the Phengodini and their Luminous Larviform Females." *Proceedings of the American Association for the Advancement of Science* 36 (1888): 362.

———. "On Some Interactions of Organisms." *American Naturalist* 15 (April 1881): 323–24.

———. "The Outlook for Applied Entomology." *Insect Life* 3 (January 1890): 209.

———. "Parasites bred from the Cotton Worm." *Canadian Entomologist* 11 (November 1879): 205.

———. "Parasites of the Cotton Worm. *Canadian Entomologist* 11 (September 1879): 161–62.

———. "Parasitic and Predacious Insects in Applied Entomology." *Scientific American* 36 Supplement No. 937 (December 16, 1893): 14978–81.

———. "Pedigree Moth-Breeding." *Entomologist's Monthly Magazine* (London) 23 (May 1889): 277–78.

———. "The Philosophy of the Movement of the Rocky Mountain Locust [delivered August 1878]." *Proceedings of the American Association for the Advancement of Science* 27 (1879): 271–77.

———. *Potato Pests: Being an illustrated account of the Colorado Potato-Beetle and the other insect foes of the potato in North America, with suggestions for their repression and methods for their destruction*. New York: Orange Judd Company, 1876.

———. "The Probabilities of Locust or 'Grasshopper' Injury in the Near Future, and a New Method of Counteracting their Injury [delivered August 1885]." *Proceedings of the American Association for the Advancement of Science* 34 (1886): 519–20.

———. *Quelques mots sur les insecticides aux Etats Unis et proposition d'un nouveau rémede contre la Phylloxera*. Montpellier: Imprimerie centrale du Midi, Hamelin Fréres, 1884.

———. "Questions and Answers." *Prairie Farmer* 46 (June 5, 1875): 180.

———. "Report of Committee on Entomology." *Transactions of the Illinois State Horticultural Society for 1864* 1 (1868): 105–17.

———. "Review of Charles Darwin, *Animals and Plants under Domestication*." *Prairie Farmer* 37 (June 6, 1868): 364–65.

———. "[*Scenopinus*]." *Proceedings of the Entomological Society of Washington* 1 (March 30, 1886): 16.

———. "Silk Culture: a new source of Wealth to the United States." *Proceedings of the American Association for the Advancement of Science* 27 (1879): 277–83.

———. "Some Interrelations of Plants and Insects." *Proceedings of the Biological Society of Washington, D. C.* 7 (1892): 81–104.

———. "Successful Management of the Insects Most Destructive to the Orange." *Scientific American* 46 (May 27, 1882): 335–36.

———. "Supplementary Notes on *Pronuba yuccasella*." *Transactions of the Academy of Science of St. Louis* 4 (1874): 178–80.

———. "The True and Bogus Yucca Moth; with Remarks on the Pollination of Yucca." *American Entomologist* 3 (1880): 142–45.

———. "The Yucca Moth and Yucca Pollination," *Report of the Missouri Botanical Garden* 3 (1892): 99–158.

Riley, C. V., ed. *Agriculture*. Vol. 5 of *Reports of the United States Commissioners to the Universal Exposition of 1889 at Paris*. Washington, DC: Government Printing Office, 1891.

Riley, C. V., and L. O. Howard. "The *Phylloxera* Problem Abroad as it Appears today." *Insect Life* 2 (April 1890): 310–12.

Ritual of the Independent Order of Good Templars, for Subordinate Lodges under the Jurisdiction of the Right Worthy Grand Lodge of North America. Adopted at Cleveland Session, May 24, 1864. Chicago: Right Worthy Grand Lodge, 1864.

Saunders, Edgar. "Chicago and St. Louis Horticulturally Considered." *Prairie Farmer* 36 (November 16, 1867): 310.

Scudder, Samuel H. "Presidential Address, Subsection of Entomology." *Proceedings of the American Association for the Advancement of Science* 29 (1881): 612–13.

Shimer, Henry. "Coccus vs. Aphis: Preliminary Notice of a New Plant-Louse Genus." *Prairie Farmer* 18 (November 3 and December 8, 1866): 290.

———. "On a new genus in Homoptera." *Proceedings of the Academy of Natural Science of Philadelphia* 19 (January 1867): 2–11.

"Shipping Intelligence." [Riley's passage from England to the US] *New York Times* (October 3, 1860).

Signoret, Victor. "Observations sur les points qui paraissent acquis à la science, au sujet des espèces connues du genre Phylloxera." *Compte rendus de l' Académie des Sciences, Paris* 79 (1874): 778.

———. "*Phylloxera vastatrix*. Hémiptère-Homoptère de la familie des Aphidiens, cause prétendue de la maladie actuelle de la vigne." *Annales de la Société entomologique de France* 9 (1869): 548–88.

Snow, F. H. "The Rocky Mountain Locust: *Caloptenus spretus* Uhler." *Transactions of the Kansas Academy of Science* 4 (1875): 26–27.

"St. Louis, Missouri Botanical Gardens." In vol. 2 of *Encyclopedia Britannica*, 183–85. 9th ed. New York: Charles Scribner, 1878.

Stowe, Harriet Beecher. "Our Florida Plantation." *Atlantic Monthly* 43 (May 1879): 648–49.

Thomas, Cyrus. "Entomological." *Prairie Farmer* 46 (July 17, 1875): 226.

Traill, H. D., ed. *Social England: A Record of the Progress of the People.* 6 vols. London: Cassell and Company, 1894–1898.

Trelease, William. Review of *The Mutual Relations Between Flowers and the Insects Which Serve to Cross Them*, by Hermann Müller. *American Naturalist* 13 (July 1879): 451–52.

Trouvelot, [É.] L[éopold]. "The American Silkworm" [part 1]. *American Naturalist* 1 (March 1867): 32. See also part 2, (April 1867): 85–94; and part 3 (May 1867): 145–49.

———. "On a Method of Stimulating Union Between Insects of Different Species [read February 27, 1867]." *Proceedings of the Boston Society of Natural History* 11 (1866–68): 136–37.

Turnbull, William W. *The Good Templars: A History of the Rise and Progress of the Independent Order of Good Templars.* [Milwaukee, WI]: B. F. Parker, 1901.

Vermorel, V. "La station viticole de Villefranche." *Revue trimestrielle de la Station Viticole de Villefranche (Rhône)* 1–2 (1890): vii–x.

Vialla, Louis, and J. É. Planchon. *Etat des vignes américaines dans le Départment de l'Herault pendant l'année 1875. Rapport adressé au nom d'une commission spéciale à la Société centrale d'agriculture du départment de l'Hérault.* Montpellier: Grollier, 1875.

———. "Etat des vignes américaines." *Messager Agricole du Midi* (December 10, 1874).

Viala, Pierre. *Une mission viticole en Amérique.* Montpellier: C. Coulet; Paris: G. Masson, 1889.

Walsh, B. D. "Grape Leaf Galls." *Practical Entomologist* 1 (August 27, 1866): 111.

———. "Grasshoppers and Locusts." *Practical Entomologist* 2 (1866): 1–5.

———. "Insects Injurious to Vegetation in Illinois." *Transactions of the Illinois State Agricultural Society for 1859–1860* [4] (1861): 335–72.

———. "Imported Insects; The Gooseberry Sawfly." *Practical Entomologist* 1 (September 29, 1866): 117–25.

———. "Notes by Benj. D. Walsh." *Practical Entomologist* 2 (November 1866): 19.

———. "On Certain Entomological Speculations of the New England School of Naturalists." *Proceedings of the Entomological Society of Philadelphia* 3 (August 1864): 207–49.

———. "On the Insects, Coleopterous, Hymenopterous, and Dipterous, inhabiting the Galls of Certain Species of Willow, Part 1, Diptera." *Proceedings of the Entomological Society of Philadelphia* 3 (December 1864): 543–644.

———. "On Phytophagic Varieties and Phytophagic Species." *Proceedings of the Entomological Society of Philadelphia* 3 (November 1864): 403–30.

Walsh, Benjamin D., and Charles V. Riley. "Imitative Butterflies." *American Entomologist* 1 (June 1869): 189–93.

"The Washington Agricultural Convention." *Prairie Farmer* 43 (March 2, 1872): 65.

Ward, Lester F. "Neo-Darwinism and Neo-Lamarckism." Presidential Address,

January 24, 1891. *Proceedings of the Biological Society of Washington* 6 (January 1891): 11–71.

[Westwood, J. O.] "New Vine Diseases." *The Gardeners' Chronicle and Agricultural Gazette* 45 (January 30, 1869): 109.

Wilder, Marshall P. "The Importance of Entomology to the Fruit Grower." *American Entomologist* 3 (January 1880): 19.

"William Saunders [obituary]," in *Yearbook of the U. S. Department of Agriculture*, 628. Washington, DC: Government Printing Office, 1900.

Whitehall, A. L. "The Seventy-Sixth Regiment Illinois," in *History of Iroquois County, Together with Historical Notes on the Northwest*, by H. W. Beckwith, 284–92. Chicago: H. H. Hill and Co., 1880.

Woodworth, C. W. "Orchard Fumigation." *University of California Agricultural Experiment Station, Bulletin 122* (January 1899), 4–6. Accessed March 20, 2018. http://catalog.hathitrust.org/Record/002137876.

Books and Articles Published Since 1900

Aldrich, J. M. "The Division of Insects in the United States National Museum." In Smithsonian Institution, *Annual Report for 1919* (Washington, DC: Government Printing Office, 1921), 367–79.

Allen, David Elliston. *The Naturalist in Britain: A Social History*. Princeton Science Library. Princeton, NJ: Princeton University Press, 1994.

Anelli, Carol M. "Darwinian Theory in Historical Context and its Defense by B. D. Walsh: What is Past is Prologue." *American Entomologist* 52 (Spring 2006): 11–18.

Anonymous. *The United States National Entomological Collections*. Washington, DC: Smithsonian Institution Press, 1976.

Anon. "Mary Esther Murtfeldt." *Journal of Economic Entomology* 6 (1913): 288–89.

Anstey, M. L., S. M. Rogers, S. R. Ott, M. Burrows, and S. J. Simpson. "Serotonin mediates behavioral gregarization underlying swarm formation in desert locusts." *Science* 323 (January 30, 2009): 627–30.

Barber, Lynn. *The Heyday of Natural History, 1820–1870*. Garden City, NY: Doubleday, 1980.

Barker, Alan. *The Civil War in America*. Garden City, NY: Doubleday and Company, Inc., Anchor Books, 1961.

Barns, Jeffrey K. *Asa Fitch and the Emergence of American Entomology*. New York State Museum, Bulletin No. 461. Albany: State Education Department, 1988.

Barrow, Mark V., Jr. *A Passion for Birds: American Ornithology after Audubon*. Princeton, NJ: Princeton University Press, 1998.

Bedall, Barbara, ed. *Wallace and Bates in the Tropics: An Introduction to the Theory of Natural Selection. Based on the Writings of Alfred Russel Wallace and Henry Walter Bates*. London: Macmillan, 1969.

Billon, Richard. "Inspiration in the Harness of Daily Labor: Darwin, Botany, and the Triumph of Evolution 1859–1868." *Isis* 102 (September 2011): 393–420.

Blaisdell, Murial Louise. "Darwinism and its Data, The Adaptive Coloration of Animals." PhD diss., Harvard University, 1976. Also available under the same title in *Harvard Dissertations in the History of Science*. New York: Garland, 1992.

Boas, Louise Schutz. "'Erasmus Perkins' and Shelley." *Modern Language Notes* 70 no. 6 (June 1955): 408–13.

Bocking, Stephen. "Alpheus Spring Packard and Cave Fauna in the Evolution Debate." *Journal of the History of Biology* 21 no. 3 (Fall 1988): 439–48.

Bogue, Allan G. *From Prairie to Corn Belt: Farming on the Illinois and Iowa Prairies in the Nineteenth Century*. Chicago: University of Chicago Press, 1963.

Bogue, Margaret B. *Patterns from the Sod; Land Use and Tenure in the Grand Prairie, 1850–1900*. Collections of the Illinois State Historical Library 34. Springfield: Illinois State Historical Library, 1959.

Bolter, Andrew, biographical sketch. *National Cyclopaedia of American Biography* 24 New York: James T. White and Company, 1935.

Brodhead, Michael J. "Elliott Coues and the Sparrow War." *New England Quarterly* 44 (September 1971): 420–32.

Brower, Lincoln P. "Understanding and Misunderstanding the Migration of the Monarch Butterfly (Nymphalidae) in North America: 1857–1995." *Journal of the Lepidopterists' Society* 49 (1995): 304–85.

Brower, Lincoln P., Orley R. Taylor, Ernest H. Williams, Daniel A. Slayback, Raul R. Zubieta, and M. Isabel Ramirez. "Decline of monarch butterflies overwintering in Mexico: is the migratory phenomenon at risk?" *Insect Conservation and Diversity* 5 (2012): 95–100.

Bruckart, William L., III. "Biological Control." In *Encyclopedia of Pest Management*, ed. David Pimentel, 373–75. Ithaca, NY: Cornell University Press, 2002.

Brunette, Donna A. "Charles Valentine Riley and the Roots of Modern Insect Control." *Missouri Historical Review* 86 (April 1992): 220–47.

Burroughs, Burt E., and Vic Johnson. *The Story of Kankakee's Earliest Pioneer Settlers*. Bradley, IL: Lindsay Publications, Inc., 1986.

Caltagirone, L. E., and R. L. Doutt. "The History of the Vedalia Beetle Importation to California and its Impact on the Development of Biological Control." *Ann. Rev. Ent.* 34 (1989): 1–16.

Campbell, Christy. *The Botanist and the Vintner: How Wine was Saved for the World*. Chapel Hill: Algonquin Books, 2005.

Carton, Yves. *Henry de Varigny, Darwinien convaincu, médecin, chercheur et journaliste (1855–1934)*. Paris: Hermann, 2008.

———. "La découvert du Phylloxera en France: Un sujet de polémique : les archives parlent (Hemiptera, Chermesidae)." *Bulletin de la Société entomologique de France* 111, no. 3 (2006): 305–16.

Carton, Yves, Conner Sorensen, Janet Smith, and Edward Smith. "Une coopération exemplaire entre entomologistes français et américains pendant la crise du Phylloxera en France (1868–1895)." *Annales de la Société entomologique de France*, n.s., 43 no. 1 (2007): 103–25.

Casagrande, R. A. "The 'Iowa' Potato Beetle, its Discovery and Spread to Potatoes." *Bulletin of the Entomological Society of America* (Summer 1985): 27–29.

Catton, Bruce. *A Stillness at Appomattox*. New York: Pocket Books, Inc., 1958. First published 1953.

Churchill, Frederick B. "August Weismann and a Break from Tradition." *Journal of the History of Biology* 1 (1968): 91–112.

———. "Hertwig, Weismann, and the Meaning of Reduction Division Circa 1890." *Isis* 61 (1970): 429–57.

———. "The Weismann-Spencer Controversy over the Inheritance of Acquired Characters." *Proceedings of the 15th International Congress of the History of Science* (1978): 451–68.

Clark, J. F. M. *Bugs and the Victorians*. New Haven and London: Yale University Press, 2009.

———. "The eyes of our Potatoes are weeping: The Rise of the Colorado Beetle as an Insect Pest." *Archives of Natural History* 34 no. 1 (April 2007): 113–18.

Claudey, G. "Victor Vermorel, connu et méconnu." *Bulletin de l'Académie de Villefranche sur Saône* (1991): 65–76.

Cockerell, T. D. A. "Alpheus Spring Packard, 1839–1905." In vol. 9 of *Biographical Memoirs of the National Academy of Sciences*, 181–236. Washington, DC: National Academy of Sciences, 1920.

Colman, Gould P. "Pioneering in Agricultural Education: Cornell University, 1867–1890." *Agricultural History* 36 (October 1962): 200–206.

Colman, Norman J. "Prof. C. V. Riley, U.S. Entomologist." *Colman's Rural World* (May 12, 1892): 1–14 (separate reprint).

Comstock, Anna B. *The Comstocks of Cornell: John Henry Comstock and Anna Botsford Comstock: An Autobiography by Anna Botsford Comstock*, ed. Glenn W. Herrick and Ruby Green Smith. Ithaca, NY: Comstock Publishing Associates, 1953.

Cooper, Matthew. "Relations of Modes of Production in Nineteenth-Century America: The Shakers and Oneida." In *American Culture: Essays on the Familiar and Unfamiliar*, ed. David Pimentel, 54–57. Pittsburgh, PA: University of Pittsburgh Press, 1990.

Corliss, Carlton Jonathan. *Main Line of Mid-America: The Story of the Illinois Central*. New York: Creative Age Press, 1950.

Cosmos Club. *The Twenty-fifth Anniversary of the Founding of the Cosmos Club*. Washington, DC: Cosmos Club, 1904.

Croker, Robert A. *Stephen Forbes and the Rise of American Ecology*. Washington and London: Smithsonian Institution Press, 2001.

Cromie, Robert. *Short History of Chicago*. San Francisco: Lexikos, 1984.

Cronon, William. *Nature's Metropolis: Chicago and the Great West*. New York: W. W. Norton & Co., 1991.

Crossette, George. *Founders of the Cosmos Club of Washington 1878: A Collection of Biographical Sketches and Likenesses of the Sixty Founders*. Washington, DC: Cosmos Club, 1966.

Davis, J. J. "Joseph Tarrigan Monell." *Journal of Economic Entomology* 8 (1915): 503.

———. "Obituary [J. T. Monell]." *Entomological News* 26 (October 1915): 380–83.

Davis, W. B., and D. S. Durrie. *An Illustrated History of Missouri, Comprising Its Early Records, and Civil, Political, and Military History from the First Exploration to the Present Time, Including . . . Biographical Sketches of Prominent Citizens*. St. Louis: A. J. Hall & Co., 1876.

Dean, G. A. "The Contribution of Kansas Academy of Science to Entomology." *Transactions of the Kansas Academy of Science* 41 (1938): 61–73.

Denis, Erving A., and John D. Jenkins. *Comprehensive Graphic Arts*. Indianapolis: Howard W. Sams & Co., Inc., 1974.

Desmond, Adrian. "The Making of Institutional Zoology in London, 1822–1836, Part 1." *History of Science* 23 (1985): 153–85.

———. "The Making of Institutional Zoology in London, 1822–1836, Part 2." *History of Science* 23 (1985): 223–50.

Dexter, Ralph W. "The Impact of Evolutionary Theories on the Salem Group of Agassiz Zoologists (Morse, Hyatt, Packard, Putnam)." *Historical Collections of the Essex Institute* 115 (1979): 144–71.

———. "The Organization and Work of the U. S. Entomological Commission (1877–82)." *Melsheimer Entomological Series* 26 (1979): 28–32.

Doutt, Richard L. "Vice, Virtue, and Vedalia." *Bulletin of the Entomological Society of America* 4 (December 1958): 119–23.

Dow, R. P. "The Rector of Barnham and His Times." *Bulletin of the Brooklyn Entomological Society* 8 (June 1913): 68–74.

Dunlap, Thomas R. *DDT: Scientists, Citizens, and Public Policy*. Princeton, NJ: Princeton University Press, 1982.

Dupree, A. Hunter. *Science in the Federal Government: A History of Policies and Activities to 1940*. New York: Arno Press, 1980. First published 1957.

Dyck, Ian. "From 'Rabble' to 'Chopsticks': The Radicalism of William Cobbett." *Albion* 21 (Spring 1989): 56–87.

Dyer, Frederick H. *A Compendium of the War of the Rebellion*. 3 vols. New York and London: Thomas Yosaloff, 1959.

"Engelmann, George." *Dictionary of American Biography*. New York: Charles Schribner's Sons, 1958.

Essig, E. O. *A History of Entomology*. New York and London: Hafner Publishing Co., 1965. First published 1931.

Evans, Mary Alice. "Mimicry and the Darwinian Heritage." *Journal of the History of Ideas* 26 (1965): 211–20.

Ewan, Joseph. "George Engelmann." In *Dictionary of Scientific Biography*, ed. Charles Coulston Gillespie, 130–31. New York: Charles Schribner's Sons, 1970.

Farber, Paul Lawrence. *Finding Order in Nature: The Naturalist Tradition from Linnaeus to E. O. Wilson*. Baltimore and London: Johns Hopkins University Press, 2000.

Filler, Martin. "Our Grand and Randy Great Architects." *New York Review of Books* 58 (May 26–June 8, 2011): 20–27.

Fisk, Catherine L. "Removing the 'Fuel of Interest' from the 'Fire of Genius:' Law and the Employee-Inventor, 1830–1930." *University of Chicago Law Review* 65 (Fall 1998): 1127–98.

Flack, J. Kirkpatrick. *Desideratum in Washington: The Intellectual Community in the Capital City 1870–1900*. Cambridge, MA: Schenkman Publishing Co., Inc., 1975.

Foster, Mike. "Ferdinand Vandiveer Hayden as Naturalist." *American Zoologist* 26 (1986): 343–49.

Friedlander, C. P. *The Biology of Insects*. New York: Pica Press, 1977.

Froude's Life of Carlyle, ed. John Clubbe. Columbus: Ohio State University Press, 1979.

Gagnon, Paul A. *France Since 1879*. New York, Evanston, and London: Harper and Row, 1964.

Gale, George. "Saving the vine from *Phylloxera*: a never-ending battle." In *Wine: A Scientific Exploration*, ed. M. Sandler and R. Pinder, 70–91. London: Taylor & Francis, 2003.

Gates, Paul Wallace. *Agriculture and the Civil War*. New York: Alfred A. Knopf, 1965.

———. *The Farmer's Age: Agriculture, 1815–1860*. New York: Holt, Reinhardt and Winston, 1960.

Gerhartz, Heinrich. "Christian Hohe: Ein Beitrag zur Geschichte der rheinischen Malerei im 19. Jahrhundert." In vol. 128 of *Annalen des historischen Vereins für den Niederrhein*, 90–120. Dusseldorf: Von L. Schwann, 1936.

Giblot-Ducray, D. R., Correll Collins, A. Nankivell, A. Downs, I. Pearce, A. C. Mckay, and K. M. Ophel-Keller. "Detection of grape phylloxera (*Daktulosphaira vitifoliae* Fitch) by real-time quantitative PCR: development of a soil sampling protocol." *Australian Journal of Grape and Wine Research* 22 (2016): 469–77.

Goetzman, William. *Exploration and Empire: The Explorer and the Scientist in the Winning of the American West*. New York: Alfred A. Knopf, 1966.

Granett, J., A. Walker, J. De Benedictis, G. Fong, and E. Weber. "California grape phylloxera more variable than expected." *California Agriculture* 50, no. 4 (July–August 1996): 387–412.

Granett, J., A. D. Omer, P. Pessereau, and M. A. Walker. "Fungal infections of grapevine roots in phylloxera-infested vineyards." *Vitis* 37, no. 1 (2015): 39–42.

Graff, Henry F. *Grover Cleveland.* New York: Times Books, 2002.

Grafton-Cardwell, Elizabeth E., and Ping Gu. "Conserving vedalia beetle, *Rodolia cardinalis* (Mulsant)(Coleoptera: Coccinellidae), in citrus: a continuing challenge as new insecticides gain registration." *Journal of Economic Entomology* 96 (2003): 1388–98.

Gravena, Santin, Winfield Sterling, and Dean Allen. *Abstracts, References, and Key Words of Publications Relating to the Cotton Leafworm, Alabama argillacea (Huebner) (Lepidoptera: Noctuidae).* Thomas Say Foundation Monographs, no. 10. College Park, MD: Entomological Society of America, 1985.

Green, Constance McLaughlin. *Washington: Village and Capital, 1800–1878.* Princeton, NJ: Princeton University Press, 1962.

Gunn, D. L. "The Biological Background of Locust Control." *Ann. Rev. Ent.* 5 (1960): 279–300.

Gurney, Ashley B. "A Short History of the Entomological Society of Washington." *Proceedings of the Entomological Society of Washington* 78 (July 1976): 225–39.

Haller, John S., Jr. *Outcasts from Evolution: Scientific Attitudes of Racial Inferiority, 1859–1900.* Urbana: University of Illinois Press, 1975.

Hawthorne, D., ed. "France Will Honor Early Texas Pioneer: Phylloxera Fighter Munson Finally Recognized." *Vinifera Wine Growers Journal* (Spring 1988): 1–6.

Hendrickson, W. B. "An Illinois Scientist Defends Darwinism: A Case Study in the Diffusion of Scientific Theory." *Transactions of the Illinois Academy of Science* 65 [314] (1972): 25–29.

Hendrickson, Walter B., and William J. Beecher. "In the Service of Science: The History of the Chicago Academy of Sciences." *Bulletin of the Chicago Academy of Sciences* 11, no. 7 (September 1972): 211–68.

Henson, Pamela. "The Comstock Research School in Evolutionary Entomology." *Osiris* 8 (1993): 158–77.

Herriott, F. I. "William Stebbins Barnard: Professor of Biology, Drake University 1886–1887." Reprint from *Annals of Iowa* (July 1936): 5–41.

Hicken, Victor. *Illinois in the Civil War.* 2nd ed. Urbana and Chicago: University of Illinois Press, 1991.

Higley, W. K. "Historical Sketches." Chicago Academy of Sciences, Special Publication No. 1 (1902), 12–13.

Hoddle, Mark S., Claudio Crespo Ramírez, Christina D. Hoddle, Jose Loayza, Maria Piedad Lincango, Roy G. Van Driesche, and Charlotte E. Causton. "Post release evaluation of *Rodolia cardinalis* (Coleoptera: Coccinellidae) for control of *Icerya purchasi* (Hemiptera: Monophlebidae) in the Galapagos Islands." *Biological Control* 67 (2013): 262–74.

Howard, Leland O. *Fighting the Insects: The Story of an Entomologist.* New York: Macmillan Company, 1933.

———. *A History of Applied Entomology (Somewhat Anecdotal).* Smithsonian Miscellaneous Collections 84. Washington, DC: Smithsonian Institution Press, 1930.

———. *The Insect Book*. New York: Doubleday, Page, and Co., 1905.

———. "The United States Department of Agriculture and Silk Culture." In *Yearbook of the United States Department of Agriculture*, 137–148. Washington, DC: Government Printing Office, 1904.

Hutchins, Ross E. *Insects*. Prism Paperback, 1972.

Ivens, J. Loreena, et al. *The Kankakee River Yesterday and Today*. Illinois State Water Survey, Miscellaneous Publication 60. Champaign: Illinois Department of Energy and Natural Resources, 1981.

Johnson, Vic. "Up 'til Now." Historical column in *The Kankakee Sunday Journal*, March 18, 1990; November 22 and 29, 1992; December 6, 13, 20, and 27, 1992; January 17, 24, and 31, 1993; and February 7 and 14 1993.

Karabell, Zachary. *Chester Alan Arthur*. New York: Henry Holt and Co., 2004.

Kenaga, William, and G. R. Letourneau. *History of Kankakee County, Seventy-Sixth Illinois Infantry*. Chicago: Middle West Publishing Co., 1906.

Kennedy, John Stoddard. "Continuous Polymorphism in Locusts." In *Insect Polymorphism*, ed. John Stoddard Kennedy, 80–90. London: Royal Entomological Society, 1961.

Khan, B. Zorina. *The Democratization of Invention: Patents and Copyrights in American Economic Development, 1790–1920*. Cambridge: Cambridge University Press, 2005.

Klasey, John, and Mary Jean Houde. *Of the People: A Popular History of Kankakee County*. Chicago: General Printing Co., 1968.

Klose, Nelson. "Sericulture in the United States." *Agricultural History* 37 (October 1963): 225–34.

Kohlstedt, Sally. *The Formation of the American Scientific Community: The American Association for the Advancement of Science, 1848–1860*. Urbana, Chicago, and London: University of Illinois Press, 1976.

Kritsky, Gene. "Darwin, Walsh and Riley: The Entomological Link." *American Entomologist* 41 (Summer 1995): 89–95.

———. "Entomological Reactions to Darwin's Theory in the Nineteenth Century." *Ann. Rev. Ent.* 53 (2008): 345–60.

Leach, William. *Butterfly People: An American Encounter with the Beauty of the World*. New York: Pantheon Books, 2013.

LeDuc, William G. *Recollections of a Civil War Quartermaster*. St. Paul, MN: North Central Publishing Co., 1963.

Lewis, David Rich. "American Indian Environmental Relations." In *A Companion to American Environmental History*, ed. Douglas Cazaux Sackman, 191–213. Chichester, UK: Wiley-Blackwell, 2010.

Lockwood, Jeffrey A. "The Fate of the Rocky Mountain locust, *Melanoplus spretus* Walsh: Implications for Conservation Biology." *Terrestrial Arthropod Reviews* 3 (2010): 129–60.

———. *Grasshopper Dreaming: Reflections on Killing and Loving*. Boston: Skinner House Books, 2002.

———. *Locust: the Devastating Rise and Mysterious Disappearance of the Insect that Shaped the American Frontier*. New York: Basic Books, 2004.

Logan, Mrs. John A. *Thirty years in Washington; or Life and scenes in our national capital: Portraying the wonderful operations in all the great departments, and describing every important function of our national government . . . With sketches of the presidents and their wives . . . from Washington's to Roosevelt's Administration*. Hartford, CT: A. D. Worthington & Co., 1901.

Luck, Robert F. "Notes on the Evolution of Citrus Pest Management in California." In *Proceedings of the Fifth California Conference on Biological Control*, ed. Mark S. Hoddle and Marshall W. Johnson, 1–7. Riverside, CA: n.p., 2006.

McKinney, H. Lewis. "Fritz Müller." In *Dictionary of Scientific Biography*, ed. Charles Coulston Gillispie, 559–600. New York: Charles Schribner's Sons, 1974.

McLeRoy, S. S., and R. E. Renfro. *Grape Man of Texas: The Life of T. V. Munson*. Austin, TX: Eakin Press, 2004.

Malcolm, S. B., and L. P. Brower. "Evolutionary and ecological implications of cardenolide sequestration in the monarch butterfly." *Experientia* 45, no. 3 (1989): 284–95.

Mallis, Arnold. *American Entomologists*. New Brunswick, NJ: Rutgers University Press, 1971.

Marcus, Alan I. *Agricultural Science and the Quest for Legitimacy: Farmers, Agricultural Colleges, and Experiment Stations, 1870–1890*. Ames: Iowa State University Press, 1985.

———. "The Ivory Silo: Farmer-Agricultural Tensions in the 1870s and 1880s." *Agricultural History* 60 (Spring 1986): 22–36.

Marlatt, C. L. "The Entomological Club." *Canadian Entomologist* 35 (March 1903): 53–54.

Miner, Craig. *West of Wichita: Settling the High Plains of Kansas, 1865–1890*. Lawrence: University Press of Kansas, 1986.

Miller, David. "Apropos C. V. Riley." *Pan-Pacific Entomologist* 22 (1946): 28–30.

Moore, David. "Fungal Control of Pests." In *Encyclopedia of Pest Management*, ed. David Pimentel, 320–24. Ithaca, NY: Cornell University Press, 2002.

Morrow, Dwight W., Jr. "The American Impressions of a French Botanist, 1873." *Agricultural History* 34 (1960): 71–76.

Muehl, Siegmar. "Isidor Bush and the Bushberg Vineyards of Jefferson County." *Missouri Historical Review* 94 (October 1999): 42–58.

Nelson, R. H., ed., "Charles Valentine Riley." *Bulletin of the Entomological Society of America* 9, no. 3 (1963): 183.

Numbers, Ronald. *Darwinism Comes to America*. Cambridge, MA, and London: Harvard University Press, 1998.

Ordish, George. *The Great Wine Blight*. London: Sidgwick & Jackson, Ltd., 1987.

Overfield, Richard. *Science with Practice: Charles E. Bessey and the Maturing of American Botany*. Ames: Iowa State University Press, 1993.

"Packet Ships." In vol. 17 of *The World Book Encyclopedia*. Chicago: World Book, Inc., 1995.

Pakenham, Simona. *Sixty Miles from England: The English at Dieppe, 1814–1914*. London: Macmillan, 1967.

Pauly, Philip J. *Biologists and the Promise of American Life: From Meriwether Lewis to Kinsey*. Princeton, NJ: Princeton University Press, 2000.

———. *Fruits and Plains: The Horticultural Transformation of America*. Cambridge, MA: Harvard University Press, 2007.

Pellmyr, O. "Yuccas, Yucca Moths, and Coevolution: a Review." *Annual Report of the Missouri Botanical Garden* 90 (2003): 35–55.

Pellmyr, Olle, and James Leebens-Mack. "Forty million years of mutualism: Evidence for Eocene origin of the yucca-yucca moth association." *Proceedings of the National Academy of Sciences* 96 (1999): 9178–83.

Perry, Matthew C., ed. *The Washington Biologists' Field Club: Its Members and Its History (1900–2006)*. Washington, DC: Washington Biologists' Field Club, 2007. Accessed March 17, 2018. https://www.pwrc.usgs.gov/resshow/perry/bios/WBFC_booksm.pdf.

Peterson, Anne E. "Hornblower & Marshall, Architects." [Washington, DC]: National Trust for Historic Preservation, The Preservation Press, 1978.

Pfeiffer, Edward J. "United States." In *The Comparative Reception of Darwinism*, ed. Thomas F. Glick, 168–226. Austin: University of Texas Press, 1974.

Pinckney, Darryl. "Elite Black and Quite Different." Review of *Negroland: A Memoir* by Margo Jefferson. *New York Review of Books* 63 (May 26–June 8, 2016): 48–50.

Pleasants, John M., and Karen S. Oberhauser. "Milkweed loss in agricultural fields because of herbicide use: effect on the monarch butterfly population." *Insect Conservation and Diversity* 6 (2013): 135–44.

Powell, Kevin S. "Grape phylloxera: An overview." In *Root Feeders, An Ecosystem Perspective*, ed. Scott N. Johnson and Phillip J. Murray, 96–114. Wallingford, UK: CAB International, 2008.

Price, Nick. "Fumigants." In *Encyclopedia of Pest Management*, ed. David Pimentel, 317–19. Ithaca, NY: Cornell University Press, 2002.

Rapp, William F., Jr. "The Andrew Bolter Insect Collection." *Entomological News* 56 (October 1945): 209.

Rasmussen, Wayne. "The Civil War: A Catalyst of Agricultural Revolution." *Agricultural History* 39 (October 1965): 187–95.

Reader, William J. *Victorian England*. New York: Putnam's Sons, 1974.

Reese, William J. "The Philosopher-King of St. Louis," in *Curriculum and Consequence*, ed. by Barry M. Franklin, 155–77. New York: Teachers College Press, 2000.

Riegert, Paul W. *From Arsenic to DDT: A History of Entomology in Western Canada.* Toronto: University of Toronto Press, 1980.

Ritland, David B., and Lincoln P. Brower. "The Viceroy Butterfly is not a Batesian Mimic." *Nature* 350 (1991): 497–98.

Ritvo, Harriett. "Animal Pleasures: Popular Zoology in Eighteenth and Nineteenth-Century England." *Harvard Library Bulletin* 33 (1985): 239–79.

Rivinus, E. F., and E. M. Youssef. *Spencer Baird of the Smithsonian.* Washington and London: Smithsonian Institution Press, 1992.

Rogers, Ben F. "William Gates LeDuc: Commissioner of Agriculture." *Minnesota History* (Autumn 1955): 287–95.

Ross, Dorothy. "The Development of the Social Sciences." In *The Organization of Knowledge in Modern America, 1860–1920*, ed. Alexandra Oleson and John Voss, 107–38. Baltimore and London: Johns Hopkins University Press, 1979.

Ross, Earle D. "The United States Department of Agriculture During the Commissionership: A Study in Politics, Administration, and Technology, 1862–1889." *Agricultural History* 20 (July 1946): 129–43.

Rossiter, Margaret. "The Organization of the Agricultural Sciences." In *The Organization of Knowledge in Modern America, 1860–1920*, ed. Alexandra Oleson and John Voss, 211–48. Baltimore and London: Johns Hopkins University Press, 1979.

Ruse, Michael. "Charles Darwin and Artificial Selection." *Journal of the History of Ideas* 36 (April–June 1975): 339–50.

Sacks, Oliver. "The Mental Life of Plants and Worms, Among Others." *New York Review of Books* 61 (April 24–May 7, 2014): 4–8.

Sandved, Kjell B., and Michael G. Emsley. *Insect Magic.* New York: Viking Press, 1978.

Saunders, William. Obituary. *Yearbook of the U. S. Department of Agriculture.* Washington, DC: Government Printing Office, 1900.

Sauer, Klaus Peter. "Hermann Schaaffhausens Beitrag zur Entwicklung des Evolutionsgedanken und des Artbegriffs vor Darwin." *Mitteilungen der Deutschen Zoologischen Gesellschaft* (2002): 39–51.

Schwarz, E. A. "A Sketch of the History of the Entomological Club of the American Association." *Canadian Entomologist* 35 (March 1903): 54–58.

Schwarz, Hermann. "Miss Mary E. Murtfeldt." *Entomological News* 24 (June 1913): 241–42.

Schlebecker, John T. *Whereby We Thrive: A History of American Farming, 1607–1972.* Ames: Iowa State University Press, 1975.

Scott, Roy V. *The Reluctant Farmer: The Rise of Agricultural Extension to 1914.* Urbana, Chicago, London: University of Illinois Press, 1970.

Semmens, Brice X., Darius J. Semmens, Wayne E. Thogmartin, Ruscena Wiederholt, Laura López-Hoffman, Jay E. Diffendorfer, John M. Pleasants, Karen S. Oberhauser, and Orley R. Taylor. "Quasi-extinction risk and population targets for the Eastern, migratory population of monarch butterflies (*Danaus plexippus*)."

Scientific Reports 6 (2016). https://www.nature.com/articles/srep23265.
"Serrell, Edward Wellman." *The Twentieth Century Biographical Dictionary of Notable Americans* 11. Boston: The Biographical Society, 1904.
Sheppard, Carol A. "Benjamin Dann Walsh: Pioneer Entomologist and Proponent of Darwinian Theory." *Ann. Rev. Ent.* 49 (2004): 1–25.
Sheppard, Carol A., and Edward H. Smith. "Entomological Heritage," *American Entomologist* 43 (Fall 1997): 142–46.
Sheppard, Carol A., and Richard A. Oliver. "Yucca Moths and Yucca Plants: Discovery of 'the most wonderful Case of Fertilization.'" *American Entomologist* 50 (Spring 2004): 32–46.
Sheppard, Carol A., and Richard A. Weinzierl. "Entomological Lucubrations: The 19th Century Spirited Conflict Concerning the Natural History of the Armyworm, *Pseudaletia unipuncta* (Haworth) (Lepidoptera: Noctuidae)." *American Entomologist* 48 (Summer 2002): 108–13.
Sherwood, Morgan B. *Exploration of Alaska, 1865–1900*. New Haven and London: Yale University Press, 1965.
Smith, Audrey Z. *A History of the Hope Entomological Collections in the University Museum Oxford: With Lists of Archives and Collections*. Oxford, UK: Clarendon Press, 1986.
Smith, Edward H. "The Grape Phylloxera: A Celebration of Its Own." *American Entomologist* 38 (Winter 1992): 212–21.
Smith, Edward, and Gene Kritsky, "Charles Valentine Riley: His Formative Years." *American Entomologist* 57 (Summer 2011): 74–80.
Smith, Edward H., and Janet R. Smith. "Charles Valentine Riley: The Making of the Man and His Achievements." *American Entomologist* 42 (Winter 1996): 228–38.
Smith, Kathryn Schneider. *Washington at Home: An Illustrated History of Neighborhoods in the Nation's Capital*. Northridge, CA: Windsor Publications, 1988.
Sorensen, W. Conner. *Brethren of the Net: American Entomology, 1840–1880*. Tuscaloosa: University of Alabama Press, 1995.
———. "Samuel Hubbard Scudder." In vol. 19 of *American National Biography*, ed. John A. Garraty and Mark C. Cornes, 542–44. Oxford and New York: Oxford University Press, 1999.
———. "Uses of Weather Data by American Entomologists 1830–1880." *Agricultural History* 63 (Spring 1989): 162–73.
Sorensen, W. Conner, and Edward H. Smith. "Charles Valentine Riley: Art Training in Bonn, 1858–1860." *American Entomologist* 43 (Summer 1997): 92–104.
Sorensen, W. Conner, Edward H. Smith, Janet Smith, and Yves Carton. "Charles V. Riley, France, and *Phylloxera*." *American Entomologist* 54 (Fall 2008): 134–49.
Spear, Robert J. *The Great Gypsy Moth War: A History of the First Campaign in Massachusetts to Eradicate the Gypsy Moth, 1890–1901*. Amherst and Boston: University of Massachusetts Press, 2005.

Sproat, John G. *"The Best Men": Liberal Reformers in the Gilded Age*. London and New York: Oxford University Press, 1968.

Stader, Karl Heinz. "Bonn und der Rhine in der englishchen Reiseliteratur." In *Aus Geschichte und Volkskunde von Stadt und Raum Bonn: Festschrift Joseph Dietz zum 80. Geburtstag am 8. April 1973*, 117–53. Veröffentlichungen des Stadtarchivs Bonn, Band 10. Bonn: Ludwig Rohrscheid Verlag, 1973.

Stanton, William. *The Leopard's Spots: Scientific Attitudes toward Race in America, 1815–59*. Chicago and London: University of Chicago Press, 1966. First published 1960.

Stebbins, Robert E. "France." In *The Comparative Reception of Darwinism*, ed. Thomas F. Glick, 117–67. Chicago and London: University of Chicago Press, 1974.

Stephens, Lester D. "The Appointment of the Commissioner of Agriculture in 1877: A Case Study in Political Ambition and Patronage." *Southern Quarterly* 15, no. 4 (1977): 371–86.

Stevenson, P. A. "The Key to Pandora's Box." *Science* 323 (January 30, 2009): 594–95.

Stiles, T. J. *Jesse James: Last Rebel of the Civil War*. New York: Alfred A. Knopf, 2002.

Stoll, Steven. *Larding the Lean Earth: Soil and Society in Nineteenth Century America*. New York: Hill and Wang, 2002.

Stork, Nigel E., James McBroom, Claire Gely, and Andrew J. Hamilton. "New approaches narrow global species estimates for beetles, insects, and terrestrial arthropods." *Proceedings of the National Academy of Sciences*, 112, no. 24 (2015): 7519–23.

Summers, Floyd G. "Norman J. Colman, First Secretary of Agriculture." *Missouri Historical Review* 19 (April 1925): 404–8.

Sutherland, Daniel E. *A Savage Conflict: The Decisive Role of Guerrillas in the Civil War*. Chapel Hill: University of North Carolina Press, 2009.

Tomalin, Claire. *Charles Dickens: A Life*. London: Viking, 2011.

Trefousse, Hans L. *Rutherford B. Hayes*. New York: Henry Holt and Co., 2002.

Uvarov, Boris. *Grasshoppers and Locusts: A Handbook of General Acridology*. 2 vols. Cambridge: Cambridge University Press, 1966.

———. "A Revision of the Genus *Locusta*, L. (= *Pachytylus*, Fieb.), with a New Theory as to the Periodicity and Migration of Locusts." *Bulletin of Entomological Research* 12 (1921): 135–63.

Van Driesche, R. G., Mark Hoddle, and Ted Center. *Control of Pests and Weeds by Natural Enemies: An Introduction to Biological Control*. Malden, MA: Blackwell Publishing, 2008.

Van Huis, Arnold, Joost Van Itterbeeck, Harmke Klunder, Esther Mertens, Afton Halloran, Giulia Muir, and Paul Vantomme. Edible insects: future prospects for food and feed security. UN Food and Agriculture Organization. FAO Forestry Paper 171. Rome, 2013. Accessed February 15, 2018. http://www.fao.org/docrep/018/i3253e/i3253e.pdf/.

Wagner, David L. "Ode to *Alabama*: The Meteoric Fall of a Once Extraordinarily Abundant Moth." *American Entomologist* 55 (Fall 2009): 170–73.

Waloff, N., and G. B. Popov. "Sir Boris Uvarov (1889–1970): The Father of Acridology." *Ann. Rev. Ent.* 35 (1990): 1–24.

Watson, Francis. *The Year of the Wombat: England, 1857*. London: Harper & Row, 1974.

Weber, Donald C. "Colorado beetle: pest on the move." *Pesticide Outlook* 14 (2003): 256–59.

Weiss, H. B. *The Pioneer Century of American Entomology*. New Brunswick, NJ: self-published, 1936.

Wheeler, Alfred G., Jr., E. Richard Hoebecke, and Edward H. Smith. "Charles Valentine Riley: Taxonomic Contributions of an Eminent Agricultural Entomologist." *American Entomologist* 56 (Spring 2010): 14–30.

Whorton, James. *Before Silent Spring: Pesticides and Public Health in Pre-DDT America*. Princeton, NJ: Princeton University Press, 1974.

Wiener, Martin J. "The Changing Image of William Cobbett." *Journal of British Studies* 13 (May 1974): 135–54.

Willard, J. T. *History of the Kansas State College of Agriculture and Applied Science*. Manhattan: Kansas State College Press, 1940.

Winter, William C. *The Civil War in St. Louis: A Guided Tour*. St. Louis: Missouri Historical Society Press, 1994.

Worster, Donald. *A River Running West: The Life of John Wesley Powell*. New York and London: Oxford University Press, 2002.

Zhang, X., Z. Tu, S. Luckhart, and D. G. Pfeiffer, "Genetic diversity of plum curculio (Coleoptera: Curculionidae) among geographical populations in the eastern United States." *Annals of the Entomological Society of America* 101, no. 5 (2008): 824–32.

Index

Page numbers in italics refer to illustrations
Entries in boldface refer to Featured Insects